Handbook of

Abiotic Stress Adaptation in Plants

Techniques and Responses

Contributors :
Huijuan Zhang,
Lei Huang, *et al.*

Volume 2

KORO PRESS

KOROS PRESS LIMITED
London, UK

Handbook of Abiotic Stress Adaptation in Plants : *Techniques and Responses*
Contributors : Lei Huang *and* Huijuan Zhang, *et al.*

Published by Koros Press Limited

www.korospress.com

United Kingdom

Printed in 2017 for Sale in the Indian Subcontinent

Handbook of Abiotic Stress Adaptation in Plants : *Techniques and Responses*

ISBN: 978-1-78163-482-0

British Library Cataloguing in Publication Data
A CIP record for this book is available from the British Library

Exclusively distributed by CBS Publishers & Distributors Pvt. Ltd.
Sales & Distribution Rights only for India, Pakistan, Bangladesh, Sri Lanka, Nepal and Bhutan.This book is not to be sold outside these territories.

PREFACE

Abiotic stress is a term used to describe nonliving factors that adversely affect living organisms. Animals can be affected by abiotic stress, but plants are more vulnerable as they are unable to move to a less stressful environment. Biotic stress factors would include insect pests and diseases, whereas plant abiotic stress results from environmental factors. These may be related to climate – drought, extremes of temperature and wind, for example – or to chemical factors in the soil or the atmosphere.

Plants show a range of responses and adaptations that help bring about abiotic stress tolerance. Some of these involve structural or chemical changes, while others involve restriction of the growing period according to conditions. In some cases, symbiotic relationships have developed as a response to stresses.

Drought is one of the commonest forms of stress encountered by plants, and those that live in arid or semi-arid areas have developed various strategies for dealing with it. One of these is succulence. Thick, fleshy stems and leaves can store large amounts of water and reduce the surface area to volume ratio of the plant, minimizing water loss by evaporation. A waxy coating on the stems and leaves also reduces evaporation. Some non-succulent plants may have long roots that extend downwards to the groundwater.

Plants can sometimes adapt to seasonal or periodic stress by simply having a short life cycle that is timed to coincide with the most favorable conditions. For example, a plant may germinate, mature, flower and produce seeds during a short rainy season within an otherwise dry year, or it may lie dormant in an underground tuber for long periods, emerging after rain to quickly complete its life cycle. In temperate areas, forest floors during the summer months may be lacking in light for photosynthesis, so low growing forest plants may quickly complete their life cycles during the spring, before the forest canopy has developed dense foliage.

Abiotic stress may result from soil factors. A high salt content can be lethal for many plants, but some – known as halophytes – have adapted to salty conditions, which are found in both in coastal areas, such as salt marshes, and inland arid areas where high evaporation tends to concentrate salts originating in the mineral content of the soil. These plants may excrete salt from their leaves or store it within their cells in bodies known as vacuoles so that it is kept separate from the cell cytoplasm. Deficiencies in nutrients or the presence of toxic substances such as heavy metals in the soil can also result in abiotic stress.

Atmospheric pollution can be another source of stress. Sulfur dioxide and nitrogen oxides from fossil fuel burning can result in acid rain, which can damage the foliage of sensitive plants. Acid rain can also reduce soil pH, harming or killing plants that are not adapted to acidic conditions.

Climate change is thought to be a major source of abiotic stress for crops. Changing patterns of temperature and rainfall are having an impact on the cultivation of food and other crops, with some formerly productive areas suffering from drought, flooding or extremes of temperature. In order to alleviate the economic impact of crop failures, research is being carried out into the development, through breeding or genetic engineering, of crop plants that are more resistant to these forms of stress.

CONTENTS

List of Contributors

Bengi (BABA) ERDAG

ADU, Faculty of Science and Arts, Department of Biology, 09010 Kepez-Aydin/TURKEY

A. Kamil YUREKLI

Ege University, Faculty of Science, Department of Biology, 35100 Bornova-Izmir/TURKEY

Lei Huang

National Key Laboratory for Rice Biology, Institute of Biotechnology, Zhejiang University, Hangzhou 310058, China; E-Mails: leihero2008@163.com (L.H.); zhanghj82@zju.edu.cn (H.Z.); yongbohong@126.com (Y.H.); liurui90913@163.com (S.L.); dyli@zju.edu.cn (D.L.)

Huijuan Zhang

National Key Laboratory for Rice Biology, Institute of Biotechnology, Zhejiang University, Hangzhou 310058, China; E-Mails: leihero2008@163.com (L.H.); zhanghj82@zju.edu.cn (H.Z.); yongbohong@126.com (Y.H.); liurui90913@163.com (S.L.); dyli@zju.edu.cn (D.L.)

Yongbo Hong

National Key Laboratory for Rice Biology, Institute of Biotechnology, Zhejiang University, Hangzhou 310058, China; E-Mails: leihero2008@163.com (L.H.); zhanghj82@zju.edu.cn (H.Z.); yongbohong@126.com (Y.H.); liurui90913@163.com (S.L.); dyli@zju.edu.cn (D.L.)

Shixia Liu

National Key Laboratory for Rice Biology, Institute of Biotechnology, Zhejiang University, Hangzhou 310058, China; E-Mails: leihero2008@163.com (L.H.); zhanghj82@zju.edu.cn (H.Z.); yongbohong@126.com (Y.H.); liurui90913@163.com (S.L.); dyli@zju.edu.cn (D.L.)

Dayong Li

National Key Laboratory for Rice Biology, Institute of Biotechnology, Zhejiang University, Hangzhou 310058, China; E-Mails: leihero2008@163.com (L.H.); zhanghj82@zju.edu.cn (H.Z.); yongbohong@126.com (Y.H.); liurui90913@163.com (S.L.); dyli@zju.edu.cn (D.L.)

Fengming Song

National Key Laboratory for Rice Biology, Institute of Biotechnology, Zhejiang University, Hangzhou 310058, China; E-Mails: leihero2008@163.com (L.H.); zhanghj82@zju.edu.cn (H.Z.); yongbohong@126.com (Y.H.); liurui90913@163.com (S.L.); dyli@zju.edu.cn (D.L.)

Chapter 18

CLIMATE CHANGE AND PLANT

EFFECT OF CLIMATE CHANGE ON PLANT BIO-DIVERSITY

Environmental conditions play a key role in defining the function and distribution of plants, in combination with other factors. Changes in long term environmental conditions that can be collectively coined climate change are known to have had enormous impacts on plant diversity patterns in the past and are seen as having significant current impacts. It is predicted that climate change will remain one of the major drivers of bio-diversity patterns in the future.

Palaeo Context

The Earth has experienced a constantly changing climate in the time since plants first evolved. In comparison to the present day, this history has seen Earth as cooler, warmer, drier and wetter, and CO_2 (carbon dioxide) concentrations have been both higher and lower. These changes have been reflected by constantly shifting vegetation, for example forest communities dominating most areas in interglacial periods, and herbaceous communities dominating during glacial periods. It has been shown that past climatic change has been a major driver of the processes of speciation and extinction. The best known example of this is the Carboniferous Rainforest Collapse which occurred 350 million years ago. This event decimated amphibian populations and spurred on the evolution of reptiles.

Modern Context

There is significant current interest and research focus on the phenomenon of recent anthropogenic climate changes, or global warming. Focus is on identifying the current impacts of climate change on bio-diversity, and predicting these effects into the future.

Changing climatic variables relevant to the function and distribution of plants include increasing CO_2 concentrations, increasing global temperatures, altered precipitation patterns, and changes in the pattern of 'extreme' weather events such as cyclones, fires or storms.

Because individual plants and therefore species can only function physiologically, and successfully complete their life cycles under specific environmental conditions (ideally within a sub-set of these), changes to climate are likely to have significant impacts on plants from the level of the individual right through to the level of the ecosystem or biome.

Effects of CO_2

Increases in atmospheric CO_2 concentration for affect how plants photosynthesise, resulting in increases in plant water use efficiency, enhanced photosynthetic capacity and increased growth. Increased CO_2 has been implicated in 'vegetation thickening' which affects plant community structure and function. Depending on environment, there are differential responses to elevated atmospheric CO_2 between major 'functional types' of plant, such as C_3 and C_4 plants, or more or less woody species; which has the potential among other things to alter competition between these groups. Increased CO_2 can also lead to increased Carbon : Nitrogen ratios in the leaves of plants or in other aspects of leaf chemistry, possibly changing herbivore nutrition.

Effects of Temperature

Increases in temperature raise the rate of many physiological processes such as photo-synthesis in plants, to an upper limit. Extreme temperatures can be harmful when beyond the physiological limits of a plant.

Effects of Water

As water supply is critical for plant growth, it plays a key role in determining the distribution of plants. Changes in precipitation are predicted to be less consistent than for temperature and more variable between regions, with predictions for some areas to become much wetter, and some much drier.

General Effects

Environmental variables will not act in isolation, but also in combination with one other, and with other pressures such as habitat degradation and loss or the introduction of exotic species. It is suggested that that these other drivers of bio-diversity change will act in synergy with climate change to increase the pressure on species to survive.

Direct Impacts of Climate Change

Changes in Distributions

If climatic factors such as temperature and precipitation change in a region beyond the tolerance of a species phenotypic plasticity, then distribution changes of the species may be inevitable. There is already evidence that plant species are shifting their ranges in altitude and latitude as a response to changing regional climates. Yet it is difficult to predict how species ranges will change in response to climate and separate these changes from all the other man-made environmental changes such as eutrophication, acid rain and habitat destruction.

When compared to the reported past migration rates of plant species, the rapid pace of current change has the potential to not only alter species distributions, but also render many species as unable to follow the climate to which they are adapted. The environmental conditions required by some species, such as those in alpine regions may disappear altogether. The result of these changes is likely to be a rapid increase in extinction risk. Adaptation to new conditions may also be of great importance in the response of plants.

Predicting the extinction risk of plant species is not easy however. Estimations from particular periods of rapid climatic change in the past have shown relatively little species extinction in some regions, for example. Knowledge of how species may adapt or persist in the face of rapid change is still relatively limited.

Changes in the suitability of a habitat for a species drive distributional changes by not only changing the area that a species can physiologically tolerate, but how effectively it can compete with other plants within this area. Changes in community composition are therefore also an expected product of climate change.

Changes in Life-cycles (Phenology)

The timing of phenological events such as flowering are often related to environmental variables such as temperature. Changing environments are therefore expected to lead to changes in life cycle events, and these have been recorded for many species of plants. These changes have the potential to lead to the asynchrony between species, or to change competition between plants. Flowering times in British plants for example have changed, leading to annual plants flowering earlier than perennials, and insect pollinated plants flowering earlier than wind pollinated plants; with potential ecological consequences. A recently published study has used data recorded by the writer and naturalist Henry David Thoreau to confirm effects of climate change on the phenology of some species in the area of Concord, Massachusetts.

Indirect Impacts of Climate Change

All species are likely to be not only directly impacted by the changes in environmental conditions discussed above, but also indirectly through their interactions with other species. While direct impacts may be easier to predict and conceptualise, it is likely that indirect impacts are be equally important in determining the response of plants to climate change.

A species whose distribution changes as a direct result of climate change may 'invade' the range of another species for example, introducing a new competitive relationship.

The range of a symbiotic fungi associated with plant roots may directly change as a result of altered climate, resulting in a change in the plants distribution.

A new grass may spread into a region, altering the fire regime and greatly changing the species composition.

A pathogen or parasite may change its interactions with a plant, such as a pathogenic fungus becoming more common in an area where rainfall increases.

Increased temperatures may allow herbivores to expand further into alpine regions, significant impacting the composition of alpine herbfields.

There are innumerable examples of how climate change could indirectly affect plant species, most of which will be extremely difficult to predict.

Higher Level Changes

Species respond in very different ways to climate change. Variation in the distribution, phenology and abundance of species will lead to inevitable changes in the relative abundance of species and their interactions. These changes will flow on to affect the structure and function of ecosystems.

Challenges of Modelling Future Impacts

Accurate predictions of the future impacts of climate change on plant diversity are critical to the development of conservation strategies. These predictions have come largely from bio-informatic strategies, involving modelling individual species, groups of species such as 'functional types', communities, ecosystems or biomes. They can also involve modelling species observed environmental niches, or observed physiological processes.

Although useful, modelling has many limitations. Firstly, there is uncertainty about the future levels of greenhouse gas emissions driving climate change and considerable uncertainty in modelling how this will affect other aspects of climate such as local rainfall or temperatures. For most species the importance of specific climatic variables in defining distribution (*e.g.* minimum rainfall or maximum temperature) is unknown. It is also difficult to know which aspects of a particular climatic variable are most biologically relevant, such as average *vs.* maximum or minimum temperatures. Ecological processes such as interactions between species and dispersal rates and distances are also inherently complex, further complicating predictions.

Improvement of models is an active area of research, with new models attempting to take factors such as life-history traits of species or processes such as migration into account when predicting distribution changes; though possible trade-offs between regional accuracy and generality are recognised.

Climate change is also predicted to interact with other drivers of biodiversity change such as habitat destruction and fragmentation, or the introduction of foreign species. These threats may possibly act in synergy to increase extinction risk from that seen in periods of rapid climate change in the past.

HOW DO INCREASED CO_2 LEVELS AFFECT PLANT GROWTH?

Carbon dioxide is the source of the food that plants make for themselves through photo-synthesis. Photo-synthesis is the process by which plants use sunlight to convert water and carbon dioxide into sugars and other biological materials that are essential to their nutrition and growth. When the concentration of carbon dioxide in the air is higher, plants become more efficient in pulling the gas out of the air. This allows the plants to process the gas more quickly. Faster conversion of carbon dioxide into sugars and other organic materials translates into faster plant growth.

There are many plants that won't benefit from higher carbon dioxide concentrations. These plants have certain tissues in their leaves that concentrate carbon dioxide over time. Since the gas is already concentrated inside the plant, a higher availability of carbon dioxide in the air won't make a difference.

Saving Water

In order to absorb carbon dioxide, plants have to open pores in their leaves. When these pores open, water can escape, either as liquid or as a gas. In regions where water is scarce, this loss of water through the pores can be a serious problem. In an environment where carbon dioxide levels are higher, plants won't have to open these pores as often. These plants will cut down on the amount of water they lose by breathing, and will thrive more in dry regions.

Other Factors Could Cancel Increased Growth

Although increased carbon dioxide levels could boost plant growth for some species, the gas will also affect the climate. As global temperatures rise, many regions of the world will get too hot for the plants that make up forests and jungles. Many plants will die off from this excess heat, even as other regions of the world see increased plant growth. Climate change induced by carbon dioxide will also increase rainfall and nitrogen deposits in soil. Although plants need both of these substances to survive, research has found that when water and nitrogen supplies and higher temperatures are accompanied by increased carbon dioxide, plants do not show increased growth. In fact, growth can slow down. Researchers aren't sure why this happens.

THE CO_2 FERTILIZATION EFFECT : HIGHER CARBOHYDRATE PRODUCTION AND RETENTION AS BIOMASS AND SEED YIELD

The rise in atmospheric carbon dioxide (CO_2) concentration from about 280 μ mol/mol before the industrial revolution to about 360 μ mol/mol currently is well documented. The consensus of many studies of the effects of elevated CO_2 on plants is that the CO_2 fertilization effect is real. However, the CO_2 fertilization effect may not be manifested under conditions where some other growth factor is severely limiting, such as low temperature. Also, plants grown in some conditions, where limitations of rooting volume, light, or other factors restrict growth, have not shown a sustained response to elevated CO_2.

The main objectives of this chapter are to assess the direct effects of rising atmospheric CO_2 and indirect effects of potential climate changes on crop growth and yield. The approach will be to (1) provide a general overview of CO_2 effects on plant growth processes; (2) analyse some specific experimental data on crop plant responses to elevated CO_2 and climate change factors; (3) summarize recent reviews of plant responses to elevated CO_2; and (4) discuss some crop modelling assessments of rising CO_2 and climate change factors on agricultural productivity based on predictions of global climate change models. Also, some adaptations for improving crop productivity in a higher CO_2 world will be suggested.

Overview of CO_2 Effects on Plant Growth Processes

Most of the following discussion of CO_2 effects on plants applies to species with the C_3 photo-synthetic pathway and not necessarily to species with the C_4 pathway. Other aerial, non-biotic environmental factors that affect plant growth and development are light and temperature. Plant photo-synthetic rates generally increase linearly with light across relatively low ranges of light intensity, and then the rates decelerate until they reach an asymptotic maximum. Because of crowding and shading of many leaves, most crop canopies do not reach light saturation at full sunlight; that is, they would be able to respond to light levels well beyond full solar irradiance. Likewise, crop photo-synthetic rates respond to increasing levels of CO_2 but then level off at higher concentrations (around 700 μ mol/mol or greater, depending upon species and other factors). However, leaf photo-synthesis usually increases with temperature up to some maximum value, and then declines. Furthermore, temperature affects not only photo-synthesis, but also respiration, growth, development phases and reproductive processes.

Elevated CO_2 may have some effects on crop phenology, although stages of development are governed primarily by temperature, time and photoperiod. If dates of planting were to be changed because of the greenhouse effect, then phenological timing of plants could be affected. For example, higher temperatures could decrease yields by decreasing the duration of the grain-filling period or changes in photoperiod could shorten or lengthen the vegetative stage.

The CO_2 fertilization effect begins with enhanced photo-synthetic CO_2 fixation. Non-structural carbohydrates tend to accumulate in leaves and other plant organs as starch, soluble carbohydrates or polyfructosans, depending on species. In some cases, there may be feedback inhibition of photo-synthesis associated with accumulation of non-structural carbohydrates. Increased carbohydrate accumulation, especially in leaves, may be evidence that crop plants grown under CO_2 enrichment may not be fully adapted to take complete advantage of elevated CO_2. This may be because the CO_2-enriched plants do not have an adequate sink (inadequate growth capacity), or lack capacity to load phloem and translocate soluble carbohydrates. Improvement of photo-assimilate utilization should be one goal of designing cultivars for the future.

In the process of growth, photo-assimilates are allocated to the vegetative shoots, root system or reproductive organs. In some cases, more photo-assimilate of CO_2-enriched plants is partitioned to the root system than to the shoots. Above ground, more photo-assimilate usually goes into stems and supporting structures than into leaves. This phenomenon may not be an inherent response to elevated CO_2, but may be a by-product of the larger size of plants often found in CO_2-enriched atmospheres, especially by species that produce branch stems along the aerial mainstems.

Reproductive biomass growth as well as vegetative biomass growth are usually increased by elevated CO_2. However, the harvest index, or the ratio of seed yield to above-ground biomass yield, is typically lower under elevated CO_2 conditions, which may also be evidence of the lack of capacity to utilize completely the more abundant photo-assimilate.

In many cases, both the amount and the carboxylation activity of ribulose 1,5-bisphosphate carboxylase-oxygenase enzyme (rubisco) is decreased in leaves of plants grown under elevated CO_2. This acclimation phenomenon may produce 'downregulation of photo-synthesis'; however, this is not universally the case. For example, there is little evidence of this downregulation response in soybean (*Glycine max* L. Merr.), a C_3 legume. In fact, photo-synthetic capacity per unit leaf area of soybean is increased under CO_2 enrichment. Leaves often develop an additional layer of mesophyll cells. Also, more structural carbohydrate may be produced in leaves, as well as stems, of CO_2-enriched plants.

We can obtain clues about reasons that some plants downregulate and others do not in response to elevated CO_2 by focusing on the capabilities of soybean. This plant has (a) symbiotic N2 fixation; (b) the capacity to form additional layers of palisade cells in the leaf tissue; (c) the capacity to shunt much of the photo-assimilate into relatively inert starch rather than soluble sugars during photo-synthesis; (d) a relatively strong leaf and stem sink during vegetative development; and (e) a strong seed-fill sink during reproductive development. Plants which lack these capacities, either inherently or because of growth in limiting environments, are more likely to demonstrate some degree of downregulation of photo-synthesis.

The carbon : nitrogen ratio of leaves of plants is usually increased under CO_2 enrichment. Plants may acclimate to elevated CO_2 by requiring less rubisco and photo-synthetic apparatus, which would lead to lower nitrogen contents. The overall change in C : N ratios is governed both by increases in structural and non-structural carbohydrates, and by decreases in protein content. However, seed nitrogen content is little affected.

Specific respiration rates may be reduced by both short-term exposure to elevated CO_2 and long-term growth at elevated CO_2. However, the long-term effect may be similar when respiration rates are reported on a per unit nitrogen basis.

In climate change scenarios, temperatures are predicted to increase following the rise of CO_2 and other greenhouse-effect gases. Carbon dioxide x temperature interactions have been observed for vegetative growth (*i.e.*, the CO_2 fertilization effect is greater at warmer temperatures than at cooler temperatures). Temperature increases in a higher CO_2 world could increase overall biomass productivity for vegetative crops (pastures and forages) both by extending the length of the growing season in temperate regions, and by the interaction of CO_2 x temperature in stimulation of vegetative growth. However, CO_2 x temperature interactions appear to be very small or negligible for reproductive processes (seed set and seed yield) although there may be more initial flowers formed by greater amounts of branching or tillering that is stimulated by CO_2 enrichment.

Precipitation changes may occur along with other climatic change effects. In general, predictions from crop models show that increased CO_2 should increase productivity of C_3 plants, but the associated predictions of temperature rise will be detrimental. Not surprisingly, changes in precipitation patterns (decreases of rainfall during growth period) could be more detrimental for crop production than changes in temperature.

Under elevated CO_2 stomatal conductance in most species will decrease which may result in less transpiration per unit leaf area. However, leaf area index of some crops may also increase. The typical 40% reduction in stomatal conductance induced by a doubling of CO_2 has generally resulted in only a 10% (or less) reduction in crop canopy water use in chamber or field experimental conditions. Actual changes in crop evapotranspiration will be governed by the crop energy balance, as mitigated by stomatal conductance, leaf area index, crop structure and any changing meteorological factors.

Water-use efficiency (WUE) (ratio of CO_2 uptake to evapotranspiration) will increase under higher CO_2 conditions. This increase is caused more by increased photo-synthesis than it is by a reduction of water loss through partially closed stomata. Thus, more biomass can be produced per unit of water used, although a crop would still require almost as much water from sowing to final harvest. If temperatures rise, however, the increased WUE caused by the CO_2 fertilization effect could be diminished or negated, unless planting dates can be changed to more favourable seasons.

Several assessments of impacts of climate change on crop productivity have been published. Progress has been made on integrating the impacts on individual countries and on economic and social interactions. For the most part, these assessments project more favourable climates for agriculture in northern latitudes and less favourable climates in the tropical and sub-tropical zones. However, the crop modelling predictions are dependent on the scenarios of outputs of General Circulation Models (GCMs) applied to the greenhouse effect. Thus, the dependency chain of assessments follows : Climate Change Scenarios → Crop Model Prediction and Agricultural Production Systems (with and without available mitigation and adaptation response strategies) → National Scenarios of Economics and Well-being of Farmers, Agricultural Commerce, and Consumers → Country-by-Country and Global Interaction Scenarios of World Trade (food and all other commodities). Population Dynamics and Economic Well-being, and Impacts on Social Systems.

As the world continues to consume fossil fuels, CO_2 concentrations will continue to rise. Other greenhouse-effect gases, such as methane, nitrous oxides, chlorofluorocarbons and chlorofluorocarbon substitutes, and perhaps tropospheric ozone, will likely rise also. The CO_2 fertilization effect on plants will increase and climate changes may occur because of the combined increase of all greenhouse-effect gases. Global agriculture could adapt to gradual regional climate changes, but sudden changes would be more serious. Adaptation and/or mitigation actions could include the following :

1. Selection of plants that can better utilize carbohydrates which are produced when plants are grown at elevated CO_2.
2. Selection of plants that produce less structural matter and more reproductive capacity under CO_2 enrichment. (This applies for seed crop plants, not necessarily vegetative biomass plants.)
3. Search for germplasms that are adapted to higher day and night temperatures, and incorporate those traits into desirable crop production cultivars to improve flowering and seed set.
4. Change planting dates and other crop management procedures to optimize yields under new climatic conditions, and select for cultivars that are adapted to these changed agricultural practices.
5. Shift to species that have more stable production under high temperatures or drought.
6. Determine whether more favourable N : C ratios can be attained in forage cultivars adapted to elevated CO_2.
7. Where needed, and where possible, develop irrigation systems for crops.

SPECIFIC RESPONSES OF CROPS TO ELEVATED CO_2

This section will focus on responses of two C_3 crops to elevated CO_2 : soybean, a symbiotic nitrogen-fixing legume representative of the pulses; and rice (*Oryza sativa* L.), a globally important cereal food crop. Findings from seven crop cycles of both soybean and rice grown in sunlit chambers at Gainesville, Florida, USA, will be emphasized. The chambers feature both large rooting volumes of real soil and light from the sun. Some studies included responses to sub-ambient as well as superambient CO_2 concentrations.

Photo-synthetic Rates

Soybean photo-synthetic rates of both leaves and the whole canopy have *always* been increased by elevated CO_2 Jones *et. al.* reported midday maximum canopy photo-synthetic rates of 60 and 90 μ mol $CO_2/m^2/s$ at 70 days after planting (DAP) for soybean grown at CO_2 concentrations of 330 and 800 μ mol/mol, respectively. The leaf area index (LAI) was 6.9 and 9.0, respectively. Midday maximum canopy photo-synthetic rates in other studies were 40 and 75 μ mol/m^2/s for soybean grown at 320 and 640 μ mol/mol CO_2 and 40 and 80 μ mol/m^2/s for soybean grown at 330 and 800 μ mol/m^2/s.

Valle *et. al.* found that midday maximum photo-synthetic CO_2 uptake rates of soybean leaves ranged from 30 to 50 μ mol/m^2/s and 15 to 25 μ mol/m^2/s on plants grown at 660 and 330 μ mol/mol CO_2 respectively. Allen *et. al.* reported that, at all light levels, leaf photo-synthetic rates increased linearly with CO_2 concentration across the range of 330 to 800 μ mol/mol.

Valle *et. al.* used a Michaelis-Menten type of rectangular hyperbola to summarize photo-synthetic responses of soybean leaves vs. CO_2 concentration. The plants had been grown at 330 and 660 μ mol/mol of CO_2 and then exposed to a wide range of CO_2 for a short period.

$$Y=(Y_{max} \times [C])/([C]+K_m)+Y_i \quad (4.1)$$

where Y is photo-synthetic rate in μ mol $CO_2/m^2/s$; $[C_3]$ is CO_2 concentration in μ mol/mol; Y_i is the y-axis intercept at zero $[C_3]$, the apparent respiration rate, in μ mol $CO_2/m^2/s$; Y_{max} is the response limit of $(Y - Y_i)$ at very high [C], the asymptotic phot`osynthetic rate, in μ mol $CO_2/m^2/s$; K_m is the value of [C] where $(Y-Y_i)=Y_{max}/2$, the apparent Michaelis-Menten constant, in (μ mol/mol; and Γ_c is the calculated [C] intercept at zero Y, the CO_2 compensation point, μ mol/mol (not shown in this equation). There was no obvious downregulation of soybean leaf photo-synthesis in response to elevated CO_2; in fact, photo-synthetic capacity was increased. Leaf quantum yield increased from 0.05 to 0.09 in the soybean leaves exposed to CO_2 of 330 and 660 μ mol/mol, respectively.

Campbell *et. al.* showed that soybean leaf photo-synthetic rates were higher for plants grown at 660 than at 330 μ mol/mol CO_2 when measured at common intercellular CO_2 concentrations. Furthermore, Campbell *et. al.* measured rubisco activity and amount in leaves of soybean grown in CO_2 concentrations of 160, 220, 280, 330, 660 and 990 μ mol/mol. They found that rubisco activity was almost constant at 1.0 μ mol CO_2/min/mg soluble protein across this CO_2 treatment range. Leaf soluble protein was nearly constant at about 2.4 g/m^2 with 55% being rubisco protein. Specific leaf weight increased across the 160 to 990 μ mol/mol CO_2 concentration range, so that the rubisco activity on a leaf dry weight basis decreased.

Campbell *et. al.* showed that photo-synthetic capacity of soybean canopies grown at 330 and 660 μ mol/mol CO_2 were similar for short exposures at concentrations below 500 μ mol/mol, but above this concentration, canopies grown at 660 μ mol/mol had a slightly higher photo-synthetic capacity. Thus, soybean did not lose photo-synthetic capacity as did some other plant species. In other studies, however, both photo-synthetic rate and rubisco activity of soybean declined during long-term CO_2 enrichment.

Respiration

Whole-canopy pre-dawn respiration rates at 40 to 60 DAP were 2-3 and 4-5 μ mol/m^2/s for soybean grown at CO_2 levels of 320 and 640 μ mol/m^2, respectively, and a constant day/night temperature of 25°C. After cross-switching two of the treatment chambers at DAP 52, the subsequent canopy photo-synthetic rates and respiration rates quickly adjusted to their new CO_2 exposure conditions.

Table 18.1 : Average asymptotic maximum photo-synthetic rate (Y_{max}) with respect to y-intercept parameter (Y_i), apparent Michaelis-Menten constant for CO_2 (K_m, and CO_2 compensation point (G_c.) for leaves grown at two CO_2 treatments and subjected to different short-term CO_2 levels. Condensed from Valle *et. al.*

Growth CO_2 treatment	Y_{max} μ mol/m^2/s	K_m μ mol/mol	Y_i μ mol/m^2/s	Γ_c μ mol/mol
330	51.8	359	-7.8	63
660	126.6	1 133	-4,6	42

Means of Y_{max} and Y_i were significantly different, p = 0.05, by a t-test.

Thus, pre-dawn respiration rates were closely connected to the previous CO_2 fixation rates.

Partitioning

Growth of plants under elevated CO_2 results in changes in partitioning of photo-assimilates to various plant organs over time. In soybean, elevated CO_2 generally promoted greater carbon (dry matter) partitioning to the supporting structure (stems, petioles and roots) than to the leaf laminae during vegetative stages of growth. During reproductive stages, there tended to be lower relative partitioning to reproductive growth (pods) by plants under elevated CO_2.

Table 18.2 : Soybean plant components as a percentage of total dry matter grown at subambient and superambient concentrations of CO_2 in 1984. Condensed from Allen *et. al.*.

Component	CO_2 concentration, μ mol/mol					
	160	220	280	330	660	990
	13 DAP [1] (V2 Stage)					
Root, %	11.3	9.8	9.3	12.4	9.8	10.5
Cotyledon, %	19.5	15.6	13.9	13.8	11.2	9.4
Stem, %	22.8	25.0	25.8	25.5	26.3	26.7
Leaf, %	46.1	49.5	50.7	48.2	52.5	53.2
	34 DAP (V8 Stage)					
Root, %	6.4	7.1	6.9	7.6	8.9	7.9
Stem, %	23.4	24.4	27.6	25.9	28.2	29.3
Petiole, %	12.2	13.1	14.6	14.4	15.7	15.7
Leaf, %	57.6	55.3	50.8	52.2	47.2	46.9
	66 DAP (R5 Stage)					
Stem, %	16.5	19.6	25.0	21.7	25.2	26.8
Petiole, %	9.9	11.3	12.9	12.7	14.1	13.6
Leaf, %	35.3	35.3	35.4	32.2	31.5	29.8
Pod, %	38.2	33.6	26.5	33.3	28.9	29.6
	94 DAP (R7 Stage)					
Stem, %	10.2	12.8	14.9	14.3	16.3	19.5
Petiole, %	5.5	6.2	6.0	6.7	7.0	6.5
Leaf, %	17.4	16.3	14.9	14.4	11.9	10.9
Pod, %	66.7	64.7	64.1	64.6	64.7	64.0

[1] Days after planting.

Growth Rates

During the linear phase of vegetative growth after full ground cover is reached, the growth rates of plants exposed to a range of CO_2 concentrations varied from 5.0 to 20.7 $g/m^2/d$ for exposures from 160 to 990 μ mol/mol. The total final dry weight ranged from 12.88 to 39.12 g/plant, and final seed weight ranged from 5.77 to 17.85 g/plant for CO_2 treatments ranging from 160 to 660 μ mol/mol.

Carbohydrates

Soybean accumulates non-structural carbohydrates, particularly starch, under CO_2 enrichment. Allen *et. al.* grew soybean under CO_2 treatments of

330, 450, 600 and 800 μ mol/mol. Sucrose, reducing sugars, and total soluble sugars of leaves remained somewhat constant throughout the day, but starch increased steadily at a rate of about 6 g/kg dry matter/hour. Average total soluble sugars increased from 24 to 36 g/kg dry weight and starch increased from 85 to 204 g/kg dry weight across the range of 330 to 800 μ mol/mol. Elevated non-structural carbohydrates in CO_2-enriched soybean plants were confirmed by Baker *et. al.* and Allen *et. al.* The concentrations also varied across the life cycle of the plants.

Nitrogen

Nitrogen content decreased from 50 to 37 g/kg dry weight over the range of 330 to 800 μ mol/mol. When the nitrogen content was adjusted to remove the effect of total non-structural carbohydrate, the relative changes were smaller, 55 to 48 g N/kg dry matter, across the 330 to 800 μ mol/mol range.

Yield

Soybean seed yield was always increased by elevated CO_2 Allen *et. al.* summarized the photo-synthetic, biomass and seed yield responses of several experiments with the equation (4.1) rectangular hyperbola model using data normalized to responses obtained at 330 μ mol/mol. The values of K_m, Y_{max} and Y_i parameters for relative photo-synthetic rates were 279 μ mol/mol, 3.08 and -0.68, respectively, for relative biomass yield were 182 μ mol/mol, 3.02 and -0.91, respectively, and for relative seed yield were 141 μ mol/mol, 2.55 and -0.76, respectively. This model was used to project yields across several ranges of atmospheric CO_2 concentration increases. For a doubling of CO_2 this model predicted a 32.2% increase in soybean grain yield and a 42.7% increase in biomass. The ratio of these two numbers, 1.322/1.427 = 0.926, gives the fraction of the harvest index expected under doubled CO_2 in comparison with ambient CO_2.

Table 18.3 : Percentage increases of soybean midday photo-synthetic rates, biomass yield, and seed yield predicted across selected carbon dioxide concentration [CO_2] ranges associated with relevant benchmark points in time. Adapted from Allen *et. al.*

Period of time (years)	[CO_2]-Midday		Biomass photosynthesis	Seed yield	Biomass yield
	Initial	Final			
	(Nmd/mol)		(% increase over initial [CO_2])		
IA-1700 [1]	200	270	38	33	24
1700-1973	270	330	19	16	12
1973-2073? [2]	330	660	50	41	31

[1] IA, the Ice Age about 13 000 to 30 000 years before present. The atmospheric CO_2 concentrations that prevailed during the last Ice Age, and from the end of the glacial melt until pre-pioneer/pre-industrial revolution times, were 200 and 270 μ mol/mol, respectively.

[2] The first world energy 'crisis' occurred in 1973 when the CO_2 concentration was 330 μ mol/mol. This CO_2 concentration is used as the basis for many CO_2 doubling studies. The CO_2 concentration is expected to double sometime within the 21st century.

Carbon Dioxide and Temperature

The CO_2 fertilization effect appears to be enhanced under elevated temperatures, at least up to a point. Idso *et. al.* and Kimball *et. al.* showed that the growth modification factor (or biomass growth modification ratio) due to a 300 μ mol/mol enrichment was 0.08 per °C (average daily temperature) across the range of 12 to 34°C. However, for soybean, Allen calculated a biomass growth modification ratio response to temperature of -0.031 per °C for seed biomass yield and -0.026 per °C for total biomass accumulation. Elevated temperatures tended to shorten the grain-filling period of this crop.

Soybean seed yield tended to decrease slightly with temperature over the day/night range of 26/19 to 36/29°C. The number of seed per plant increased slightly with increase of both CO_2 and temperature. Mass per seed decreased sharply with increasing temperature. Although CO_2 enrichment resulted in increased seed yield and above-ground biomass, harvest index was decreased with both CO_2 and temperature. The data show no tendency for the growth modification factor to increase with temperature for either seed yield or biomass accumulation.

Subsequent experiments with soybean substantiate these data although it was found that seed yields dropped sharply for day/night temperatures of 40/30°C and above, and biomass yields were maintained up to 44/34°C before rapidly failing.

The sensitivity of the growth modification factor data of Idso *et. al.* and Kimball *et. al.* may have been affected by other factors at Phoenix, Arizona. Apparently, the growth *vs.* temperature data were obtained throughout the year, and the findings may have been impacted by factors such as total solar radiation, photoperiod, stages of development, or other conditions caused by the changing seasons. Of course, these other environmental factors are all part of the complex of plant responses to climatic conditions. However, a better test of temperature effects alone would be CO_2 enrichment throughout the season or life cycle of plants under natural conditions, but with consistent temperature differences (cooler and warmer).

Table 18.4 : Seed yield, components of yield, total above-ground biomass and harvest index of soybean grown at two CO_2 concentrations and three temperatures in 1987.

CO_2 conc. (m mol/mol)	Day/night temperature (°C)	Grain yield (g/plant)	Seed/plant (no./plant)	Seed mass (mg/seed)	Above-ground biomass (g/plant)	Harvest index
330	26/19	9.0	44.7	202	17.1	0.53
330	31/24	10.1	52.1	195	19.8	0.51
330	36/29	10.1	58.9	172	22.2	0.45
660	26/19	13.1	58.8	223	26.6	0.49
660	31/24	12.5	63.2	198	27.6	0.45
660	36/29	11.6	70.1	165	26.5	0.44

F-values					
CO_2 conc.	12.3**	11.4**	2.5*	NA	NA
Temperature	0.0 NS	8.4**	106.2**	NA	NA
CO_2 x Temperature	2.0 NS	0.1 NS	11.2**	NA	NA

*, ** Significant at the 0.05 and 0.01 probability levels, respectively.
NS = not significant; NA = *not* available.

Evapotranspiration and Water-use Efficiency

The direct effect of increasing temperatures across the range of 28 to 35°C appears to increase transpiration rate about 4 to 5% per °C, based on both experimental and modelling studies. This is in close agreement with the rise in saturation vapour pressure of about 6% per °C. Allen *et. al.* showed that the increase in WUE of soybean contributed by increased photo-synthesis under elevated CO_2 was much greater than the increase in WUE contributed by decreased transpiration. The fractional increase in WUE attributable to increased photo-synthesis and decreased transpiration were about 0.8 and 0.2, respectively.

RICE

Leaf and Canopy Photo-synthesis

Leaf photo-synthetic rates, in high light, of cv. IR72 rice, determined in 1992 for plants grown at three day/night temperature regimes, 32/23, 35/26 and 38/29°C (but measured at ambient temperature), averaged 18.8 and 30.4 μ mol $CO_2/m^2/s$ for 330 and 660 μ mol/mol CO_2 treatments, respectively.

There were essentially no differences in A between the two CO_2 treatments at each specific C. level. Similar pairs of curves were obtained for the other temperature treatments. Each pair of leaf A/C_i curves were similar within each temperature regime. Thus, the leaf A/C_i curves gave no indication of a change in photo-synthetic capacity. Extrapolations of the response curves to zero values of A gave an intercept (leaf CO_2 compensation point) of about 60 μ mol/mol CO_2.

Following canopy closure, the rice canopy net photo-synthetic rate (Pn) *vs.* photo-synthetic photon flux density (PPFD) responses were linear and did not approach light saturation in any of the experiments, probably because of the erect leaf orientation of cv. IR30 and high plant populations. Canopy Pn *vs.* PPFD at 60 DAP for the six CO_2 treatments of the RICE II experiment gave values of about 34, 50, 60, 80, 85 and 90 μ mol $CO_2/m^2/s$ at high light (1600 μ mol photons/m^2/s) for treatments of 160, 250, 330, 500, 660 and 900 μ mol/mol.

Linear canopy Pn responses to PPFD on day 60 of the RICE IV experiment were similar among all temperature treatments (25/18, 28/21, 31/24, 34/27 and 37/30°C) that were exposed to 660 μ mol/mol CO_2 responses of the 330 μ mol/mol treatment were about 25% less. Other studies of canopy Pn have shown only small differences across wide ranges of temperature; *e.g.*, cotton and soybean. There may be two reasons for the lack of a clear canopy photo-synthetic response to air temperature across the 25 to 37°C range. Evaporative cooling may lower foliage temperature below air temperature increasingly with increasing air temperature and increasing vapour pressure deficit. The summed response of the photo-synthetic rates of leaves to temperature at all exposures of light may broaden the temperature response for the whole canopy photo-synthetic rates.

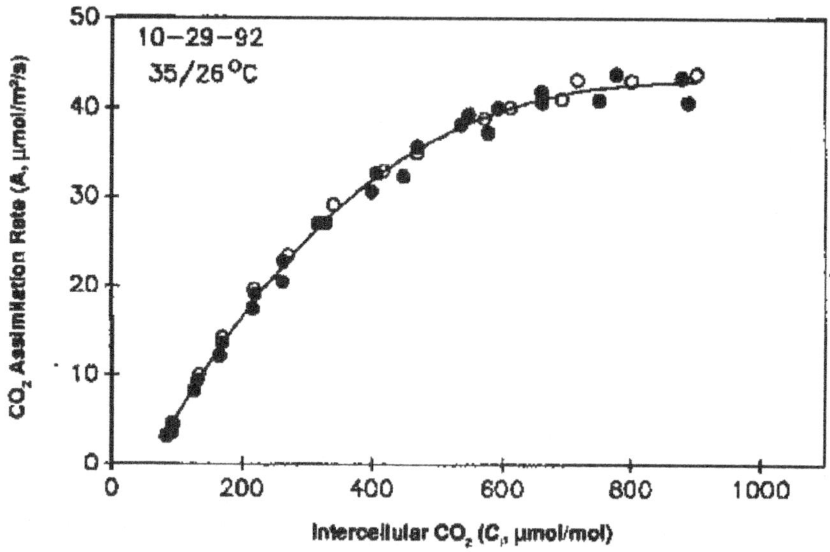

Figure 18.1 : Photo-synthesis CO_2 assimilation rate (A) as a function of intercellular CO_2 (C_i) for single, attached, fully expanded leaves of rice plants grown at CO_2 concentrations of 330 (open symbols) or 660 (filled symbols) μ mol/mol and dry bulb air temperatures (day/night) of 35/26°C. Measurements were made on 29 October 1992 at an irradiance of 1 200 to 1 300 μ mol (photons)/m^2/s.

Baker and Allen fitted whole-day canopy Pn data to a rectangular hyperbola of the form of equation (4.1) for DAP 61 of the RICE II experiment. The parameters were 70.8 μ mol/mol, 3.96 and -2.21, for K_m, Y_{max} and Y_i, respectively, with a relative Pn response ceiling of 1.75 at infinite [C]. The calculated percentage increase in response of whole-day canopy Pn at 660 *vs.* 330 μ mol/mol was 36%.

Photo-synthetic Acclimation to CO_2 at the Canopy Level

To test for acclimation of canopy photo-synthetic capacity, Baker *et. al.* used rice that was grown at 160, 250, 330, 500, 660 and 900 μ mol/mol CO_2. For half-day periods (mornings) on 62,63 and 64 DAP, common CO_2 setpoints of 160, 330 or 660 μ mol/mol were imposed on each chamber on those respective days. Within each of these short-term CO_2 exposure comparisons, Baker *et. al.* and Baker and Allen showed that the short-term canopy net photosynthetic rate decreased with increasing long-term CO_2 treatment. The relative effects of the short-term CO_2 exposure were greatest for the lowest short-term CO_2 concentration. For example, when compared at a common short-term CO_2 exposure of 160 μ mol/mol, the canopy photo-synthetic rate of the chamber containing the 900 μ mol/mol long-term CO_2 growth treatment was only about one-third of that of the 160 μ mol/mol long-term CO_2 growth treatment. However, a large part of this apparent acclimation effect may be attributed to the greater respiration rates of the plants that had been grown under elevated CO_2.

For rice, rubisco activity expressed on a leaf area basis decreased by 66% across the 160 to 900 μ mol/mol long-term CO_2 treatments.

Although Rowland-Bamford *et. al.* reported decreases in both amount and activity of rubisco with increasing CO_2 in rice plants grown across the range of 160 to 900 μ mol/mol, the leaf A/C. response curves gave no indication of a downward acclimation of photo-synthetic capacity across the smaller range of 330 to 660 μ mol/mol CO_2. One possibility would be that rubisco is not as limiting for photo-synthesis under CO_2 enrichment as would be expected. More work is needed, under a range of CO_2 treatments, to explore the interaction effects of sink capacity, nitrogen nutrition, and other internal CO_2-fixation processes on photo-synthetic behaviour and crop yield.

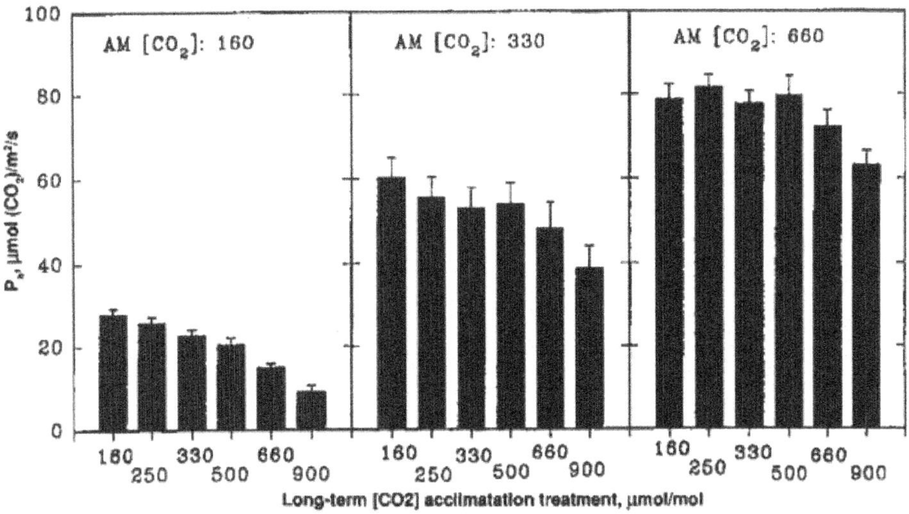

Figure 18.2 : Comparison of rice canopy net photo-synthetic rate (Pn) *vs.* long-term $[CO_2]$ acclimation treatment for rice canopies grown at sub-ambient (160, 250), ambient (330), and superambient (500, 660 and 900 μ mol CO_2/mol air) $[CO_2]$ treatments.

Table 18.5 : For soybean, leaf blade soluble protein expressed on a leaf blade area basis and percentage rubisco protein expressed on a leaf blade soluble protein basis for 34-day-old soybean plants grown under a wide range of CO_2 concentrations. Adapted from Campbell *et. al.* and Baker and Allen. For rice, leaf nitrogen content expressed on a leaf area basis and percentage rubisco protein expressed on a leaf soluble protein basis for 75-day-old rice plants grown under a wide range of CO_2 concentrations. Adapted from Rowland-Bamford *et. al.* (1991) and Baker and Allen.

Soybean			Rice		
CO_2 growth concentration (μ mol/mol)	Leaf soluble protein. (g/m²)	Rubisco protein (%)	CO_2 growth concentration (μ mol/mol)	Leaf nitrogen protein (μ mol/m²)	Rubisco protein (%)
160	2.5	56	160	95	62
220	3.2	54	250	90	59
280	2.6	-	330	81	54
330	2.3	57	500	62	49
660	2.3	54	660	78	43
990	2.3	55	900	64	42

Development Stages

In rice, two distinctly different rates of leaf appearance for vegetative and reproductive phases of growth were found, as has been observed many times before. Leaf appearance rates (leaves per day) are about twice as great in the vegetative phase as in the reproductive phase.

In the 'late' planted RICE II experiment, the number of mainstem leaves at panicle initiation and the final number of mainstem leaves decreased across CO_2 treatments of 160 to 500 μ mol/mol and remained similar from 500 to 900 μ mol/mol. Panicle initiation and boot stage occurred about 12 days earlier in the superambient CO_2 treatments compared with the 160 μ mol/mol treatment. Therefore, plant developmental rate was clearly accelerated with increasing CO_2 up to about 500 μ mol/mol.

Temperature effect data were assembled from treatments ranging from 25/18/21 to 40/33/37°C (day/night/paddy water temperatures) in RICE I to RICE V experiments. Air temperature greatly influenced leaf appearance rate, developmental rate, and total growth duration, whereas, across the limited range of 330 to 660 μ mol/mol the effects were comparatively small. No consistent differences in phyllochron interval (days per leaf appearance) between 330 and 660 μ mol/mol were observed, whereas phyllochron interval increased with increasing temperature across the range of 25/18/21 to 40/33/37°C, especially for the reproductive phase. In RICE III, IV and V experiments, anthesis ranged from 0 to 6 days earlier in the 660 *vs.* 330 μ mol/mol treatments. Also, the number of days to anthesis was shortened by approximately 10 days across the temperature treatment range from 25/18/21 to 34/27/31°C for the RICE IV experiment.

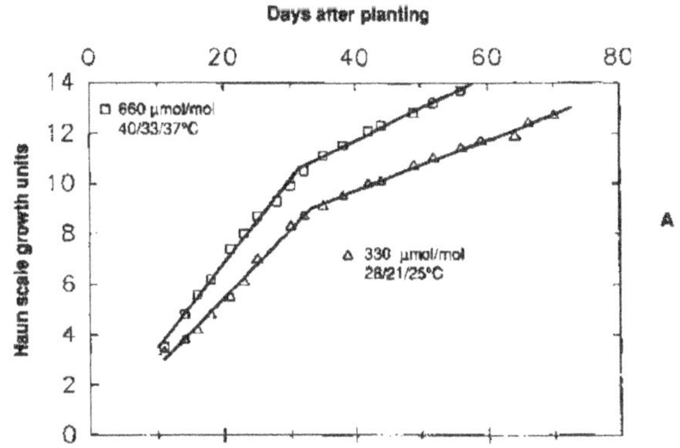

Figure 18.3 : (A) Mainstem Haun scale growth units *vs.* days after planting for rice plants in two different CO_2 and temperature regimes from the RICE III experiment.

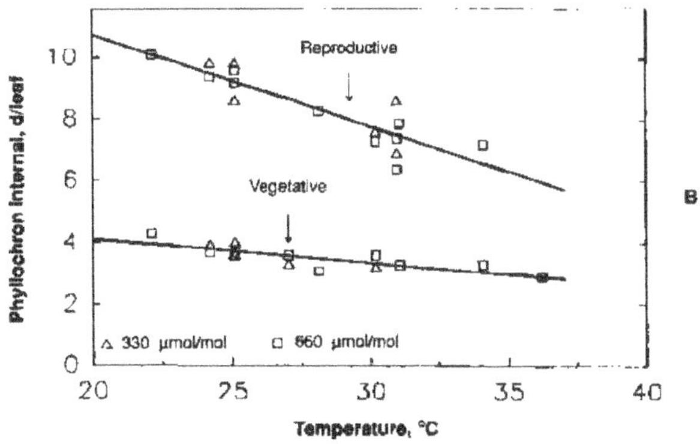

Figure 18.3. (B) Phyllochron interval *vs.* temperature treatment for all five rice experiments. 'Vegetative' data are plotted against paddy water temperature and 'reproductive' data are plotted against average day/night temperature adjusted for thermoperiod.

Growth and Yield

A rectangular hyperbola (equation (18.1)) was fitted to the seed yield data. Values of K = 284 μ mol/mol, Y_{max} =2.24 and Y_i= -0.13 were obtained for relative grain yield with asymptotic relative yield response ceiling (Y_{max} + Y_i) = 2.11 at infinite [C]. The calculated percentage increase in response at 660 *vs.* 330 μ mol/mol was 44% for grain yield.

Across a relatively wide temperature range from 25/18/21 to 37/30/34°C, Baker *et. al.* found a broad temperature optimum for biomass production in the mid-temperature ranges. Plants grown at 40/33/37°C were near the upper temperature limit for survival. High temperature spikelet sterility of rice is induced almost exclusively on the day of anthesis when temperatures greater than 35 °C for more than one hour induce a high percentage of sterility. In the 40/33/37°C treatments, plants in the 330 μ mol/mol CO_2 chamber died during internode elongation whereas plants in the 660 μ mol/mol chamber produced small, abnormally shaped panicles that were sterile. Therefore, elevated CO_2 may slightly increase the maximum temperature at which rice plants can survive.

At both CO_2 levels, grain yield was highest in the 28/21/25°C treatment followed by a decline to zero yield in the 40/33/37°C treatment. The CO_2 enrichment from 330 to 660 μ mol/mol increased yield by increasing the number of panicles per plant, whereas the number of filled grains per panicle and individual seed mass were less affected. Temperature effects on yield and yield components were highly significant. The number of panicles per plant increased while the number of filled grains per panicle decreased sharply with increasing temperature treatment. Individual seed mass was stable at moderate temperatures but tended to decline at temperature treatments above 34/27/31°C. Final above-ground biomass and harvest index were increased by CO_2 enrichment while harvest index declined sharply with increasing temperature. Notably, there were no significant CO_2 x temperature interaction effects on yield, yield components, or final above-ground biomass.

At each CO_2 concentration, polynomial regression equations were fitted to the rice seed yield (Y) vs. temperature (X) data of Table 18.7. A third-degree polynomial [$Y = -239.226 + 25.557 * X - 0.848 * X^2 + 0.00896 * X^3$ ($R^2 = 0.91$, $S_{y.x} = 1.36$)] provided the best fit for the 660 μ mol/mol CO_2 treatments. However, a second-degree polynomial [$Y = -1.196 + 1.042 * X - 0.0278 * X^2$ ($R^2 = 0.91$, $S = 1.36$)] fits the data best for the 330 μ mol/mol treatments. While future increases in atmospheric CO_2 should benefit rice yields, large negative effects are likely if temperatures also rise. The figures show that vegetative productivity is maintained at higher temperatures than is reproductive growth.

Table 18.6. Grain yield, components of yield, total above-ground biomass and harvest index of rice sub-ambient and superambient CO_2 concentration experiments conducted in 1987.

CO_2	Temperature [1]	Grain yield	Panicle/plant	Filled grain	Grain mass	Biomass	Harvest index	Experiment RICE No.
μmol/mol	°C	Mg/ha	no./plant	no./panicle	mg/seed	g/plant		
160	31/31/27	3.4c[2]	3.6c	24.8a	17.0a	4.0c	0.36a	I & II
250	31/31/27	4.1c	4.8bc	20.8a	18.2a	5.1bc	0.34a	I & II
330	31/31/27	4.8bc	5.7ab	21.0a	17.6a	6.3ab	0.34a	I & II
500	31/31/27	6.8[3]	7.3	23.0	18.1	9.8	0.30	II only
660	31/31/27	6.6ab	6.5ab	25.0a	17.8a	8.4a	0.35a[4]	I & II
900	31/31/27	7.3a	7.4a[4]	24.8a	17.8a	8.2a	0.39a	I & II

[1] Daytime dry bulb air temperature/night-time dry bulb air temperature/paddy water temperature.
[2] Values followed by the same letter in the same column are not significantly different by Duncan's Multiple Range Test (p - 0.05).
[3] Plants in 500 μ mol/mol CO_2 treatment of the RICE I experiment experienced hydrogen sulphide toxicity in the root zone which inhibited seed fill and reduced grain yield to levels below those measured in sub-ambient treatments. These data were not used in this analysis. The values in the 500 μ mol/mol CO_2 treatment row are from the RICE II experiment only.
[4] Indicates significant difference between planting dates as determined by the t-test at the 0.05 level of confidence.

Table 18.7. Grain yield, components of yield, total above-ground biomass and harvest index for five separate rice experiments. Adapted from Baker and Allen and Baker *et. al.*.

CO_2	Temperature [1]	Mean air temperature [2]	Grain yield	Panicle/plant	Filled grain	Grain mass	Biomass	Harvest index	Experiment No.
μ mol/mol	°C	°C	Mg/ha	no./plant	no./panicle	mg/seed	g/plant		
330	28/21/25	24.2	7.9	5.1	34.5	17.4	7.3	0.47	RICE III
	28/21/25	25.1	6.6	3.9	39.6	17.5	6.5	0.44	RICE IV
	28/21/25	25.1	8.0	4.0	47.5	18.5	8.1	0.43	RICE V
	31/31/27	31.0	5.2	5.9	23.0	17.1	5.5	0.42	RICE I
	31/31/27	31.0	4.3	5.4	19.0	17.9	7.2	0.26	RICE II
	34/27/31	30.2	4.2	7.7	15.2	16.2	5.6	0.43	RICE III
	40/33/37	36.2	0.0	-.-	-.-	-.-	-.-	-.-	RICE III
660	25/18/21	22.1	8.4	4.4	46.0	18.2	7.8	0.47	RICE IV
	28/21/25	24.2	8.4	5.0	37.7	18.0	7.9	0.46	RICE III
	28/21/25	25.1	10.4	4.2	58.3	18.3	8.9	0.50	RICE IV
	28/21/25	25.1	10.1	4.4	54.2	19.0	9.3	0.47	RICE V
	31/31/27	31.0	6.8	6.9	25.1	17.2	7.5	0.40	RICE I
	31/31/27	31.0	6.4	6.0	24.8	18.4	9.3	0.29	RICE II
	34/27/31	30.2	4.8	6.5	18.5	16.7	6.3	0.32	RICE III
	34/27/31	31.1	3.4	7.5	12.9	16.3	8.1	0.18	RICE IV
	37/30/34	34.1	1.0	8.0	3.0	14.2	7.1	0.06	RICE IV
	40/33/37	36.2	0.0	-.-	-.-	-.-	-.-	-.-	RICE III
F-values									
CO_2			4.2*	4.3*	0.4 NS	0.1 NS	13.1 **	4,6*	
Temperature			51.8**	22.6 **	51.1 **	27.0 **	3.2*	32,8 **	
CO_2 x temperature			1.9 NS	1.7 NS	1.3 NS	0.3 NS	0.2 NS	1.9 NS	

[1] Day time dry bulb air temperature/nighttime dry bulb air temperature/paddy water temperature.

[2] Mean air temperature is the average of day and night temperature adjusted for thermoperiod.

** * Significant at the 0.01 and 0.05 probability levels, respectively. NS = Not significant.

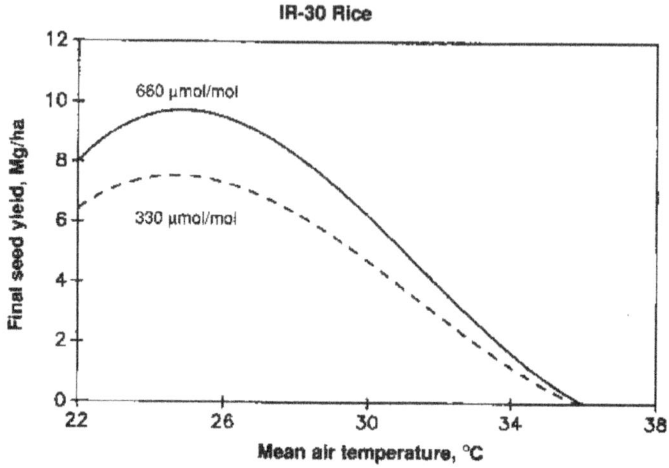

Figure 18.4 : Rice seed yield *vs.* weighted mean day/night air temperature for plants grown to maturity in CO_2 concentrations of 330 and 660 μ mol/mol in five separate experiments.

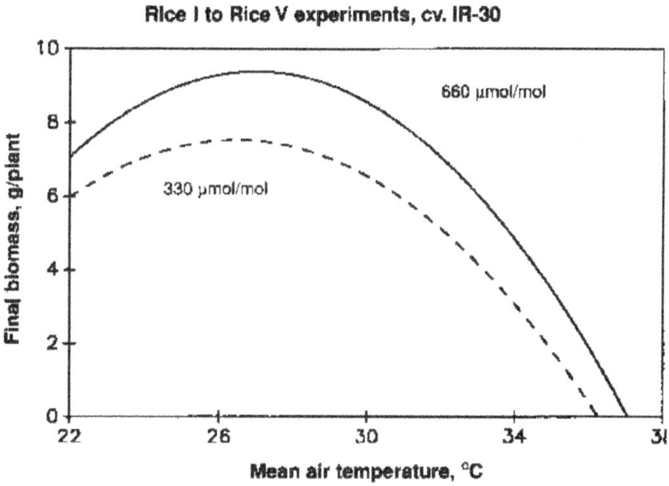

Figure 18.5 : Second-degree polynomial fit of rice final biomass yield *vs.* weighted mean day/night air temperature for plants grown to maturity in CO_2 concentrations of 330 and 660 μ mol/mol in five separate experiments

Canopy Dark Respiration Rates

Plant respiration rates may be decreased by both short- and long-term exposure to high CO_2 concentrations; however Baker *et. al.* found that night-time canopy dark respiration rates [R_d, μ mol (CO_2)/m^2 (ground area)/s] increased for rice exposed to day time CO_2 ranging from 160 to 900 μ mol/mol. Similar to photo-synthetic rates, R_d increased with CO_2 exposure from

160 to 500 μ mol/mol but levelled off somewhat across the 500 to 900 μ mol/mol range. The R_d of the ambient and super-ambient CO_2 treatments reached a broad maximum around 30 to 50 DAP whereas the broad maximum of the sub-ambient treatments occurred later around 50 to 70 DAP. The maximum values of R_d were about 6, 8, and 9 μ mol/m²/s for the 160, 250 and 330 μ mol/mol CO_2 treatments and about 11 to 12 μ mol/m²/s for the three superambient CO_2 treatments.

Specific respiration rate [R_{dw}, μ mol (CO_2)/s/kg (total above-ground dry matter)] decreased exponentially with DAP at all CO_2 exposures and was higher in the sub-ambient (160 and 250 μ mol/mol) than in the ambient (330 μ mol/mol) and super-ambient (500, 660 and 900 μ mol/mol treatments). At each CO_2 exposure level, the patterns of R_{dw} with DAP were very similar to the patterns of plant tissue nitrogen concentration with DAP. Furthermore, R_{dw} was linearly related to total above-ground plant tissue nitrogen concentration [mg (N)/g (DW)] across the range of CO_2 exposures and the six dates of plant sampling (R_{dw} = -13.0 + 0.952 * [N]; r2=0.9, P=0.01). Baker *et. al.* concluded that the CO_2 treatments affected R_{dw} mainly by altering the protein content of the plant tissue. Another explanation is that elevated CO_2 increased the amount of structural and non-structural carbohydrates in the plant tissues, so that a larger proportion of dry matter was sequestered in non-protein materials. This study showed no respiration acclimation to long-term CO_2 enrichment (indirect acclimation) of rice that could not be explained by nitrogen concentration of the plant tissues.

Evapotranspiration and Water-use Efficiency

Both plant transpiration and direct evaporation from the flood-water surface contribute to water use (evapotranspiration, ET) of rice growing in the SPAR chambers. Diurnal trends of ET followed diurnal patterns of solar irradiance. After canopy closure, maximum ET rates and total day time water losses were about 35 and 30% greater, respectively, from rice grown at 160 μ mol/mol compared to rice grown at 900 μ mol/mol CO_2. The ET rates were similar when all chambers were exposed for one-half day to the same CO_2 concentration, which demonstrates the effect of the exposure level of CO_2 on stomatal control of transpiration.

Temperature also has a large effect on diurnal ET rates and day time water use, mediated primarily through vapour pressure deficit of the air. Solar irradiance also has a large effect, probably through both the energy inputs to the canopy and through the stomatal opening response to light (directly or indirectly).

Water-use Efficiency : Effects of CO_2

Day time totals of CO_2 uptake, ET, and calculations of WUE *vs.* CO_2 treatment are shown in Table 18.8. Stomatal conductance decreases with increasing CO_2 concentration which can cause a reduction of both leaf and whole canopy transpiration. However, CO_2 enrichment may also increase canopy leaf surface area for transpiration, thereby off-setting some of the water savings. The leaf area index of the RICE II experiment ranged from 7.6 to 10.8 across the CO_2 treatments from 160 to 900 μ mol/mol. Calculations of WUE increased with increasing CO_2 due to the decline in ET across this CO_2 range and the increase in Pn with CO_2 up to the 500 μ mol/mol treatment.

Table 18.8 : Comparison of total daytime responses of rice canopies grown season-long in various CO_2 concentration (RICE II, 23 August, 1987, 61 days after planting) and temperature (RICE IV, 10 September 1989, 58 days after planting) treatment regimes. Day time dry bulb air temperature/night time dry bulb air temperature/paddy water temperature : dewpoint temperature treatments are given in the second column. Adapted from Baker *et. al.* and Baker and Allen.

CO_2	Temperature [1]	Total daytime photons	Total daytime CO_2 uptake	Total daytime H_2O loss	Water-use efficiency [2]
m mol/mol	°C	mol(Photons)/m²	mol(CO_2)/m²	mol(H_2O)/m²	μ mol/mol
	Effects of carbon dioxide : RICE II, 23 August 1987, 61 DAP				
160	31/31/27:18	39.4	0.74	608.5	1.22
250	31/31/27:18		1.13	643.4	1.76
330	31/31/27:18		1.34	611.0	2.19
500	31/31/27:18		1.80	553.6	3.25
660	31/31/27:18		1.79	536.2	3.34
900	31/31/27:18		1.83	469.5	3.90
	Effects of temperature : RICE IV, 10 September, 1989, 58 DAP				
330	28/21/25:12.0	26.4	0.87	612.7	1.42
660	25/18/21:10.5		0.83	359.3	2.30
660	28/21/25:12.0		1.13	475.8	2.37
660	31/24/28:13.5		0.83	491.3	1.69
660	34/27/31:15.0		0.89	736.4	1.20
660	37/30/34:16.5		0.97	909.4	1.06

[1] Daytime dry bulb air temperature/night time dry bulb air temperature/paddy water temperature : dewpoint temperature.
[2] mmol CO_2/mol (H_2O).

Water-use Efficiency : Effects of Temperature

In the 28°C air temperature treatment, CO_2 enrichment from 330 to 660 μ mol/mol resulted in 30% increase in day time total CO_2 uptake and 22% decrease in ET with a concomitant increase of WUE from 1.42 to 2.37 μ mol (CO_2)/mmol (H_2O), an increase of 67%. Daytime ET was more than doubled from the 25 to the 37°C daytime temperature treatment. WUE declined from the 28 to the 37°C daytime treatment due to the sharp increase in ET and the relatively stable Pn across this temperature range. If WUE were based on grain yield, it would decrease drastically with increasing temperature not only because of increasing ET, but also because of sharply decreasing seed production.

Rising atmospheric CO_2 is likely to benefit rice production by increasing photo-synthesis, growth and grain yield while reducing water use and increasing WUE. In warm areas of the world, however, possible future global warming may result in substantial yield decreases because of the sensitivity of flowering and seed set to high temperatures and the possibility of water shortages that may result from increased evapotranspiration.

Summary of Comprehensive Reviews

Several recent symposia proceedings and reviews leave little doubt that crop plants can respond well to elevated CO_2. Poorter compiled information from 156 plant species and found that doubling CO_2 provided an average growth increase of 37%. Furthermore, the fast growing wild species responded more strongly than slow-growing wild species (54 *vs.* 23%).

Poorter imposed two restrictions on his compilations that may have led to larger than expected responses to elevated CO_2. Firstly, plants grown in competition were not included. Secondly, only vegetative stages of plants were compared since compiled data were selected prior to flowering.

A number of studies have shown that vegetative growth responses may be greater than reproductive (seed yield) responses. Therefore, the compiled data of Poorter may give an impression of greater response than would be observed throughout the life cycle. Secondly, crops in field conditions usually are grown in dense populations where they compete for space and light. Under more realistic field conditions, crop plants are likely to respond as a community rather than individual plants, wherein light (solar radiation) becomes a limiting factor for growth. Under these conditions, elevated CO_2 cannot promote horizontal expansion and greater light capture. Although the actual field responses may be less, the CO_2 fertilization effect is clearly well-established.

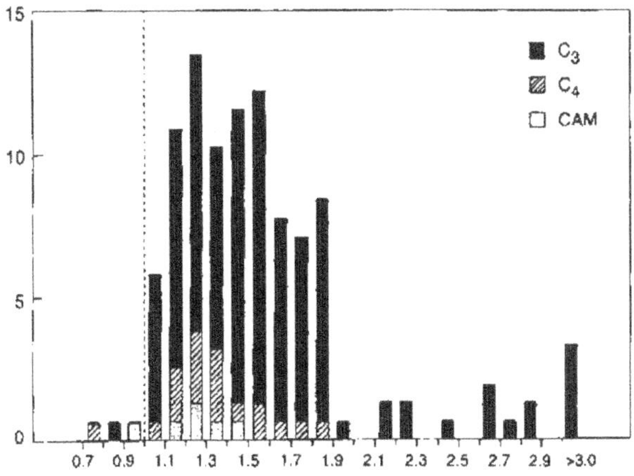

Figure 18.6 : Distribution of the biomass growth modification ratio (weight ratio) of CO_2-enriched plants (600 to 720 μ mol/mol) in comparison with control treatments (300 to 360 μ mol/mol).

The CO_2 fertilization effect for forest species has also become firmly established. Wullschleger *et. al.* estimated the biotic growth factor for 58 controller-exposure studies of forest tree species which included 398 observations. They also found a mean response ratio of 1.32. Of the 398 observations, 51 showed a relative growth response less than 1, and 31 showed responses greater than 2. However, under competitive conditions, tree response may be much less.

Kimball *et. al.* discussed the data of Idso *et. al.* which related the growth modification factor (or biomass growth modification ratio) caused by a 300 μ mol/mol increase in CO_2 concentration above ambient. Thus, the *vegetative* response to CO_2 should be enhanced at increased temperatures. As discussed before, this growth modification factor may not apply for reproductive growth responses (seed yield) of crops (*e.g.*, rice).

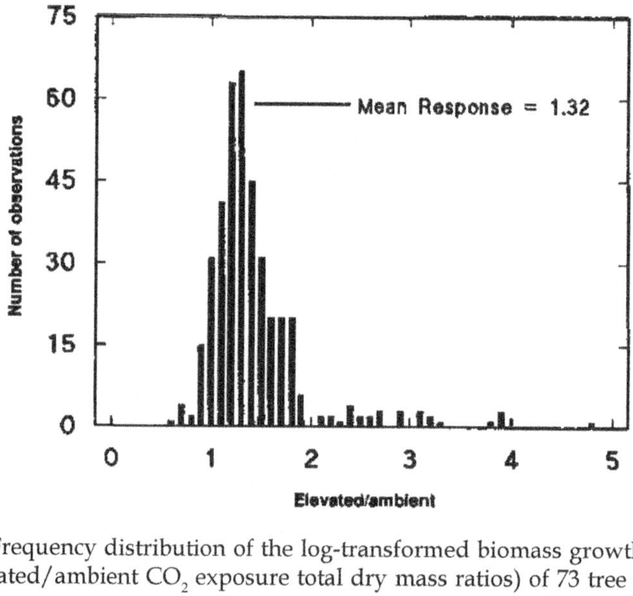

Figure 18.7 : Frequency distribution of the log-transformed biomass growth modification ratio (log elevated/ambient CO_2 exposure total dry mass ratios) of 73 tree species.

CROP MODELLING : PREDICTIONS FOR THE FUTURE

Modelling Crop Responses to CO_2 and Climate Changes

Peart *et. al.* and Curry *et. al.* predicted growth and yield responses of soybean and maize to doubled-CO_2 climate change scenarios of the southeastern USA. Their simulations used 30 years of baseline weather data from 19 sites in 11 states. Predicted climate changes of the Goddard Institute for Space Studies (GISS) model and the Geophysical Fluid Dynamics Laboratory (GFDL) model were used to change temperatures, precipitation and solar radiation, month-by-month, for the 30 years of baseline data at each site.

These modified baseline weather data provided GISS and GFDL climatic change scenarios. At each site, prevailing cropped soil types, planting dates, and cultivars were used in the simulations.

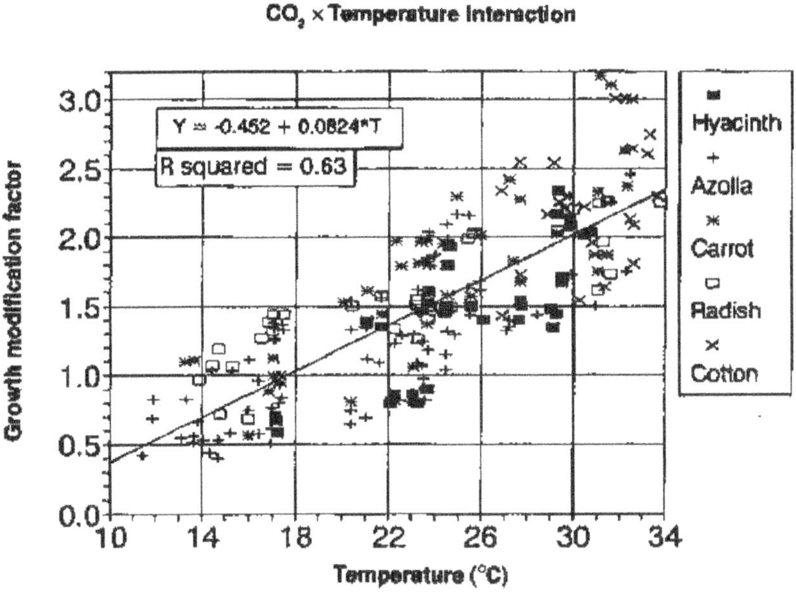

Figure 18.8 : Biomass growth modification ratio (growth modification factor or relative increase in biomass growth) resulting from a 300 μ mol/mol increase in CO_2 concentration above ambient (almost doubled) versus mean daily air temperature for the plants indicated in the legend.

Baseline, GISS and GFDL scenario temperature and rainfall data that were used by Peart *et. al.* were condensed for two representative locations; Columbia, South Carolina, and Memphis, Tennessee. Crop yield responses to climate change were simulated under four conditions : with or without direct CO_2 fertilization effects, and under rainfed or optimum irrigation culture. The Penman-Monteith equation, which contains a term for canopy conductance, was used to compute the effects of elevated CO_2 on canopy transpiration. Simulations were run for doubled-CO_2 conditions with a crop photosynthetic enhancement factor of 1.35 for soybean (a C_3 plant) and 1.10 for maize (a C_4 plant). No simulations were run with CO_2 fertilization effects only (without climate change effects).

Simulated soybean seed yields were averaged over 30 years and 19 sites. Under rainfed conditions with climate change effects only, simulated soybean yields under the GFDL scenario were reduced 71 % compared to the baseline climate, whereas the average yields were reduced only 23% under the GISS scenario. Yields under the GFDL scenario were severely impacted because of the rainfall reductions predicted by this model.

Table 18.9 : Summary of 30-year mean (1951-1980) baseline (BASE) weather data of two locations with GISS and GFDL scenarios. Condensed from Peart *et. al.*

	BASE			GISS			GFDL		
	Prec. (mm)	T_{max} (°C)	T_{min} (°C)	Prec. (mm)	T_{max} (°C)	T_{min} (°C)	Prec. (mm)	T_{max} (°C)	T_{min} (°C)
Columbia. SC									
APR-SEP; Jul	684	33.3	21.2	817	35.2	23.1	471	38.2	26.1
OCT-MAR; Jan	561	13.5	0.7	602	15.8	3.0	571	15.6	2.8
TOTAL; Mean	1245	24.1	10.7	1419	26.6	13.2	1042	27.3	13.9
Memphis. TN									
APR-SEP; Jul	656	33.1	22.6	748	35.4	24.9	549	36.0	25.5
OCT-MAR; Jan	654	9.0	-0.6	565	12.0	2.4	758	11.3	1.7
TOTAL; Mean	1 310	22.0	11.1	1313	25.4	14.5	1307	24.8	13.9

Under optimum irrigation conditions, average yields under both the GISS and GFDL scenarios were reduced 18 to 19%. However, in spite of higher temperatures, the irrigated yields under the GISS and GFDL scenarios were about 25% greater than the baseline climate scenario without irrigation.

When CO_2 fertilization effects were included with climate change effects, simulated average yields of the GISS climate scenario were increased 11% under rainfed conditions, whereas yields under the GFDL climate scenario were still decreased (-52%). Under optimum irrigation with CO_2 fertilization effects, yields under both GISS and GFDL scenarios were increased 13 to 14%.

Yields of maize were simulated for doubled-CO_2 fertilization effects at only four weather station locations (Charlotte, North Carolina; Macon, Georgia; Memphis, Tennessee; Meridian, Mississippi). For climatic change effects alone, predicted maize yields declined only 6% in the GISS scenario, but declined 73% in the GFDL scenario. Although irrigation increased predicted crop yields, the GISS and GFDL climate scenarios gave yield decreases of 18 and 27%, respectively, with respect to the irrigated baseline weather. The yield reduction of the GFDL scenario with respect to the GISS scenario was 10%, attributable to slightly higher temperatures. Including the CO_2 fertilization effects with climatic change scenarios had little effect on the predicted yields of maize because it is a C_4 plant.

In the Great Lakes and Corn Belt Area, simulations by Ritchie *et. al.* showed that higher temperature would have the greatest effect of the climate change factors on predicted yields of soybean and maize, mainly through decreases in the crop life cycle. For the southern part of this region, yield reductions were greater for the GFDL scenario. Predicted yields increased for the northernmost stations because temperatures and growing season duration became more favourable. Average irrigation water requirements in this region increased about 90% for the GISS and GFDL scenarios.

Rosenzweig modelled maize and wheat yields in the Great Plains under GISS and GFDL scenarios. Yields decreased in the Southern Great Plains because higher temperatures shortened the life cycle of the crops. Where precipitation was predicted to decrease, irrigation requirements increased. In a separate modelling study, Allen and Gichuki predicted a 15% increased requirement for irrigation for this region, with greater requirements for alfalfa because its growing season was increased, and lower requirements for maize and winter wheat because their growing seasons were decreased. The CO_2 fertilization effect off-set the adverse effects of climate change at some locations of Ritchie *et. al.* and Rosenzweig.

Table 18.10 : Doubled CO_2 soybean yield simulations (SOYGRO) for the south-eastern USA.

BASE	GISS Model		GFDL Model		Model
Yield (kg/ha)	Yield (kg/ha)	% Diff.	Yield (kg/ha)	% Diff.	% Diff.
Climate Change Effects Only, Rainfed					
2497	1 929	-23	733	-71	-62
Climate Change Effects Only, Irrigated					
3 837	3 158	-18	3092	-19	- 2
CO_2 Fertilization plus Climate Change Effects, Rainfed					
2497	2780	+11	1 206	-52	-57
CO_2 Fertilization plus Climate Change Effects. Irrigated					
3837	4350	+13	4393	+14	+1

Adapted from Peart *et. al.*.

Table 18.11 : Doubled CO_2 maize yield simulations (CERES-Maize) for Charlotte NC, Macon GA, Meridian MS, and Memphis TN.

Base	GISS model		GFDL model		Model
Yield (kg/ha)	Yield (kg/ha)	% Diff.	Yield (kg/ha)	% Diff.	% Diff.
Climate Change Effects Only, Rainfed					
8468	7926	-6	2289	-73	-71
Climate Change Effects Only. Irrigated					
13,899	11 455	-18	10257	-26	-10
CO_2 Fertilization plus Climate Change Effects. Rainfed					
8577	8 136	-5	2224	-74	-73
CO_2 Fertilization plus Climate Change Effects. Irrigated					
14052	11 545	-18	10363	-26	-10

Adapted from Peart *et. al.*.

Dudek predicted productivity changes of several Californian vegetable, fruit and nut crops in response to GISS and GFDL scenarios for doubled CO_2. Without CO_2 fertilization effects, statewide average yield changes, depending on the crop, would be -8 to -34% for the GISS scenario and -6 to -31 % for the GFDL scenario. With CO_2 fertilization effects, predicted state-wide yield changes ranged from about +17 to -12% for the GISS scenario and about +21 to -8% for the GFDL scenario.

Dudek used a California Agriculture and Resource Model (CARM) to further predict economic and market impact of the productivity changes. Production declined, in general, under the scenarios of climate change without CO_2 fertilization effects; however, commodity prices generally increased. Under the CO_2 fertilization plus climate change effects scenarios, predicted impacts on commodity prices were much less.

The effect of temperature on the phenology of crop plants plays a critical role. One crucial need is for more detailed research on responses of plants to temperature, and temperature x CO_2 interactions, as inputs to crop models. More modelling studies are also needed on different planting dates as an adaptive strategy. Cultivars need to be designed for future climatic conditions (Hall and Allen, 1993). Factors that should be considered are : extension of the grain-filling period and perhaps shortening the duration of vegetative growth (which would also improve harvest index); ability to flower and set seed at higher temperatures; photoperiod and thermoperiod adaptive interactions; selection for positive photo-synthetic acclimation where negative photo-synthetic acclimation has been observed; and capability of utilizing photo-assimilates (storage carbohydrates) more effectively. These factors need to be integrated into whole-plant physiology under real-world conditions.

ADAPTATIONS AND EVAPOTRANSPIRATION REQUIREMENTS

The simulated crop yield responses to climatic changes provided by Peart *et. al.* and Curry *et. al.* manifest two main points : (a) the serious adverse impact of inadequate rainfall scenarios on crop production coupled with rising temperature scenarios, and (b) the importance of beneficial CO_2 fertilization effects in the face of elevated temperatures. However, the climate change for an effective doubling of CO_2 may occur at CO_2 concentrations less than those used in this simulation, if radiatively active trace gases other than CO_2 play a large role in the greenhouse effect. All of the simulations assumed that climatic changes would occur simultaneously with increasing concentrations of CO_2 and other trace greenhouse-effect gases. If global warming lags the increases of atmospheric CO_2, then some beneficial effects of CO_2 fertilization are likely to occur before the full impact of climate change is manifested. However, Broecker and others caution that climate changes have not always been gradual during interglacial periods of the Pleistocene Era. There is clear evidence of relatively rapid climatic oscillations in the northern hemisphere during the previous interglacial period 110000 to 140000 years before present based on Greenland Ice-core Project (GRIP) records. These oscillations produced cold periods that were as severe as the proceeding glacial period.

Much of the reduction in soybean yields reported by Peart *et. al.* and Curry *et. al.* was caused by decreases in the length of the grain-filling period under higher temperatures. Changes in management practices, such as changing planting dates or selection of other cultivars, may help to prevent some of the potential reductions in yield. In the future, plant breeders may need to adapt combinations of temperature tolerance and photo-period responses into new germplasm. In situations where non-structural carbohydrates accumulate as a CO_2 fertilization effect response, new germplasm needs to be developed that can make better use of the photo-assimilate source.

Irrigation is not likely to be a panacea for climate change. Predicted irrigation requirements for soybean in the south-eastern USA were increased 33 and 134% under the GISS and GFDL scenarios, respectively, based on simulations of Peart *et. al.* and Curry *et. al.*. However, under the GFDL scenario, water resources would become scarce, and may not be readily available for crops. Some areas of the USA may have to adapt by irrigating less land area. Increasing temperatures and decreasing precipitation for the USA as predicted by the GFDL model would have a serious negative impact overall on agricultural productivity although producers in favourable regions may benefit from scarcity-mediated higher prices.

ASSESSING INTERNATIONAL IMPACTS OF RISING CO_2 AND CLIMATE CHANGE

Several assessments have been conducted on the impacts of rising CO_2 and global climatic changes on crop production patterns and economics responses within national or regional zones and in various countries around the world. Furthermore, progress has been made on predicting impacts of climate change using world trade models.

The studies conducted by Rosenzweig and colleagues used existing crop models and climate change scenarios from three GCMs to predict changes in crop yields for a number of countries around the world. The doubled-CO_2 climate change scenarios (temperature, rainfall and evaporation changes) were based on the climate change potential expected from increases of all greenhouse-effect gases. These climate changes are expected to occur well before CO_2 concentration has actually doubled. Therefore, the CO_2 levels for fertilization effects were estimated to be 555 μ mol/mol rather than 660 μ mol/mol. Thus, the CO_2 fertilization effect photo-synthetic ratios for four crops (soybean, wheat, rice and maize) were taken as 1.21, 1.17, 1.17 and 1.06, respectively.

The impact of climate change scenarios was more severe in the tropical latitudes than in the mid- or high-latitudes. For example, averaged over all three GCM scenarios, wheat production changes predicted in Brazil, India,

China and Canada were -47, -43, -11 and -20%, respectively, without CO_2 effects, and -28, -13, +8 and +16% with CO_2 fertilization effects.

Simulations of soybean yields throughout the USA were a part of this international study. They found that aggregated yields were 2.42, 2.80, 2.37 and 1.31 Mg/ha for the BASE, GISS, GFDL and UKMO scenarios, respectively. Temperatures were about 4.0°C higher than BASE for the GISS and GFDL scenarios, but were about 5.2°C higher for the UKMO model.

Refinements and improvements in prediction methodology will continue, but these assessments provide the best currently available insights into the CO_2 fertilization and climate change effects on global crop productivity.

THE EFFECTS OF ELEVATED CO_2 AND TEMPERATURE CHANGE ON TRANSPIRATION AND CROP WATER USE

The projected climatic effects of the continuously increasing concentrations of CO_2 and other radiatively active trace gasses in the atmosphere have caused concern over the last decades and increasingly attracted scientists' and policy-makers' attention. The expected changes at a global level will be reflected in changed weather conditions in the growing season at regional and local levels that directly affect agriculture and natural vegetation. The exchange of water and energy is determined both by the climate and by the gas exchange properties of the vegetation. Feedback on land cover, changed vegetation properties and local and regional climate is therefore of prime importance.

So far, the General Circulation Models (GCMs) are not able to give accurate predictions of climate change with great geographical detail. Also, feedback processes by the biosphere are still poorly represented in the models. These shortcomings limit predictions at the regional level that are needed in impact studies for agricultural production. Qualitative changes in potential production may well be estimated on the basis of the scenarios, but quantitative changes in the actual production, including the limitations caused by the changed variability of weather conditions (daily and seasonal amplitudes of temperatures, seasonal patterns in precipitation, cloudiness) remain difficult to estimate.

In addition to changes in precipitation and total water availability for irrigation, that directly affect agricultural production, changes in the pattern of water use by crop plants throughout the season may affect the outcome. Of special concern is the change in the physiological functioning of the vegetation as a consequence of the changed atmospheric composition. Most plants react to the changed atmospheric CO_2 concentration with changed stomatal response, and not only is growth affected but also the transpiration. The complex nature of the physiological response in interaction with micrometeorological processes at the leaf and canopy level requires further attention.

This paper gives an indication of the state of the art regarding the effects of elevated CO_2 on stomatal behaviour, transpiration, water-use efficiency and total water-use. It briefly touches upon the interaction between vegetation and atmosphere at a larger scale. Aspects related to these effects at the level characteristic for a doubling of the present atmospheric CO_2 concentration [CO_2] are discussed. Changes in atmospheric concentration will not occur overnight but are expected to lag behind relative to changes in greenhouse gas concentration. Therefore uncertainties remain not only on the expected time course of modifications and adaptations, but also on the level at which new threats and opportunities arise in future conditions.

WATER USE AND WATER-USE EFFICIENCY

Definitions

It should be realized that various expressions for water-use efficiency are used, and that the physiological and agronomic definitions may differ. Time-averaged water use and water-use efficiency (seasonal or periodic whole-plant transpiration rate (WPTR), and water-use efficiency (WUE)) are often confounded with instantaneous transpiration rate and transpiration efficiency (ITE). The latter expression is best described as the ratio between the instantaneous CO_2 fixation (actual net photo-synthesis) and water loss by transpiration (both measured with a leaf or plant chamber). Water-use efficiency is the ratio of the net gain in dry matter over a given period, divided by the water loss (from the vegetation alone or from soil and vegetation together) over the same period. Mostly the ITE, normally only measured during the light period, is higher than WUE, both because of variations in ITE over day and season and because of carbon losses by respiration in the dark. WUE seems to be the more relevant parameter to the question of impact on agricultural production.

Interactions with Experimental Conditions

Most available data on water use and water-use efficiency have been collected from single plants grown in pots or hydroponically, and using porometer measurements supplemented with periodic weight determinations. Values for water-use efficiency and total water use in (semi-) field conditions over the full season, or an extended period, are scarce. Moreover, a complication for the correct estimation of the effect of elevated [CO_2] and temperature changes on transpiration is that almost all experiments in this field have been performed in environmentally controlled and generally well mixed and ventilated experimental set-ups (enclosures or Open-Top Chambers (OTCs)), where the indirect effects may not show up so prominently as in a real and outside future climate. Similar interpretation problems have been met in the evaluation of air pollution research.

Instantaneous transpiration efficiency (ITE)
calculated from photosynthesis/gas exchange

$$ITE = \frac{\text{mmol } CO_2}{\text{mol } H_2O} \quad \text{instantaneous}$$

(at leaf or canopy level)

Water-use efficiency (from gas exchange data)
integrated at leaf or canopy level, including respiration loss

$$WUE = \frac{(\text{Net mg } CO_2)}{(\text{g } H_2O)} \quad \text{per day or extended period}$$

(per m^2 leaf or soil surface)

Crop water-use efficiency / Seasonal water use efficiency
integrated over (part of) the season

$$(C)WUE = \frac{(\text{g DM})}{(\text{g H2O})}$$

(of canopy per m^2, with or without soil evaporation)

Crop / Plant water use
integrated over (part of) the season

$$WU = \text{g } H_2O \text{ per plant or per } m^2$$

(incl. or excl. soil evaporation)

Figure 18.9 : Different expressions for the relation water use - carbon assimilation and growth.

RESPONSE AT THE LEVEL OF STOMATA

STOMATAL DENSITY

The plant leaves lose water primarily by evaporation through the stomata. The stomatal density depends upon plant species, and can be related to the plant-ecotype (between 300 and 800 stomata/mm. Woodward correlated the decrease in the stomatal density over time, observed in herbarium leaves collected over the last centuries, with the rising CO_2 concentrations and concluded from the shift in d C that the water-use efficiency concomitantly has improved. The nitrogen content in the leaves had dropped, in line with most data from elevated CO_2 experiments. Experimentally, an increase in $[CO_2]$ up to about 310 µ l/l decreased the stomatal density, but sometimes no effect is found above this $[CO_2]$. This matter is still under dispute although such a correlation has also been confirmed for palaeo-records. Among species large differences in response of stomatal density to elevated $[CO_2]$ seem to exist. Experiments with a range of CO_2 concentrations (160-900 µ l/l) showed an increase of stomatal density of rice and bean leaves, with a differential effect at abaxial (increasing) and adaxial sides. At subambient CO_2 concentrations stomatal density dropped. This is in contrast with the findings of Oberbauer *et. al.* for tropical trees. The relative effect of changes in stomatal density and decreasing stomatal aperture at elevated [CO_2] for the water relations, however, have not been evaluated. A gradual change in [CO_2] over the next century may lead to a natural selection that favours cultivars having a lower stomatal density, especially for water-limited growth conditions. It should, however, be realized that other environmental factors like salt stress can also modify stomatal density. Whatever the net effect, the resulting stomatal conductance is primarily determined by stomatal functioning, and much less by density.

Stomatal Functioning

In the pathway from the stomatal cavity to the leaf surface, and from ambient air to the photo-synthetic machinery in the mesophyll, the stomata are a major resistance for gas transport between the leaf and the surrounding air. A change in gas exchange resistance of the stomatal pores therefore affects the entrance of CO_2 and even more the exit of water vapour. The opening status of the stomata is a compromise between water loss and uptake of CO_2 from ambient air. In line with this proposition, stomatal response to elevated concentrations of CO_2 (C_a) in general is reflected in partial stomatal closure. The mechanism behind this stomatal closure is not yet clear. The observations are in line with the notion that plants tend to regulate the internal CO_2 concentration (C_i) in the substomatal cavity such that for a given

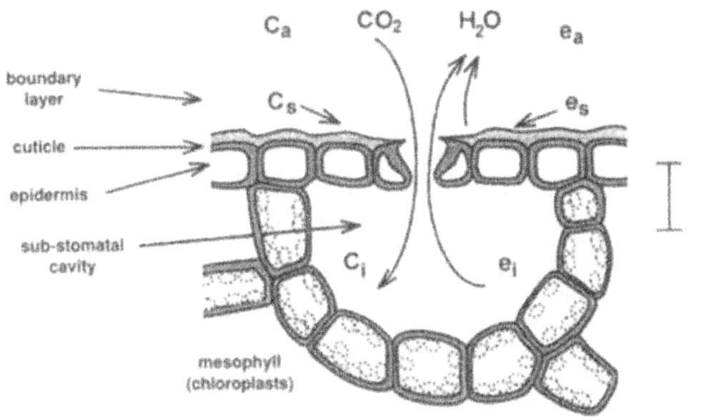

Figure 18.10 : Schematic cross-section of a stoma of a leaf showing the pathway of CO_2, and H_2O in the light. C_i, C_s, C_a : internal, surface and ambient CO_2 concentration; e_i, e_s, e_a : internal, surface and ambient air humidity. Bar indicates 100 μ m.

water vapour deficit there is a constant ratio (C_i/C_a) with the atmospheric concentration. Such a regulation would directly lead to the partial closure at elevated CO_2, as observed in many studies using porometers. Jackson *et. al.* measured photo-synthesis and water relations of native grassland species and calculated C_i/C_a for C_3 and C_4 plants. They confirmed the conservation of the value, with only a small (not significant) tendency to rise with increasing [CO_2]. The rate of photo-synthesis and therefore the required supply of carbon dioxide is directly coupled to light intensity. In line with this, the conductance of stomata is also highly correlated to light. This relation can be modified by environmental conditions like drought or stress by air pollution. The ratio C_i/C_a in stationary conditions is about two-thirds for C_3 but about one-third for C_4 plants. The lower value for C_4 plants reflects the higher affinity for CO_2 of the C_4. photo-synthetic pathway, and reflects the more efficient water use for these plants.

Acclimation of Stomatal Movement

Little is known about the acclimation of the stomatal movement to long-term exposure to elevated $[CO_2]$. Some studies suggest that the lower stomatal conductance not only persists over extended exposure periods to elevated $[CO_2]$, but is also conserved after subsequent lowering of the $[CO_2]$ (Gorissen, pers. comm.). For glasshouse horticulture it was reported that the stomatal conductance of high-$[CO_2]$ tomato plants was less sensitive to short-term $[CO_2]$ fluctuations, and higher than that of plants grown at ambient concentrations and measured at double present $[CO_2]$. This may indicate that the sensitivity of stomatal movement of high $[CO_2]$ plants to changes in $[CO_2]$ is reduced, but still a lower stomatal conductivity exists when compared to leaves growing and measured in present $[CO_2]$ conditions.

EFFECTS AT THE LEVEL OF THE LEAF

The partial closure of stomata is reflected in the reduced conductance at the leaf level. At a CO_2 concentration double the present, conductance is reduced by 30-40%, although large differences among species exist.

Leaf Temperature and Vapour Pressure Deficit

Water loss by transpiration is not only affected by the conductivity of the stomata, but also by the driving forces for exchange of the water vapour from the leaf surface to the surrounding atmosphere. Therefore the gradient in partial pressure of the water vapour at the leaf surface is also of importance. All other factors being equal, the existing vapour pressure deficit (VPD) between stomatal cavity and surrounding air, the boundary layer, will increase at a reduced transpiration rate, and feed back to stimulate transpiration. The reduced transpiration will cool the leafless, and consequently a rise in temperature in the stomatal cavity and at the leaf surface may occur. Thus, in addition to the global greenhouse effect on air temperature, the temperature at the leaf surface may rise by 0.5 to 1.5 °C. In their FACE experiments (Free Air Carbon dioxide Enrichment) in Arizona, Kimball *et. al.* measured an average rise in canopy temperature of 0.56°C over the growing season. Such a higher leaf temperature may also have important consequences for the longevity and photo-synthetic capacity of the individual leaves and at the canopy level, as ageing may be accelerated. It is presently not known how different crops respond to the increase in leaf and canopy temperature.

The mostly negative feedback on water vapour exchange will at least partly counteract the effect of the reduced stomatal conductance even without rise in ambient temperature at the global scale. A rise of ambient growing season temperatures due to the greenhouse effect will also tend to stimulate transpiration. Depending on changes in air humidity it may affect VPD at the leaf surface determining the fluxes of sensible and latent heat and thereby the energy balance. Increased water availability may, however, offset the relative warming at the leaf surface, especially if, through preceding water saving and lower instantaneous requirements, partial stomatal closure alleviates water stress during periods of high water requirements in the diurnal sequence. It requires detailed micro-climatological data and complex calculations to integrate these various mechanisms in operation at the leaf level.

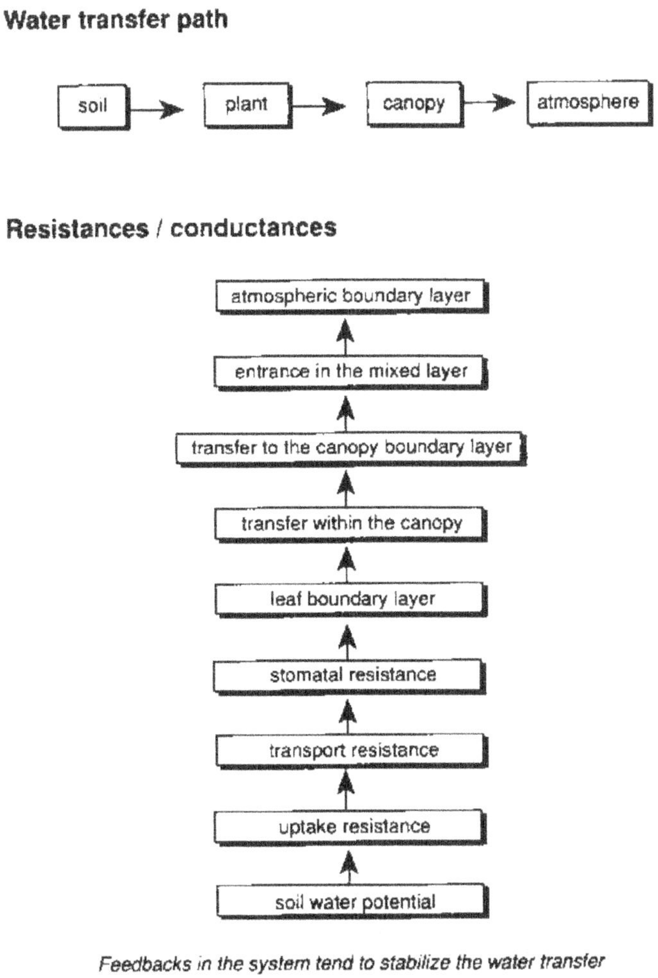

Figure 18.11 : Stepwise transfer of water from the root environment to the atmosphere.

Leaf Area and Weight

Changes in area of individual leaves also have a direct effect on water loss at a per leaf basis. In general a substantial increase in individual leaf weight is found. This is primarily due to an increased leaf thickness and additional accumulation of non-structural carbohydrates. Leith *et. al.* found for soybeans that a small but significant increase in leaf area with CO_2 concentration may be expected (about 5% increase with a doubling [CO_2]). This is consistent with trends reported in other papers. In general, however, weight gain through increased leaf thickness is more important than that by the increase in leaf area. The relative gain in leaf area may also depend on growing conditions such as water shortage.

EFFECTS AT THE PLANT AND CANOPY LEVEL

Aspects of the water economy at the leaf level also apply at the canopy level. Moreover, light distribution in the canopy, leaf age, humidity gradients and coupling with the atmosphere for different canopy layers and canopy structures all have their effect. The direct effects on vegetation properties discussed here are linked to the change in CO_2 concentration and temperature, but indirect effects of a climate change related with changes in ambient temperature and air humidity, coupling to the lower atmosphere and albedo changes, may even dominate. Mismanagement or over-exploitation of natural resources may lead to changes in the vegetation and even to desertification, and via a positive feedback aggravate changes in local and regional climate as studied in the EU-EFEDA-programme. This field of research is still under development and the subject of international programmes like the IGBP-BAHC.

Leaf Area Index

The leaf area index (leaf surface area per unit soil surface area) of a crop with adequate water supply at elevated [CO_2] increases, especially early in the season, as a result of earlier and more rapid leaf production in the vegetative growth phase. This applies especially for indeterminate growing species and under non-limiting supply of nutrients. Early development of the canopy will lead to an earlier full ground cover, and may thus limit water loss from direct soil evaporation. Depending on the local precipitation and available soil water reserves, such an early enhanced canopy development may also be favourable for the full utilization of water resources. The higher transpiration early in the season may also lead to an earlier depletion of water reserves in the soil.

In their FACE experiments Hendrey *et. al.* found that for cotton leaf area increased by 25% in dry plots, but by only 11% in irrigated plots as compared to the respective controls. Such enhancements in leaf area and biomass formation in dry conditions could improve the duration of the vegetative soil cover, and help counteract erosion and other forms of land degradation.

Plant Water Status

It has been shown that plant water status is generally improved at elevated CO_2. Part of the effect can be ascribed to a reduced transpiration demand (per unit leaf area), and therefore a partial alleviation of the water stress. Also for plants with a normal water supply, pre-dawn and midday water potentials have been found to be less negative in high [CO_2]. Even in saline conditions the effect of partial stomatal closure is reflected in a higher salt tolerance. As a consequence, the restriction of photo-synthesis under low water supply may be less severe, or be delayed.

Changes in the Rooting Pattern

Several studies have reported an increased input of carbon in below-ground processes. An important aspect to be considered in reviewing the literature at this point is that most experiments concern plant growth over a short period and with a soil volume limited by pot or container size. In a study with container-grown grasses and clover Nijs *et. al.* showed that in a rapidly developing terminal drought period water-use efficiency is about doubled, and stress is developing later under elevated [CO_2]. These conditions are thus improving the possibility to escape drought stress. Even though the trend is clear, elevated [CO_2] plants grown with restricted soil volume may not take full advantage of a larger root system. In (semi-) field experiments the larger root system indeed leads to a better exploration of the soil volume, and an earlier, or higher, rooting density at the larger depths. Experiments with potted plants can therefore not be expected to show the full extent to which elevated [CO_2] plants may profit in a (transient) drought period. The observation of Morison and Gifford that the pattern of depletion of soil water reserves is rather similar may partly be an under-estimation of the real moisture availability in field conditions because of a better access to deeper soil layers. It has also been observed that a faster recovery of the crop is possible when grown at elevated [CO_2]. The explanation of this observation has not yet been given, but could be related to deeper and more extensive rooting patterns.

Temperature Effects

The importance of temperature at the canopy level is two-fold. Firstly higher temperatures increase transpiration by changing the VPD at the leaf surface, and secondly the higher canopy temperature may lead to an accelerated ageing of the foliage, and a shortening of the growing season, or, for example, grain-filling period. The latter type of effect is discussed extensively in crop growth modelling studies that describe changes in crop productivity due to climate change.

On theoretical grounds it can be expected that transpiration losses will increase with higher air temperatures. It is likely that the evaporative demand as determined by the vapour pressure deficit would increase by about 5 to 6% per degree warming. The overall effect of a higher temperature alone is an intensification of the hydrological cycle. It should be realized that the combined result of a higher ambient temperature, leading to a higher evaporative demand, and partial stomatal closure, counteracting this, could be overridden by changes in other environmental and atmospheric conditions like soil water availability, precipitation patterns, cloudiness and air humidity.

EFFECTS OVER THE GROWING SEASON

The effects of elevated CO_2 alone on transpiration integrated over the season show that any of the cases (higher water use; no change; lower water use) can be found. Especially in the case of non-limiting water supply a higher water use can be expected even with a lower stomatal conductance as leaf area will increase with about the same factor as total biomass. This will compensate for the reduction in water exchange per leaf area. This applies especially in early season. In conditions with transient water shortage, profit from partial stomatal closure will be highest, as both the depletion of soil water reserves is delayed and plant sensitivity to midday water shortage (suppressed photo-synthesis) is lessened.

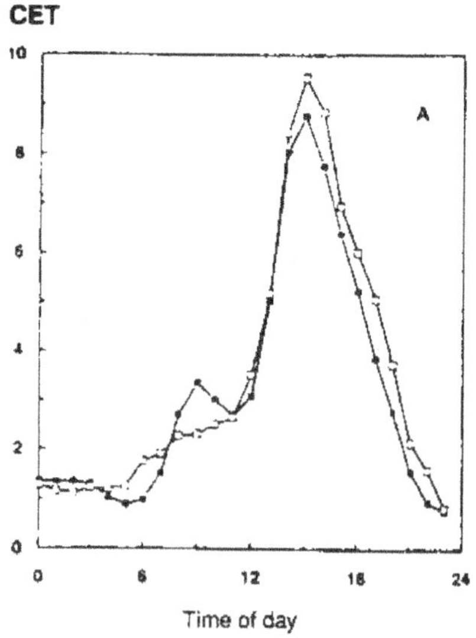

Figure 18.12 : Effect of ambient (□) and elevated (■) [CO_2] on CET (Canopy Evapotranspiration rate, mmol/m^2/s) on spring wheat (1992).

In conditions with adequate water supply, water use over the whole season, and especially in periods in the season with a closed canopy (*e.g.*, with leaf area index over 3), is little different if at all. ITE and WUE (24h basis) are both higher (about 40-50%) at elevated [CO_2], but transpiration per unit soil surface is not changed significantly. Here treatment temperatures have been kept unchanged, tracking outside conditions. The improvement in seasonal water-use efficiency (SWUE) ranges from 23% (wheat) to 54% (faba bean), similar to the results reported by Dijkstra *et. al.*

Similarly, Baker and Allen presented results of canopy gas exchange for soybean and citrus at a range of temperatures. Although day time water-use efficiency (DWUE) fell with temperature, they measured at 800 μ l/l [CO_2] a doubling (soybean) or more of the DWUE, independent of temperature (28-35°C). Seasonal water-use efficiency varies with crop and growth conditions.

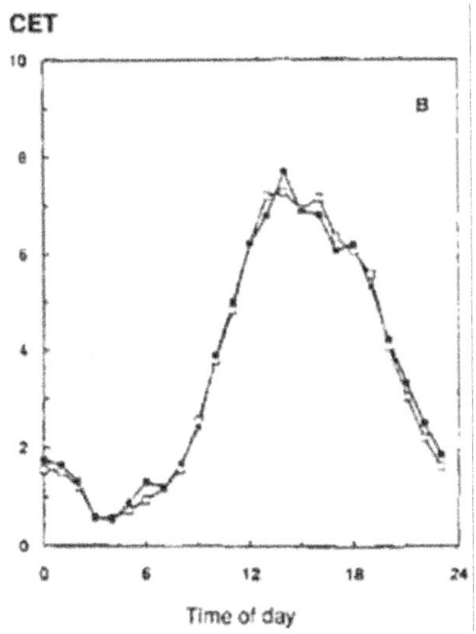

Figure 18.13 : Effect of ambient (□) and elevated (■) [CO_2] on CET (Canopy Evapotranspiration rate, mmol/m^2/s) on faba bean on Julian day 188.

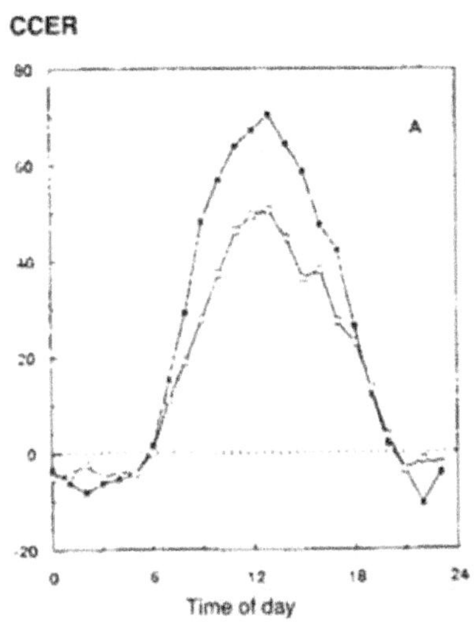

Figure 18.14 : Effect of ambient (□) and elevated (■) [CO_2] on CCER (Canopy Carbon Exchange Rate, μ mol/m^2/s) on spring wheat.

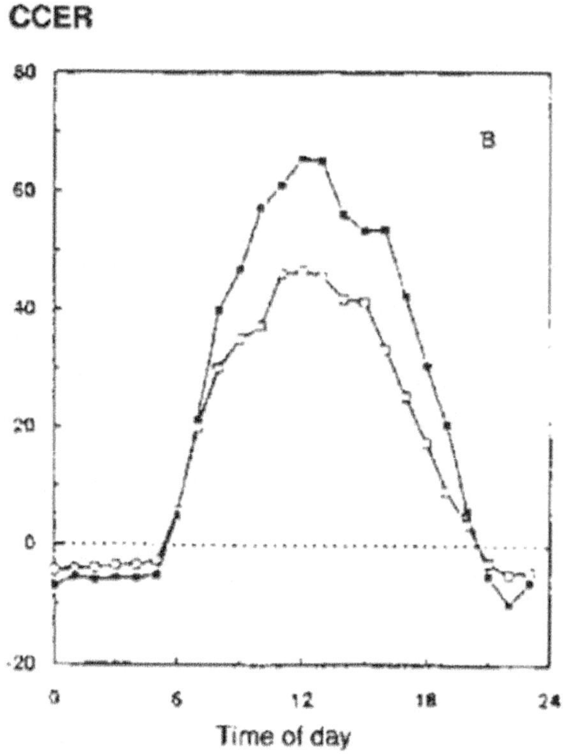

Figure 18.15 : Effect of ambient (□) and elevated (■) [CO_2] on CCER (Canopy Carbon Exchange Rate, μ mol/m²/s) on faba bean on Julian day 188.

Water-use efficiency rises especially in conditions with water limitations. In their study of the potential changes of productivity of cool-season legumes, Grashoff *et. al.* concluded that for rainfed faba bean crops in the Netherlands, Syria and Israel productivity would increase, especially under water limitation. Faba bean is a special case, because crop development rate is accelerated by drought. A small increase in temperature (1.7°C) would decrease yield, but the lower water requirement at higher [CO_2] (1.7°C and 460 μ l/l for the climate in 2030) more than compensates for this, raising yields from 2.8 to 4.7 t/ha (Syria), 3.9 to 5.0 t/ha (Israel) and 5.1 to 5.7 t/ha (Netherlands). In fully irrigated conditions yields would be higher, but the relative CO_2 effect would be lower. Interestingly the standard deviation of the predicted yield (>10 year simulated average) shows a tendency to decrease in the changed climate scenario.

Table 18.12 : Effect of [CO_2] on full canopy gas exchange parameters for spring wheat and faba bean both measured on Julian day 188.

	Spring wheat			Faba bean		
[CO_2] µ mol/mol	**350**	**700**	**ratio**	**350**	**700**	**ratio**
CCERmax [1]	50.2	70.4	1.40	45.6	66.8	1.47
CDR [2]	-3.8	-6.0	1.59	-3.85	-6.50	1.69
CCER light period [7]	67.5	95.6	1.42	67.7	100.7	1.49
CDR dark period [7]	-5.4	-8.6	1.59	-5.5	-9.3	1.69
Net CCER shoot-24h [7]	62.1	87.0	1.40	62.2	91.4	1.47
CETmax [3]	9.55	8.77	0.92	7.29	7.71	1.06
CET-24h [4]	5.24	4.90	0.94	5.50	5.60	1.02
CCERmax/CETmax [5]	5.26	8.03	1.53	6.26	8.66	1.39
CCER-24h/CET-24h [6]	11.85	17.75	1.49	11.31	16.32	1.44

[1] CCERmax = maximum canopy CO_2 exchange rate on a day (µ mol/m^2/s)
[2] CDR = mean canopy dark (night) respiration rate (µ mol/m^2/s)
[3] CETmax = maximum canopy evapotranspiration rate on a day (µ mol/m^2/s)
[4] CET-24h = integrated canopy evapotranspiration (mm/d)
[5] Or maximum ITE in mmol CO_2/mol H20
[6] Or daily WUE in g CO_2/kg H_2O
[7] g CO_2/m^2/d

Table 18.13 : Day time canopy water-use efficiency (D-WUE : photo-synthesis per unit water transpired : mmol CO_2/mol H_2O)

	[CO_2] µ mol/mol	Temp. (°C)	DWUE	High/low ratio
Soybean	330	28	2.95	
		31	2.63	
		33	2.37	
		35	2.33	
	800	28	6.00	2.03
		31	5.11	1.94
		33	4.95	2.09
		35	4.71	2.02
Citrus	330	25	2.30	
		34	1.25	
	840	25	5.40	2.34
		34	3.52	2.81

Table 18.14 : Seasonal crop water-use efficiency ((C)WUE in g DM/kg water)

	Ambient [CO_2]	Double [CO_2]	Ratio	
Sorghum	3.08	4.13	1.34	
Wheat well-watered	5.1	6.3	1.23	
Wheat water shortage	6.2	8.9	1.43	
Wheat	2.62	3.45	1.31	
Wheat well-watered	1.58	2.14	1.35	grain only
Wheat water shortage	1.27	1.86	1.46	grain only
Faba beans	4.91	7.82	1.54	
Water hyacinth	1.4	2.6	1.85	

Table 18.15 : Average simulated seed yield in t/ha of rainfed and fully irrigated faba bean crops in Wageningen (Netherlands), Tel Hadya (Syria) and Migda (Israel) under present and changed climate conditions (year 2030 : + 1.7°C and [CO_2] at 460 μ mol/mol; year 2080 : +3.0°C and [CO_2] at 700 μ mol/mol; no precipitation change).

Location and Scenario	Rainfed			Fully irrigated		
	yield	change (%)	sd (%)	yield	change (%)	sd (%)
Wageningen						
Current climate	5.1	-	32	6.1	-	9
Year 2030	5.7	12	29	6.4	5	7
Year 2080	6.7	31	22	7.2	18	6
Tel Hadya						
Current climate	2.8	-	41	6.9	-	11
Year 2030	4.7	68	35	8.0	16	8
Year 2080	7.4	164	22	9.5	38	6
Migda						
Current climate	3.9	-	39	6.4	-	9
Year 2030	5.0	28	41	7.2	12	10
Year 2080	6.7	72	35	8.5	33	11

EFFECTS ON THE REGIONAL VEGETATION-ATMOSPHERE WATER VAPOUR EXCHANGE

One of the more complex aspects of the estimation of changes in the water use of arable crops is the extrapolation from leaf and canopy to the field and region. This has long been recognized by the international scientific community, and has given rise to the establishment of several large-scale measuring campaigns (HAPEX : Hydrologic Atmospheric Pilot Experiment (Sahel and Spain), GEWEX : Global Energy and Water Cycle Experiment (World Climate Research Programme)) and to the start of the IGBP-BAHC programme. A major part of the efforts will go into the development of the SVAT-models (Soil-Vegetation-Atmosphere-Transfer models). The extrapolation from the stomatal and leaf level to the region have been the subject of experimental and theoretical studies. In general the structure of the canopy determines to a large extent the transfer of heat and water vapour to the lower atmosphere. Most models to date neglect or oversimplify the feedback between the vegetated surface and the lower atmosphere, the planetary boundary layer (PBL). Jacobs showed in a study using coupled models (one-dimensional model for the PBL coupled to a modified Penman-Monteith big-leaf model) that in such a model the sensitivity of regional transpiration to changes in surface resistance is reduced (about halved) but that the sensitivity to changes in albedo is increased by between 25 and 250% relative to the values obtained without PBL feedback.

In addition to the sensitivity of stomata to photo-synthesis and [CO_2], Jacobs introduced an additional correlative relationship between the humidity deficit at the leaf surface and the C_i/C_a ratio. The changes in transpiration caused by changes in surface resistance are damped. However, within the canopy the changes in the specific humidity lead to a positive feedback, *e.g.*, the CO_2 effects on surface resistance are enhanced. The inclusion of PBL feedback and stomatal response to humidity deficit leads in his study to an estimated overall decrease of the regional transpiration by 10 to 30%. Modifying factors are surface roughness determined by the vegetation type, temperature and air humidity. In particular, information about the last factor, extrapolated from regionalized GCM output, is virtually absent or highly speculative.

Arable crops, especially in the temperate regions in general, have a low surface roughness, and in that case the transpiration is primarily determined by the radiation energy. This is also true for pastures. In rangelands with sparse higher vegetation the higher roughness leads to a stronger coupling, a situation comparable to forests and mixed vegetation.

CONCLUDING REMARKS

Present knowledge does not allow a firm statement to be made concerning a future situation with respect to water-limited agricultural production. At a regional production level the available quantitative data from regionalized GCM runs are insufficient to be used as direct inputs for crop growth models. At a global level the simultaneous changes in rainfall pattern, air humidity and possible shifts in vegetation zones add to uncertainty, as many and often non-linear feedbacks are expected to operate.

The limited amount of available experimental data has as a trend that water use per unit soil surface area will change little (-10 to +10%). As, however, the general trend towards an improved water-use efficiency is clear, the productivity per unit of available water is expected to rise by 20-40%, probably much less than the value (100%) calculated from a reduced stomatal conductivity and an increased photo-synthesis. Some studies show that in situations with marginal water availability the threshold for a successful crop may shift to lower values. Whether at otherwise unchanged water availability this would open up possibilities to reverse existing trends towards desertification in certain areas is very questionable, as apart from precipitation falling short to maintain existing vegetation, other aspects like over-exploitation might dominate. It should, however, be emphasized that present knowledge on feedback among vegetation characteristics, gas exchange and albedo and the regional climate is insufficient to draw firm conclusions.

Some related problems have not been touched upon. For example, changes in precipitation distribution throughout the season may lead to shifts in the accessibility of fields for farm operations in early spring or late autumn. In addition they may cause changes in soil moisture and thereby modify the rate of mineralization of nutrients (nitrogen, phosphorus). Such interactions, combined with changes in (air and soil) temperatures, may markedly change soil fertility and thereby local farming systems and crop productivity. Depending on the aim of the study, such impacts have to be considered more or less relevant to the present subject.

In many studies the impact of climate change on crop growth and yield is analysed using crop simulation models. A proper analysis of the performance of such models should be made to verify their reliability in the projected conditions of changed atmospheric composition and changed climate. The absolute values of the regionally and globally aggregated predicted crop yields, often calculated using a proportionality factor for the CO_2 response, are at present probably less reliable than the predicted sensitivity of the yield to various climate and management factors.

The above considerations relate to a projected world, changed primarily in terms of climate and atmospheric composition. It should be emphasized that at many points agricultural practice is very dynamic, and will respond to changed conditions by adaptation. Crop and cultivar choice will, in most instances and in the most productive areas, change over time and gradually incorporate the traits necessary for adapted performance, or change to better adapted species. A rising temperature will for most current cultivars, for instance, accelerate crop development. This would in itself lead to a reduced water use over the shortened growth period, but also to a loss of potential yield. One may argue that farmers will repair such a loss of production potential by a proper choice of adapted cultivars or crop species, unless temperatures exceed the appropriate temperature window.

These developments require, however, that new technologies and genetic resources be practically and economically accessible for all farmers, a situation that is not reached at present for farmers in arid and semi-arid regions in developing countries where the risks but probably also opportunities for agricultural production may be greatest.

EFFECTS OF HIGHER DAY AND NIGHT TEMPERATURES ON GROWTH AND YIELDS OF SOME CROP PLANTS

Gaseous emissions from human activities are substantially increasing the concentrations of atmospheric greenhouse gases, particularly carbon dioxide, methane, chlorofluorocarbons and nitrous oxides. Global circulation models predict that these increased concentrations of greenhouse gases will increase average world temperature. Under the business-as-usual scenario of the Inter-governmental Panel on Climate Change (IPCC), global mean temperatures will rise 0.3°C per decade during the next century with an uncertainty of 0.2 to 0.5%. Thus global mean temperatures should be 1°C above the present values by 2025 and 3°C above the present value by 2100. Although global circulation models do not all agree as to the magnitude, most predict greenhouse warming. There is also general agreement that global warming will be greater at higher latitudes than in the tropics. Different global circulation models have predicted that global warming effects will vary diurnally, seasonally and with altitude.

It is also possible that there will be an autocatalytic component to global warming. Photo-synthesis and respiration of plants and microbes increase with temperature, especially in temperate latitudes. As respiration increases more with increased temperature than does photo-synthesis, global warming is likely to increase the flux of carbon dioxide to the atmosphere which would constitute a positive feedback to global warming.

This paper describes the effects of higher day and night temperatures on crop growth and yield. Temperature effects at different levels of organization - biochemical, physiological, morphological, agronomic and systems - are considered. This is followed by identification of options for germplasm improvement and crop management that may mitigate the adverse effects of higher day and night temperatures. The main focus is on wheat (*Triticum aestivum* L.) and rice (*Oryza saliva* L.).

Mechanisms for Heat Tolerance

Crop plants are immobile. They must adapt to prevalent soil and weather conditions. Except for transpirational cooling, plants are unable to adjust their tissue temperatures to any significant extent. On the other hand, plants have evolved several mechanisms that enable them to tolerate higher temperatures. These adaptive thermotolerant mechanisms reflect the environment in which a species has evolved and they largely dictate the environment where a crop may be grown.

Four major aspects of thermotolerance have been studied : (1) thermal dependence at the biochemical and metabolic levels; (2) thermal tolerance in relation to membrane stability; (3) induced thermotolerance through gradual temperature increase *vis-a-vis* production of heat shock proteins; and (4) photosynthesis and productivity during high temperature stress.

Biochemical Processes

Temperature effects on the rates of bio-chemical reactions may be modelled as the product of two functions, an exponentially increasing rate of the forward reaction and an exponential decay resulting from enzyme denaturation as temperatures increase. The greatest concern is whether it is possible to increase the upper limit of enzyme stability to prevent denaturation.

Failure of only one critical enzyme system can cause death of an organism. This fact may explain why most crop species survive sustained high temperatures up to a relatively narrow range, 40 to 45°C. The relationship between the thermal environment for an organism and the thermal dependence of enzymes has been well established.

The shape of this function also describes temperature effects on most biological functions, including plant growth and development. The function can be categorized by the three cardinal temperatures - minimum, optimum and maximum. Modellers frequently simplify the relationship into a stepwise linear function. The stepwise linear function has a plateau rather than an optimum temperature.

The thermal dependence of the apparent reaction rate for selected enzymes may indicate the optimal thermal range for a plant. The range over which the apparent Michaelis-Menten constant for CO_2 (K_m) is minimal and stable is termed the thermal kinetic window. For crop plants, the thermal kinetic window (TKW) is generally established as a result of thermally induced lipid phase changes, rubisco activity and the starch synthesis pathway in leaves and reproductive organs.

In cotton and wheat, the time during which foliage temperature remained within the TKW was related to dry matter accumulation. The cumulative time that rainfed crop foliage is outside the TKW provides an index of the degree of extreme temperature stress of the environment. Irrigation is one management option to reduce crop exposure to heat stress.

Temperatures that inhibit cellular metabolism and growth for a cool season C_3 species such as wheat may not inhibit warm-season C_3 species such as rice (*Oryza sativa* L.) and C_4 species such as sorghum, maize (*Zea mays* L.) and sugar cane (*Saccharum spontaneum* spp.). The identification of TKWs for different species can aid in the interpretation of the differential temperature stress responses for crop growth and development among species.

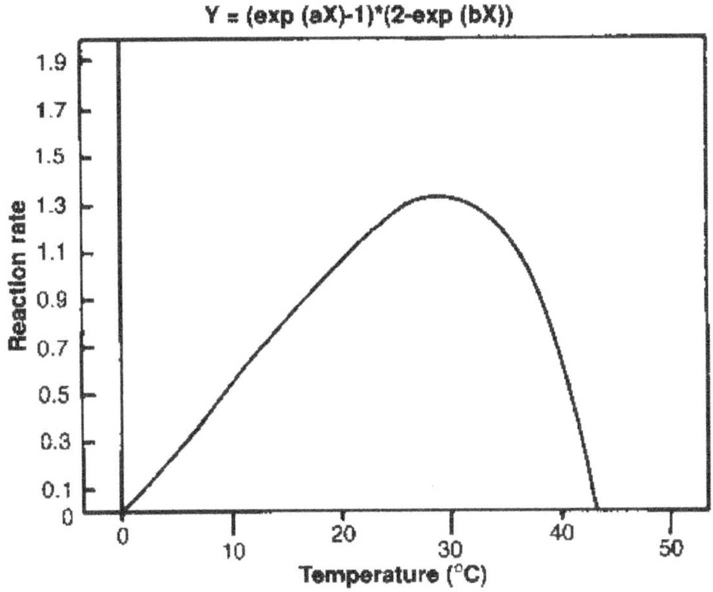

Figure 18.16 : (a) Exponential rate of reaction as a function of temperature.

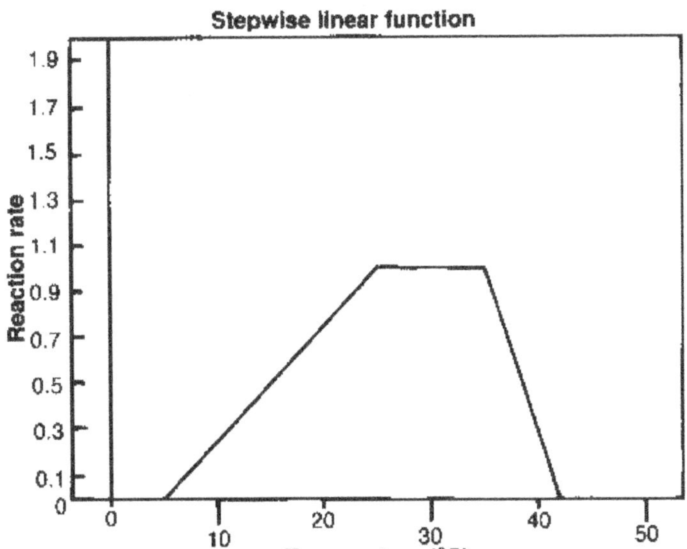

Figure 18.17 : (b) Step-wise increase in reaction rate as a function of temperature.

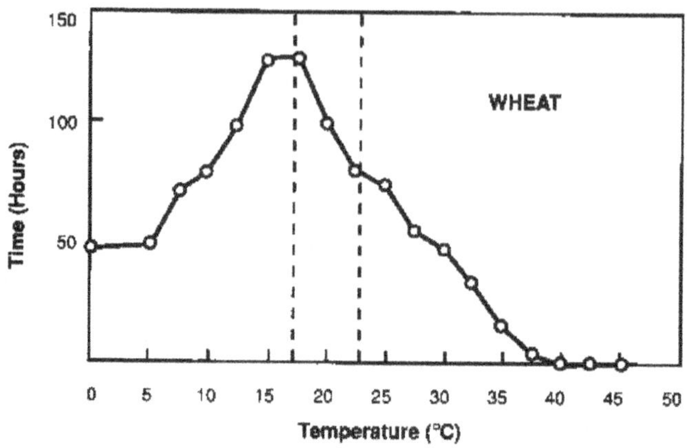

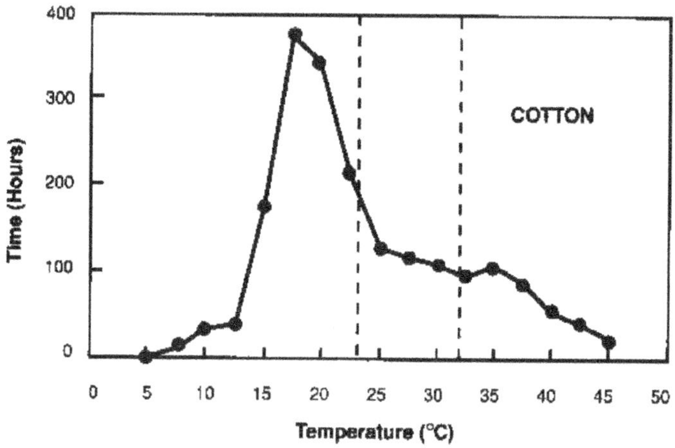

Figure 18.18 : Seasonal foliage temperatures of wheat (cv. Kanking) and cotton grown at Lubbock, Texas. The vertical lines represent the temperature range that comprises the species-specific thermal kinetic window as determined from the changes in the apparent K_m of purified enzymes with temperature. Foliage temperatures were measured with a 50° field-of-view Teletemp Model 50 infrared thermometer (Teletemp Corp., Fullerton, CA) positioned at 1.5 μ above the crop. Instruments were scanned at 1 min intervals with a 15 min average computed and stored. The infrared thermometer viewed an area of 0.75 m^2, with the same area continuously sampled

Thermal Stability of Cell Membranes

The plasmalemma and membranes of cell organelles play a vital role in the functioning of cells. Any adverse effect of temperature stress on the membranes leads to disruption of cellular activity or death. Heat injury to the plasmalemma may be measured by ion leakage (Chaisompongpan *et. al.,* 1990; Hall, 1993). Injury to membranes from a sudden heat stress event may result from either denaturation of the membrane proteins or from melting of membrane lipids which leads to membrane rupture and loss of cellular contents.

Heat stress may be an oxidative stress. Peroxidation of membrane lipids has been observed at high temperatures, which is a symptom of cellular injury. Enhanced synthesis of an anti-oxidant by plant tissues may increase cell tolerance to heat but no such anti-oxidant has been positively identified.

A relationship between lipid composition and incubation temperature has been shown for algae, fungi and higher plants. In *Arabiodopsis,* exposed to high temperatures, total lipid content decreases to about one-half and the ratio of unsaturated to saturated fatty acids decreases to one-third of the levels at temperatures within the TKW. Increase in saturated fatty acids of membranes increases their melting temperature and thus confers heat tolerance. An *Arabiodopsis* mutant, deficient in activity of chloroplast fatty acid W-9 desaturase, accumulates large amounts of 16 : 0 fatty acids, resulting in greater saturation of chloroplast lipids. This increases the optimum growth temperature.

In cotton, however, heat tolerance does not correlate with degree of lipid saturation and similar differences in genotypic differences in heat tolerance have been unrelated to membrane lipid saturation in other species. In such species, a factor other than membrane stability may be limiting growth at high temperature.

Heat Shock Proteins

Synthesis and accumulation of proteins were ascertained during a rapid heat stress. These were designated as 'Heat Shock Proteins' (HSPs). Subsequently it was shown that increased production of these proteins also occurs when plants experience a gradual increase in temperature more typical of that experienced in a natural environment.

Three classes of proteins as distinguished by molecular weight account for most HSPs, namely HSP90, HSP70, and low molecular weight proteins of 15 to 30 kDa (LMW HSP). The proportions of the three classes differ among species. In general, heat shock proteins are induced by heat stress at any stage of development. Under maximum heat stress conditions, HSP70 and HSP90 mRNAs can increase ten-fold and LMW HSP increase as much as 200-fold. Three other proteins, though less important, are also considered to be heat shock proteins *viz.* 110 kDa polypeptides, ubiquitin, and GroEL proteins.

In arid and semi-arid regions, dryland crops may synthesize and accumulate substantial levels of heat shock proteins in response to elevated leaf temperatures. The induction temperature for synthesis and accumulation of heat shock proteins in laboratory-grown cotton ranged from 38 to 41°C. Soil water deficits resulting in midday canopy temperature of 40°C or greater for two to three weeks were used to study heat shock proteins in field-grown cotton. A comparison of polypeptide patterns of dryland and irrigated cotton leaves showed that at least eight new polypeptides accumulated in about half of the dryland leaves analysed. The polypeptides that accumulated in the dryland leaves but not irrigated cotton leaves had molecular weights of 100, 94, 89, 75, 60, 58 and 21 kDa. In a similar experiment with field-grown soybean (*Glycine max* (L.) Merr.), several heat shock proteins were observed in both irrigated and dryland treatments, although levels were greater in the non-irrigated treatments (Kimpel and Key, 1985).

Correlation between synthesis and accumulation of heat shock proteins and heat tolerance suggests, but does not prove, that the two are causally related. Further evidence for a causal relationship is that some cultivar differences in heat shock protein expression correlate with differences in thermotolerance. In genetic experiments, heat shock protein expression co-segregates with heat tolerance. Another evidence for the protective role of heat shock protein is that mutants unable to synthesize heat shock proteins, and cells in which HSP70 synthesis is blocked or inactivated, are more susceptible to heat injury.

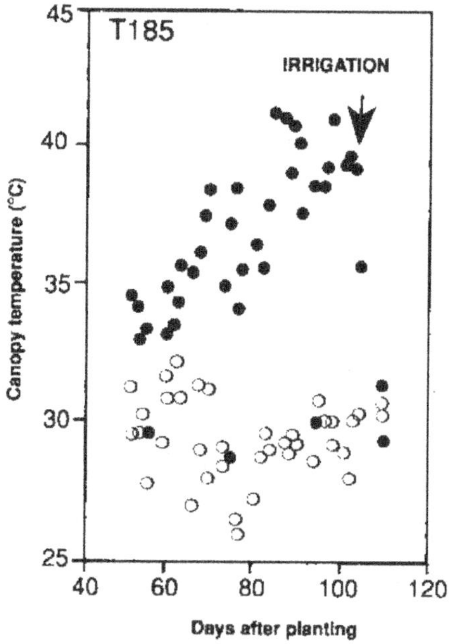

Figure 18.19 : Seasonal changes in the midday canopy temperatures of irrigated (○) and dryland (●) cotton. Arrow indicates the day on which dryland plots were irrigated with 10 to 12.5 cm of water (106 DAP). The cotton strain is T185.

The mechanism by which heat shock proteins contribute to heat tolerance is still not certain. One hypothesis is that HSP70 participates in ATP-dependent protein unfolding or assembly/disassembly reactions and that they prevent protein denaturation during stress. If this mechanism is true, then heat shock proteins may provide a significant basis for increasing heat tolerance of crop plants in a global warming situation. The LMW HSPs may play a structural role in maintaining cell membrane integrity during stress. Other heat shock proteins have been associated with particular organelles such as chloroplasts, ribosomes and mitochondria. In tomato (*Lycopersicon esculentum* L.), heat shock proteins aggregate into a granular structure in the cytoplasm, possibly protecting the machinery of protein synthesis.

HSPs provide a significant opportunity to increase heat tolerance of crops. To elucidate their mechanisms of action and to exploit their potential contribution to increasing heat tolerance, four lines of investigations are suggested :

1. Establish the bio-chemical activities of individual HSPs as a preliminary step.
2. Characterize the genetic variability of specific heat shock proteins across a wide range of germplasm. Develop iso-population and near isogenic lines selected for production of low and high levels of HSP synthesis.
3. As HSPs appear to participate in maintaining the conformation or assembly of other protein structures, analyse the molecular details of these processes and establish all participating protein substrates. Such bio-chemical studies are needed to understand how these processes protect or allow recovery from heat stress.
4. Identify specific HSP mutants or create transgenic mutant plants to complement molecular and biochemical understanding with genetic approaches.

PHOTO-SYNTHESIS AND HIGH TEMPERATURE STRESS

Variability in leaf photo-synthetic rates within or between species is often unrelated to differences in productivity. Similarly, high photo-synthetic rates at high temperatures do not necessarily support high rates of crop dry matter accumulation. The temperature optimum for photo-synthesis is broad, presumably because crop plants have adapted to a relatively wide range of thermal environments. A 1 to 2°C increase in average temperature is not likely to have a substantial impact on leaf photo-synthetic rates. Further, there is a possibility that photo-synthesis of crop plants can adapt to a slow increase in global average temperatures. Thus, global warming is not likely to affect photo-synthetic rates per unit leaf area gradually or on a closed canopy basis over the next century.

While photo-synthetic rates were found to be temperature-sensitive in other crops, wheat and rice appear to be different. In wheat, no measurable differences were found in photo-synthetic rates per unit flag leaf area or on a whole-plant basis in the temperature range from 15 to 35°C. In rice, there is little temperature effect on leaf carbon dioxide assimilation from 20 to 40°C.

Recent research has shown significant variation among wheat cultivars with respect to reduction in photo-synthesis at very high temperature. Photosynthesis of germplasm adapted to higher temperature environments was less sensitive to high temperature than was germplasm from cooler environments. When this germplasm was grown under moderate (22/17°C) and high

(32/27°C) temperatures in the seedling stage or from anthesis to maturity, there was a highly significant correlation between photo-synthesis rate and either seedling biomass (r=0.943***) or grain yield of mature plants (r=0.807**). Genotypes most tolerant to high temperatures had the most stable leaf photo-synthetic rates across temperature regimes or they had the longest duration of leaf photo-synthetic activity after anthesis and high grain weights. The above relationship was exemplified by 'Ventnor' from the high temperature area of Australia and 'Lancero' from the high altitude area of Chile. See Al-Khatib and Paulsen.

Despite observed negative effects of high temperature on leaf photo-synthesis, the temperature optimum for net photo-synthesis is likely to increase with elevated levels of atmospheric carbon dioxide. Several studies have concluded that CO_2-induced increases in crop yields are much more probable in warm than in cool environments. Thus, global warming may not greatly affect overall net photo-synthesis.

Night Respiration Response to CO_2 and Temperature

In tomato (*Lycopersicon esculentum*) and cotton (*Gossypium hirsutum*), elevated CO_2 increased dark respiration possibly because of increased carbohydrate accumulation in tissues. The latter has been shown to increase alternative pathway respiration as well.

Apparent dark respiration may decline under elevated CO_2 if there is dark CO_2 fixation or if elevated CO_2 directly inhibits or inactivates respiratory enzymes as may occur through increased formation or carbamate.

Table 18.16 : Mean weekly photo-synthetic rate (fmol *$CO_2/m^2/s$*) and duration of photo-synthetic activity (weeks, in parentheses), and grain biomass of two wheat genotypes grown at two temperature regimes

Genotype	Seedlings after two weeks of treatment		From anthesis to maturity		Grain biomass (g/tiller)	
	22/17 °C	32/27 °C	22/17 °C	32/27°C	22/17 °C	32/27 °C
Ventor	8.6	7.0	6.4(7)	5.3 (7)	0.93	0.8
Lancero	7.9	4.6	4.1 (10)	3.1(7)	0.43	0.28

Figures in parentheses give duration of photo-synthetic activity in weeks from anthesis to physiological maturity.

Few studies have successfully partitioned the effects of elevated CO_2 on growth and maintenance respiration. Both components appear to decline, probably because of decrease in leaf protein levels which results in reduced construction and maintenance costs.

Elevated CO_2 reduced maintenance respiration of *Medicago sativa* and *Dactylis glomerata* at lower temperatures (15 to 20°C), whereas elevated CO_2 reduced growth respiration of *M. sativa* at 20 to 30°C and *D. glomerata* at 15 to 25°C.

Crop Growth and Development

Wheat development is customarily divided into vegetative and reproductive phases, with either ear emergence or anthesis as the event that separates the two phases. In the past 30 to 40 years, the sequence of pre-anthesis phenological events has been critically assessed with respect to grain yield potential and sensitivity to weather variables, particularly prevailing temperature and day length. Several systems are used to classify the sequence of phenological events. We identify five such developmental stages :

(i) germination - seeding to seedling emergence;

(ii) canopy development - emergence to first spikelet initiation, the double ridge stage;

(iii) spikelet production - first spikelet initiation to terminal spikelet formation;

(iv) spikelet development - terminal spikelet formation to anthesis;

(v) grain development - anthesis to maturity.

These stages are generally based on early recognized features of the apical meristem. They mark significant changes in morphology or physiology of different crop organs. Numbers of leaf and tiller primordia are determined before spikelet initiation but their subsequent growth and development are controlled by temperature and day length during the differentiation of spikes into spikelets. Similarly, floret number within each spikelet is established by anthesis, at which time the potential grain number per spike is established.

Productivity of wheat and other crop species falls markedly at high temperatures. Wheat in India is invariably exposed to extreme temperatures during some stages of development. In Australia, wheat is usually exposed to brief periods of heat stress during grain development.

All stages of development are sensitive to temperature. It is the main factor controlling the rate of crop development. Development generally accelerates as temperature increases, a phenomenon that is often described as a linear function of daily average temperature. The growing degree day concept is a common example of a linear model of developmental response to temperature.

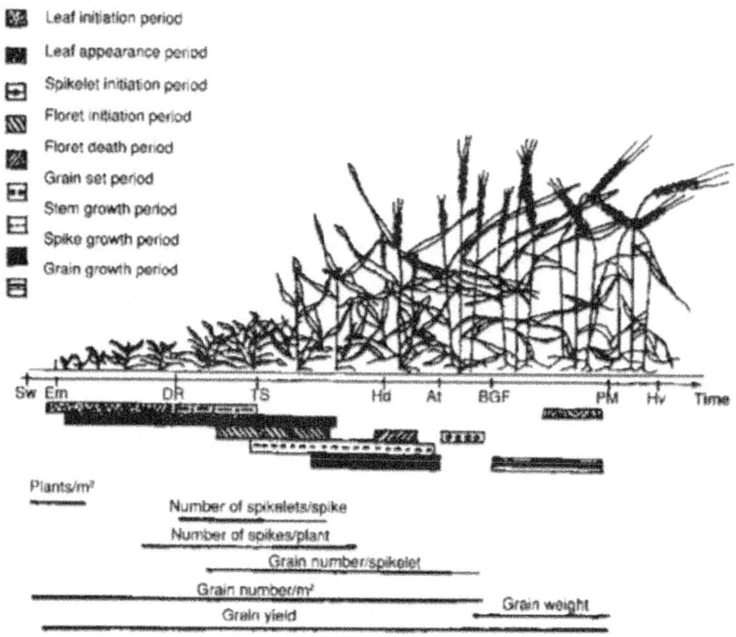

Figure 18.20 : Schematic diagram of wheat growth and development, showing the stages of sowing (Sw), emergence (Em), first double ridge appearance (DR), terminal spikelet appearance (TS), heading (Hd), anthesis (At), beginning of the grain-filling period (BGF), physiological maturity (PM), and harvest (Hv). Patterned boxes indicate the period of differentiation or growth of specific organs. Bars represent the periods of development when different components of grain yield are produced (heavy lines refer to main shoots and light lines represent extension associated with tillers.

Vegetative Phase

Several experiments have observed the effects of temperature on the duration from sowing or emergence to heading under controlled environment and field conditions. Unfortunately, however, few experiments have been conducted with enough cultivars to assess the genetic variability in this trait. The major conclusions from these studies are :

1. All genotypes are sensitive to temperature at one stage or another. Temperature sensitivity, however, varies greatly with genotype.
2. Phenological stages differ in sensitivity to temperature.
3. The duration of phase from sowing to first spikelet initiation is less sensitive to change in temperature than are other phases, although genotypes do differ in thermotolerance during this phase.
4. The stages during which environment has the greatest impact on yield are from first spikelet initiation or terminal spikelet formation until anthesis. Spikelet number and floral number (potential grain

number), both dominant yield contributing attributes, are established during these phases. Grain weight, on the other hand, appears to be much less sensitive to heat stress than is grain number.

Table 18.17 : Response of phasic development to temperature photoperiod and vernalization.

Developmental phase	Temperature	Photoperiod	Vernalization
Sowing-emergence	++++/+++++	0	0
Emergence-double ridges	+++/+++++	0/+++	0/+++++
Double ridges-terminal spikelet	+/+++	0/+++++	0/+++++
Terminal spikelet-heading	+++/+++++	0/+++	0/+(?)
Heading-anthesis	+++/+++++	0/+(?)	0
Anthesis-maturity	+++/+++++	0	0

For the estimation of sensitivity, the total life span was divided into the stages. An arbitrary scale was used to show when the effects are strong (+++++), moderate (+++) or slight (+). 0 denotes that the factor does not affect the process and question marks refer to uncertainties in the literature. For each factor, genetic variation in response was considered.

GRAIN DEVELOPMENT PHASE

In experiments under controlled conditions from 25 to 35°C, mean grain weight declined 16% for each 5°C increase in temperature. In pot experiments, grain yield decreased by 17% for each 5°C rise. For every 1°C rise in temperature, there is a depression in grain yield by 8 to 10%, mediated through 5 to 6% fewer grains and 3 to 4% smaller grain weight.

To elucidate the causal factor for reduced grain filling in wheat because of higher temperatures, Wardlaw studied the three main components of the plant system. The three components are : (a) source – flag leaf blade; (b) sink – ear; and (c) transport pathway – peduncle. He observed that photo-synthesis had a broad temperature optimum from 20 to 30°C with photo-synthesis declining rapidly at temperatures above 30°C. The rate of 14C assimilate movement out of the flag leaf, phloem loading, was optimum around 30°C; the rate of 14C assimilate movement through the stem was independent of temperature from 1 to 50°C. Thus, in wheat, temperature effects on translocation result indirectly from direct temperature effects on source and sink activities.

In a subsequent experiment with source-sink relationships altered through grain excision, defoliation and shading treatments, heat stress still reduced grain weight (Wardlaw *et. al.,* 1980). This result supports the earlier findings that temperature effects on grain weight are direct effects rather than assimilate availability. Furthermore, respiration effects do not appear to be the direct cause of decreased grain size in heat-stressed wheat.

Reduction of grain weight by heat stress may be explained mostly by effects of temperature on rate and duration of grain growth. As temperature increased from 15/10°C to 21/16°C, duration of grain filling was reduced from 60 to 36 days and grain growth rate increased from 0.73 to 1.49 mg/grain/day with a result of minimal influence on grain weight at maturity. Further increase in temperature from 21/16°C to 30/25°C resulted in decline in grain filling during 36 to 22 days with a minimal increase in grain growth rate from 1.49 to 1.51 mg/grain/day. Thus, mature grain weight was significantly reduced at the highest temperature.

Research on the effects of brief periods of ear warming after anthesis on ear metabolism have identified differential responses of starch and nitrogen accumulation in grain of four wheat cultivars. Warming increased the rate of dry matter accumulation in all the cultivars but the increase was less in cv. Aus 22645 than in the other cultivars studied. Rate of increase in nitrogen accumulation was, however, higher than the increase in total dry matter accumulation. Under long-term exposure to heat stress, increased grain nitrogen concentration is almost entirely as a result of decreased starch content rather than a change in total grain quality. The conversion of sucrose to starch within the endosperm is decreased by elevated temperatures. Furthermore, heat stress effects on final grain weight were associated with reduced levels of soluble starch synthetase activity.

In summary, high temperature reduction of grain yield results from : (a) reduced numbers of grains formed; (b) shorter grain growth duration; and (c) inhibition of sucrose assimilation in grains.

Extreme Temperature Effects on Crops

There are two major forms of extreme temperature stress on crops - heat and cold. An increase in global temperatures may have either or both of these two acute effects : more frequent high temperature stress and less frequent cold temperature stress.

Increase in temperature will lengthen the effective growing season in areas where agricultural potential is currently limited by cold temperature stress. Thus, increased temperature will cause a poleward shift of the thermal limits to agriculture. This poleward shift will be especially important for crops such as rice that have tropical centres of origin and adaptation but are also grown in temperate latitudes during warm seasons. Global warming impact will be greater in the northern than southern hemisphere because there is more high-latitude area cultivated in the northern hemisphere.

Table 18.18 : Effect of whole-plant warming on the rate of total dry matter and nitrogen accumulation, between days 10 and 20 after anthesis, in the grains of four cultivars of wheat.

Cultivar	Treatment	Rate of increase (mg/grain/day)	
		Total dry matter	N content
AUS 22645	C	1.94±09	0.03±002
	W	2.07±07(107)	0.48+003(160)
Kite	C	1.72+15	0.027±004
	W	2.28±17(133)	0.043±003(159)
Sonora	C	1.65±18	0.034±009
	W	2.06±19(125)	0.051±009(150)
WW15	C	1.89±20	0.037±005
	W	2.37±20(125)	0.053+005(143)

Plants were grown at 21/16°C and some (W) were warmed, between 10 and 20 days after anthesis, to 33/25°C and then returned to 21/16°C where they stayed until maturity. Control plants (C) were grown at 21/16°C throughout. Values given are the means ±.e; values in parentheses are percentage of cotton values.

Increased temperature would also affect the crop calendar in tropical regions. In the tropics, however, global warming, though predicted to be of only small magnitude, is likely to reduce the length of the effective growing season, particularly where more than one crop per year is grown. In semi-arid regions and other agro-ecological zones where there is wide diurnal temperature variation, relatively small changes in mean annual temperatures could markedly increase the frequency of highest temperature injury. For example, canopy temperature is 10 to 15°C higher in dryland cotton (*Gossypium hirsutum* L.) than in irrigated cotton. Thus, global warming would reduce dry matter accumulation in dryland cotton because of increased respiration, and reduced photo-synthesis and cellular energy.

In India, the growing season for wheat is limited by high temperatures at sowing and during maturation. As wheat is grown over a wide range of latitudes, it is frequently exposed to temperatures above the threshold for heat stress. For example, rainfed wheat depends on soil moisture remaining after the monsoon rains recede in September. High maximum and minimum temperatures in September (about 34/20°C), which adversely affect seedling establishment, accelerate early vegetative development, reduce canopy cover, tillering, spike size and yield. Hence, sowing is typically delayed until after mid-October when seedbeds have cooled, though much of the residual soil moisture may be lost. High temperatures in the second half of February (25/10°C), March (30/13°C) and April (30/20°C) reduce the numbers of viable florets and the grain-filling duration. High temperature stress particularly reduces yield of wheat sown in December/January which is necessitated in some regions because of the multiple cropping system.

The situation is similar for sorghum (*Sorghum bicolor* (L.) Moench) and pearl millet (*Pennisetum glaucum* (L.) R. Br.) which are exposed to extreme high temperatures in Rajasthan, India. After sowing, air and soil temperatures often exceed 40°C and midday soil surface temperatures above 50°C are common.

Acute effects of high temperature are most striking when heat stress occurs during anthesis. In rice, heat stress at anthesis prevents anther dehiscence and pollen shed, to reduce pollination and grain numbers.

Clearly, many crops in tropical areas are already subjected to heat stress. If temperatures increase further, crop failure in some traditional areas would become more commonplace.

Long-term Effects of High Temperatures on Crops

More important than acute effects of extreme temperature stress are the chronic effects of continuously warmer temperatures on crop growth and development. Chronic effects of high temperature include effects on grain growth discussed above. Record crop yields clearly reflect the importance of season-long effects on crop yields : crops generally yield the most where temperatures are cool during growth of the harvested component.

Crop growth simulations show that rice yields decrease 9% for each 1°C increase in seasonal average temperature. This chronic effect of high temperature differs significantly from the acute effect of short-term temperature events, because seasonal temperature effects are mostly a result of effects on crop development. For most grain crops, there is much greater genotypic variation in thermal requirements for vegetative than for reproductive development. As long-term temperatures increase, grain-filling periods decrease, and there appears to be little scope to manipulate this effect through existing genetic variation within species.

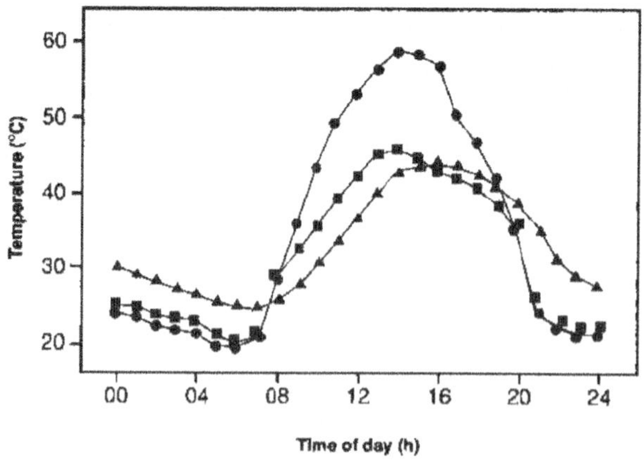

Figure 18.21 : Diurnal temperature data recorded in Fatehpur, Rajasthan, India. (Latitude 27°C 37'N) in June 1989). Each measurement is the mean value from three thermocouples placed at either 5 cm depth of soil (▲) : 0.5 cm depth of soil (●) : or 150 cm above the soil surface (■).

COMBINED EFFECTS OF CHANGING CO_2 TEMPERATURE, UV-B RADIATION AND O_3 ON CROP GROWTH

Our understanding of the relationships between crop growth and the atmospheric environment has developed substantially in the past few decades. Improvements in technology have enabled crop physiologists to study processes such as transpiration and carbon dioxide exchange in the field. Mechanistic understanding of the factors controlling photo-synthesis and respiration has improved greatly, but we still have no mechanistic understanding of the ways in which many abiotic stresses affect these processes. Our knowledge of factors that control crop development and the partitioning of assimilate between organs is still at the descriptive stage, and poses one of the main limitations on our ability to develop fully mechanistic mathematical models of crop growth. Nevertheless, crop science has reached a stage where reliable semi-empirical simulation models of some of the major world crops have been constructed and used to explore the potential for crop growth and yield in different environments. For the reasons discussed above, these models are restricted in their application to situations where the empirical relationships in them are well defined, but they are often useful in the design of particular experiments to investigate new environments, for example, to apply them to scenarios of climatic change. It should be emphasized that, although we have a much improved understanding of the growth and yield potential of major agricultural crops, this knowledge needs to be extended to include other crops, especially those important for developing countries.

However, as we steadily improve our understanding of how crop yield is influenced by the atmospheric environment, that environment is constantly changing as a result of human activities, and the challenge today is to predict how crops will respond to the changed environments of tomorrow. Traditionally, crop scientists have used controlled environments and field-based facilities to investigate crop responses to individual factors, such as radiation, temperature, humidity and water availability. More recently, manipulation of air quality has been used to study crop responses to air pollutants, such as SO_2, O_3 NO_x, and some of the technology developed for these experiments has been successfully applied to study responses to increased CO_2. Although this 'single component' approach is a highly effective way of discovering how one atmospheric factor influences crops, in practice the atmosphere in the future will alter in several ways simultaneously.

Some of the impending changes are more certain than others. Without doubt, CO_2 concentration is increasing worldwide at about 1.8 μ mol/mol per year and under all likely scenarios will continue to do so. This increase will enhance the global greenhouse effect, and the consensus of scientific opinion is that it will result in global warming and in changed global distributions of rainfall and other weather components. The amount of climatic change is, however, very uncertain, because even the best computer simulation models do not adequately describe features such as ocean-atmosphere linkages, and feedbacks associated with changes in atmospheric aerosols or cloud amount and distribution. At present, expert opinion is that, when the CO_2 concentration reaches double its present value of about 350 μ mol/mol, global temperatures will be about 2.5°C warmer.

In contrast to the uncertainty associated with computer models, there is clear evidence from satellite and from surface monitoring that stratospheric ozone is being depleted, probably because of emissions of chlorine- and bromine-containing substances, so that globally the column of ozone from the surface through the atmosphere is thinning at about 3% per decade, a rate that seems to be increasing. Because the chemicals believed to be causing this thinning are long-lived in the stratosphere, this thinning process is likely to continue for several decades even though the chemicals are being phased out of production. The consequence of this thinning is that ultra-violet (UV) radiation at the ground, normally severely attenuated by stratospheric O_3 will increase. There is, however, an inadequate network of surface monitoring stations to define the UV climate at the ground (*i.e.*, taking into consideration effects of cloud, dust, etc.) or to determine whether UV has increased significantly. Even if annual mean UV exposure has not changed, there are indications that thinning of stratospheric ozone, similar but less extreme than that observed in the Antarctic, can occur episodically at temperate latitudes.

Air quality in the troposphere is also changing. In particular, in many highly populated regions, emissions of hydrocarbons and exhaust gases from motor vehicles provide the precursors for photo-chemical production of tropospheric ozone. There is evidence that background tropospheric ozone concentrations have increased by about 10% per decade in Europe over the past 20 years, and that the European background has approximately doubled since the beginning of the century. Although there are no equivalent long-term records, it is likely that the tropospheric O_3, concentration in the United States and Southeast Asia has also increased. Perhaps more important than the mean background concentration of O_3 is the regular occurrence on a regional scale of high concentrations of tropospheric O_3 (typically 5 to 10 times background) associated with stagnant weather systems in summer. These concentrations have been demonstrated to reduce the growth and yield of many agricultural crops.

The atmospheric environment of many world crops of the future will undoubtedly contain more CO_2 than present, is likely to be somewhat warmer, and rainfall (and hence humidity) and cloudiness will probably also change, though in a direction unknown at present. Whether UV will increase much on a seasonal basis remains unclear, but episodes of increased UV at the ground, associated with patches of thinned stratospheric O_3, may occur, especially in temperate latitudes. Without doubt, in regions with large population densities, episodes of high concentrations of tropospheric O_3 and of other pollutants associated with fossil-fuel burning will become more common.

In this paper we review our knowledge of some of the interacting effects of these changing atmospheric conditions on agricultural crops, and we draw attention to particular gaps where further research is needed. In an attempt to structure what could otherwise be a very disparate summary, we will concentrate on two major gases, CO_2 and O_3, and will describe work that has investigated the effects of other stresses combined with them. In each case, we first briefly review responses to these gases alone.

Carbon Dioxide

Several recent reviews of responses of crops to CO_2 have been published (Goudriaan and Unsworth, 1990; Lawlor and Mitchell, 1991; van de Geijn *et. al.*, 1993; chapters 4 and 5 of this publication), and there is no advantage in repeating the detail of these here. At the leaf level, the two most well-known responses to elevated CO_2 are an increase in the rate of net photo-synthesis, P_N, and a decrease in stomatal conductance, g_s. The increase in P is much larger in C_3 species (50 to 100% when CO_2 concentration doubles) than in C_4 species (10%), but a substantial decrease in g (30 to 40%) is observed in both types of species. There is some evidence of photo-synthetic acclimation to elevated CO_2, so that increased rates of P_N do not persist at the leaf level in long-term studies with some species, but there is no evidence of acclimation in g. It is not clear whether increased CO_2 also indirectly affects plant growth, *e.g.*, by changing rates of leaf initiation, expansion or longevity, but such effects could have a large influence on crop productivity.

Much of our understanding of effects of CO_2 on plants has been gained from studies with individual leaves. There have been far fewer studies of long-term effects of elevated CO_2 on canopy-scale photo-synthesis and transpiration. Most of those reported have been in closed- or open-top chambers, though a few reports of field-scale exposures (FACE) are now becoming available. Drake and Leadley summarized the data available at that time and concluded that : (1) canopy photo-synthesis increases in elevated CO_2 when there is a sink available for the carbon; (2) the relative effect of CO_2 is greatest at highest temperatures; and (3) elevated CO_2 alters many interacting factors, such as canopy architecture and partitioning of assimilates, that mediate gas exchange of canopies and ecosystems.

As a consequence of increased net photo-synthesis, dry matter production and yield are substantially increased by elevated CO_2. Several authors have reviewed the literature and concluded that, provided there is adequate water, nutrients and pest control, yields of C_3 and C_4 crops growing in about 700 μ mol/mol CO_2 would be about 30 to 40% and 9%, respectively, greater than present yields (350 μ mol/mol CO_2) if all other climatic factors were unchanged.

Although harvest index (seed or pod yield per plant) was not altered by CO_2, yield increases were consistent with other reports for unstressed C_3 and C_4 crops.

However, in much of the world, ideal crop growing conditions, with adequate water and nutrition, are wishful thinking, and the influence of elevated CO_2 on aspects such as water-use may be much more relevant. Reviews by Morison and Eamus make it clear that, on a leaf area basis, elevated CO_2 improves the water-use efficiency (WUE – the ratio of weight of dry matter produced to weight of water transpired) of plants. On a whole-plant basis, the increase in leaf area resulting from enhanced photo-synthesis may cause the amount of water-used per plant to remain close to that of plants growing at current levels of CO_2, although the WUE as defined here increases. We return to this topic later.

CO_2 and Temperature

As discussed earlier, the atmospheric environment of the future is likely to include correlated increases of atmospheric CO_2 concentration and temperature. Long reviewed the mechanisms by which temperature and CO_2 affect photo-synthesis in C_3 species, and developed models of the response of leaf canopy carbon exchange to changes in these variables. He showed that the interaction of carbon dioxide concentration and temperature causes the temperature optimum of the light-saturated rate of CO_2 uptake to increase as CO_2 increases. Without this interaction, an increase in global temperature of 2.5°C would increase the frequency of temperatures in temperate regions that are supra-optimal for light-saturated photo-synthesis. However, if warming on this scale was coincident with a CO_2 concentration increase to above 500 μ mol/mol, the temperature optimum would increase, and this inhibition would not occur. Long went on to develop a simple model of canopy CO_2 uptake (Ac) in response to temperature, taking into account the substantial fraction of the canopy leaf area that would be shaded and therefore not photo-synthesizing at the saturated rate. Using a simple model of the diurnal variation of radiation and temperature at different latitudes. These calculations are in good agreement with observations from long-term studies reported by Oechel and Strain, Drake and Leadley, and Kimball.

Table 18.19 : Harvest indices (HI, seed or pod yield per plant/total dry mass per plant) and seed or pod yields for stands of groundnut *(Arachis hypogaea* L.) and sorghum *(Sorghum bicolour* L.) grown in controlled-environment glasshouses at 350 μ mol/mol or 700 μ mol/mol CO_2

Crop	HI		Pod/seed yield (g/plant)		Percentage yield increase at elevated CO_2
	350 μ mol/mol CO_2	700 μ mol/mol CO_2	350 μ mol/mol CO_2	700 μ mol/mol CO_2	
Groundnut	0.195	0.212	9.06	13.29	30.9
	±0.019	±0.038	±0.06	±1.70	±13.3
Sorghum	0.333	0.334	25.78	27.37	6.2
	±0.016	±0.015	±0.11	±0.85	±3.3

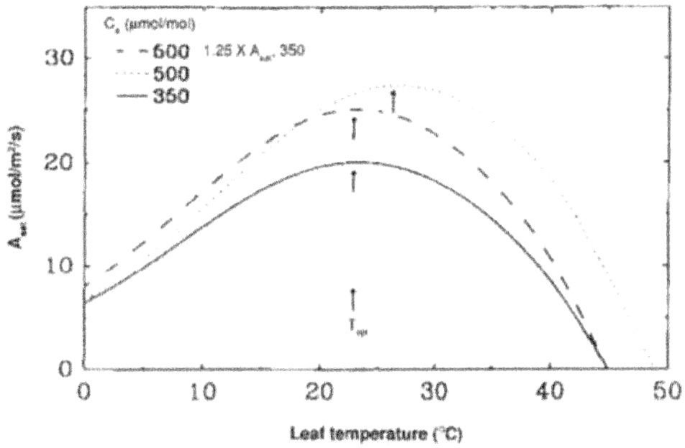

Figure 18.22 : Dependence of light-saturated rates of leaf CO_2 uptake (A_{sat}) on leaf temperature for three atmospheric CO_2 concentrations (C_a, µ mol/mol of CO_2 in air). Arrows indicate T_{opt}, the temperature at which A_{sat} is maximal for each value of C_a

Idso grew crops in field chambers in Phoenix, Arizona, at 300 and 600 µ mol/mol CO_2.

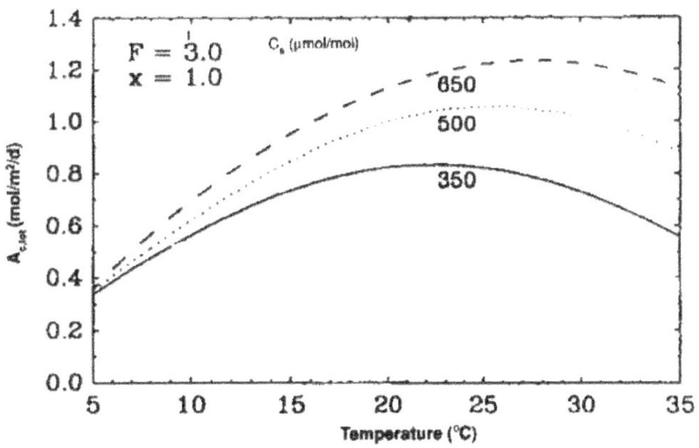

Figure 18.23 : Simulated dependence of net canopy CO_2 uptake ($A_{c,tot}$), integrated over 24 h, on temperature at three atmospheric CO_2 concentrations (C_a). The simulation is for a canopy with a leaf area index (F) of 3.0 and a rate of horizontally to vertically projected leaf area (x) of 1.0, on 9 July, latitude 52°N and assuming cloudless conditions.

Interactions between temperature and CO_2 also need to account for effect of temperature on development. Squire and Unsworth pointed out that determinate crops with discrete elements to their life cycle develop faster in higher temperatures, and so the stage of seed filling is shortened, limiting the benefits of elevated CO_2. When CO_2 concentration doubled, but daily weather data from a typical year were used, the potential grain yield was 27% larger than in a control run at 340 µ mol/mol CO_2 (11.5 and 9.0 t/ha, respectively). The date of maturity was unchanged. When daily temperatures were increased by 3°C and CO_2 was doubled, more rapid development of the crop shortened the growing season, and the potential grain yield (10.4 t/ ha) was only 15% larger than in the control, but maturity occurred 30 days earlier. The wheat model used in these simulations contains a number of elements that are sensitive to temperature. Recently, Kocabas completed a detailed investigation of the model sensitivity to changes in mean temperature at various stages of crop development, and to changes in variability of temperature in various developmental phases and over the whole season. The results indicated the particularly strong dependence of yield on temperature in the phase from emergence to double ridges in wheat, but suggested that changes in temperature variability of ±20% would not have any consistent effects on yield variability. It would be useful to incorporate Long's analysis into the wheat model to allow more precise simulation of the interactive effects of temperature and CO_2 on carbon assimilation and crop development.

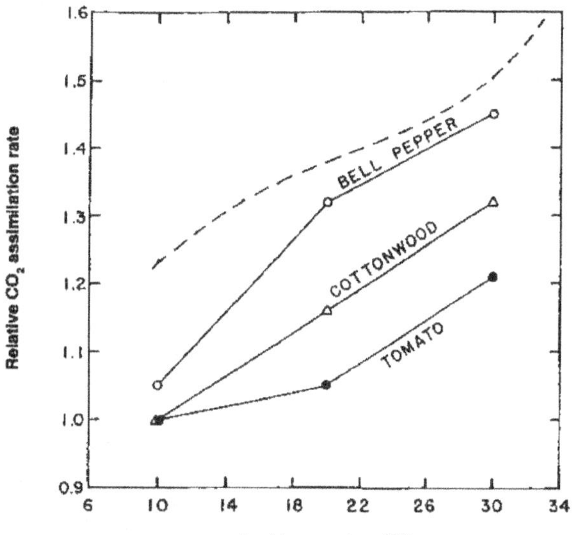

Figure 18.24 : Variation of relative CO_2 assimilation rate (net photo-synthesis at an inter-cellular CO_2 concentration of 600 μ mol/mol divided by net photo-synthesis at 300 μ mol/ mol) with leaf temperature for bell pepper, tomato and cottonwood tree. The dotted line is the relationship predicted by the analysis of Long.

Indeterminate crops, such as grass and sugar beet, continue to grow and produce yield as long as the temperature is above a minimum threshold, provided that other factors such as drought and nutrition do not limit growth. Consequently, simultaneous increases in temperature and CO_2 concentration increase the potential yield of such crops in temperate environments, where low temperatures at the beginning and end of the season usually limit productivity.

Perennial tree crops, which yield fruit, nuts and wood, pose complex problems in assessing the effects of CO_2 and temperature interactions, because their yield depends on a phased sequence of development over at least two years. Climatic warming could disrupt this sequence, for example, by providing insufficient winter chilling to synchronize spring bud break, or by advancing ovule development and leading to poor fruit set in spring. Cannell and Smith pointed out that climatic warming could either advance or delay bud burst in apples depending on the extent to which chilling requirements were met. The long-term influence of CO_2 on tree growth, and its interaction with temperature responses, is very uncertain at present.

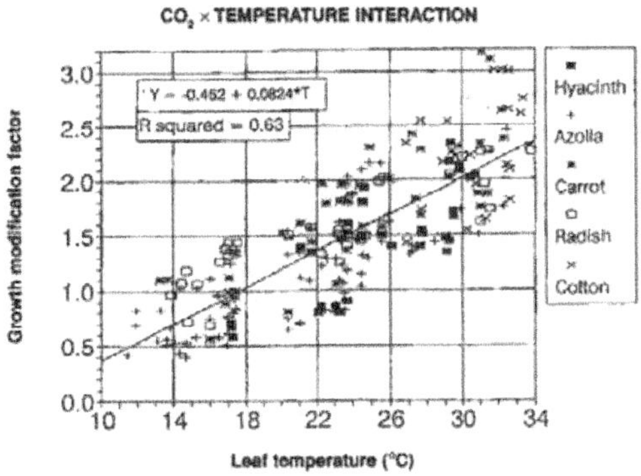

Figure 18.25 : Influence of mean air temperature on the growth modification factor (weekly dry matter production (W) at 650 μ mol/mol CO_2 divided by W at 350 μ mol/mol CO_2) for water hyacinth, azolla (water fern), carrot, radish and cotton.

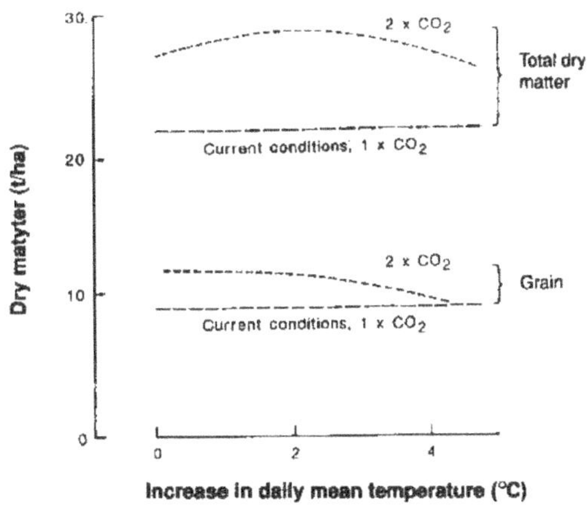

Figure 18.26 : Calculations using the model ARCWHEAT, modified for CO_2 responses to illustrate potential production of winter wheat based on weather data from Rothamsted, England. The straight lines correspond to total dry matter production and grain yield predicted at 340 μ mol/mol CO_2 and observed weather. The curves are predicted production and yield at 680 μ mol/mol CO_2 and increases of daily mean temperature up to 4°C.

CO_2 and Water aVailability

The influence of CO_2 on the water-use of crops may well prove to be the most important benefit of increased CO_2 concentrations for agriculture. Morison reviewed the recent literature and discussed the effect at a range of scales, pointing out that there are now many studies, including annual cereal crops, legumes and woody perennials, which show that, contrary to earlier assumptions, although the absolute amount of dry matter produced decreases at all CO_2 concentrations as water availability is reduced, the *relative* effect of increasing CO_2 on plant growth increases with decreasing water supply. There are at least two mechanisms responsible for this relative enhancement : (1) the reduction in stomatal conductance, with a consequent beneficial effect on leaf water potential and hence on leaf expansion; (2) greater allocation of carbon to roots in elevated CO_2, and hence improved potential for exploring the soil volume for water and nutrients. In contrast, the increased production of leaf area in elevated CO_2 acts to increase the water-use per unit ground area of crops.

It is less clear whether increasing CO_2 will decrease water-use of crop canopies in the field over periods of days or months. This is not only because of the various mechanisms described above, but also because soil evaporation must be accounted for, and feedbacks between leaf temperature, atmospheric humidity and canopy evaporation apply. Consequently, the very rough estimate that a 33% increase in dry matter production combined with a 33% decrease in transpiration in response to doubling CO_2 concentration would increase water-use efficiency by a factor of 2 is unlikely to apply. Goudriaan and Unsworth used a crop model with feedbacks between vegetation and the atmosphere to estimate that daily plant canopy water-use efficiency would increase by about 35% when carbon dioxide concentration doubled. In a simpler model, but including soil evaporation, Morison calculated that seasonal water-use by crops with leaf area indices between 2 and 6 would decrease by only 10 to 15% in doubled CO_2 concentration. Nijs *et. al.* grew ryegrass in 350 μ mol/mol CO_2 and 600 μ mol/mol CO_2. While the total water-use of the canopy was unaffected by growth in elevated CO_2, WUE was increased by 25% on a canopy basis or 87% on a leaf area basis. Results from this experiment indicated that, because the influence of elevated CO_2 on P_N and g_s was greatest early in the season, before the canopy closed, WUE probably varied seasonally, and this may be a general feature for annual crops.

Azam Ali (pers. comm.) used two methods to assess transpiration of crops of sorghum and groundnut growing at 350 and 700 μ mol/mol CO_2. Seasonal monitoring of soil water content with a neutron probe, combined with metered irrigation and subsidiary measurements of soil evaporation, enabled cumulative crop transpiration to be calculated. Porometry on individual leaves, and measurements of leaf area distribution, enabled transpiration per unit leaf area to be calculated. In groundnut but not in sorghum, elevated CO_2 reduced transpiration per unit leaf area throughout the season, but more so in the earlier stages. For the C_3 groundnut, seasonal WUE was significantly larger at 700 μ mol/mol CO_2 (4.4 ± 0.2 g/kg) than for 350 μ mol/mol CO_2 (3.0 ± 0.2 g/kg); equivalent values for sorghum (4.9 ± 0.3 and 4.2 ± 0.3 g/kg, respectively) did not differ significantly. The figure also shows that, in spite of the increased WUE in groundnut, the total amount of water-used in the season was about the same (450 mm) in both CO_2 treatments.

On an even larger scale, de Bruin and Jacob estimated the influence of doubling CO_2 concentration on regional-scale transpiration, allowing for interactions between transpiration, leaf temperature and the planetary boundary layer. They concluded that, for vegetation where stomatal conductance decreased by 34% in response to elevated CO_2, but leaf area was unchanged, regional transpiration on a typical summer day would decrease by about 11 % for short crops such as cultivated grass and 17% for tall crops such as plantation forests. If negative feedback from the planetary boundary layer had been ignored, these values would have been over-estimated (15 and 24%, respectively).

CO_2 and Ozone

When Krupa and Kickert attempted to review the interactive responses of crops to CO_2 and O_3 they found no publications describing experimental work. We summarize here the few papers that have been published since then, and comment on other work known to be in progress. We briefly review the effects of ozone alone later in this paper.

Allen made an estimate of the possible scale by which CO_2 might influence responses of soybeans to O_3. He used published results relating O_3 exposure to yield and a model of the sensitivity of stomatal conductance for soybean to CO_2. According to these models, doubling CO_2 concentration would reduce g_s by about 30% and thus decrease the stomatal uptake of O_3 by the crop. Allen estimated that this would result in a yield decrease of about 15% compared with yield in ambient air. This estimate takes no account of the large direct effects of CO_2 on soybean growth and assumes no interactions between the two gases.

Scientists at the USDA Beltsville Research Center have studied soybeans growing in open-top chambers in the field and exposed to CO_2 and O_3. Kramer *et. al.*, in a preliminary study, showed that the yield loss attributed to ozone at the site was about 12%, but when CO_2 concentrations were increased to 500 μ mol/mol the yield loss was only 6.7%. Mulchi *et. al.* reported a more complex design with three O_3 treatments (charcoal-filtered, CF, non-filtered, NF, NF + 40 μ mol/mol O_3) and three CO_2 treatments (ambient, + 50, + 150 μ mol/mol). Increasing CO_2 concentration increased grain yield and grain oil content, and decreased protein content. Increasing O_3 reduced grain yield but did not significantly alter oil or protein content. Ozone caused a significant decrease in yield compared to the charcoal-filtered controls. When CO_2 was added to the ozone treatments, it partially counteracted the negative effect of O_3 so that yields in chambers with plus 150 μ mol/mol CO_2 and NF air were similar to those in ambient CO_2 and CF air. The results suggested that most effects of CO_2 and O_3 were additive.

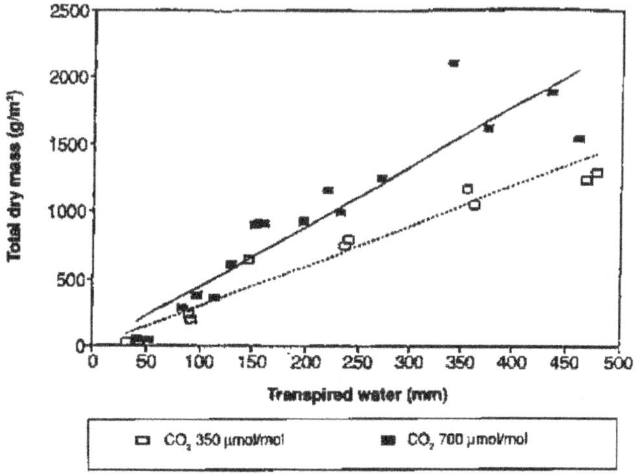

Figure 18.27 : Seasonal relationship between dry mass and transpired water for crops of groundnut growing in controlled-environment glasshouses at 350 μ mol/mol CO_2 and 700 μ mol/mol CO_2. Regression lines are shown dotted for 350 μ mol/mol and solid for 700 μ mol/mol. The slopes of the regression line are the water-use efficiencies (WUEs).

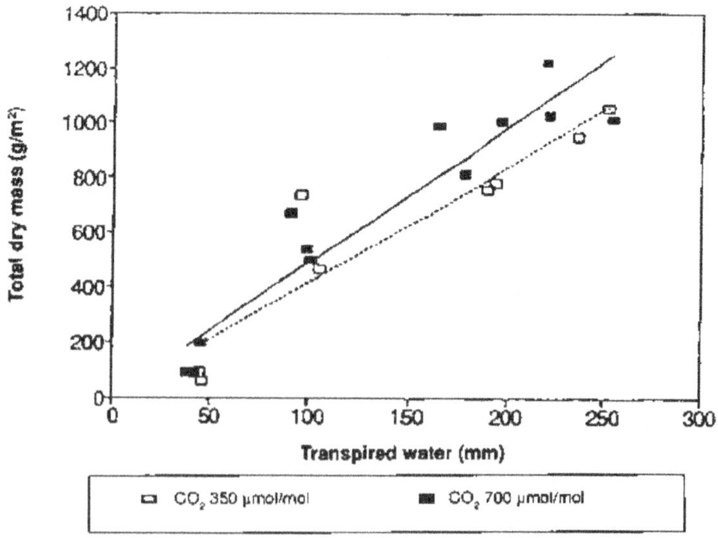

Figure 18.28 : As 27 but for a crop of sorghum.

Barnes and Pfirrman grew radishes in a phytotron in Munich, Germany, at two CO_2 concentrations (385 and 765 μ mol/mol) and two O_3 concentrations (20 μ mol/mol and 73 μ mol/mol), and studied gas exchange, growth and productivity, and mineral composition of the crop. In clean air the typical effects of elevated CO_2 on photo-synthesis (increased) and stomatal conductance (decreased) were observed; the extra carbon assimilation stimulated root growth rate by 43% but there was no significant effect on shoot growth or leaf area. At ambient CO_2, the high O_3 treatment depressed photosynthesis by 26% (compared to the low treatment) and induced a slight reduction in g_s, with the net result that water-use efficiency declined. The reduction in carbon uptake was reflected in reduced growth, with roots more affected than shoots. Interactive effects of ozone and CO_2 were generally consistent with a reduction in ozone-induced responses in elevated CO_2. Early in the experiment, elevated CO_2 reduced the effect of O_3 on photo-synthesis, but later this benefit disappeared, perhaps because long-term exposure to O_3 has an overall limiting effect on plant response to CO_2. The effects of CO_2 and O_3 on g_s appeared to be additive. Work is in progress in Munich on responses of grass swards to O_3, CO_2 and water availability. Preliminary results indicate that the growth depression which is observed as water is withheld in ambient CO_2 is alleviated at higher CO_2 levels. Responses of yield and water-use efficiency to the mixtures of CO_2 and ozone seem to be additive rather than synergistic.

In Newcastle, England, work is in progress exposing winter and spring cultivars of wheat to two CO_2 concentrations and two O_3 concentrations. Early results indicate that after 50 days' exposure there was evidence of down-regulation of photo-synthesis in response to prolonged exposure to elevated CO_2; this was most pronounced in the winter cultivars. Long-term exposure to O_3 resulted in a decrease in the light-saturated rate of CO_2 assimilation, partial stomatal closure, and accumulation of water soluble carbohydrate and starch in leaves. These resulted in decreased growth, with root growth more severely affected than shoot growth. In plants exposed to elevated CO_2 and O_3, effects on stomatal conductance were less than additive, and CO_2 enhancement effects on photo-synthesis and growth were reduced by O_3.

In summary, from the very limited experimental results available at the time of writing, interactions between exposures to elevated O_3 and CO_2 seem to result in approximately additive effects on growth and yield. Both gases decrease stomatal conductance, and this results in less uptake of O_3 in elevated CO_2 treatments than in ambient air. Several authors have noted that this reduces (or at least delays) the damaging effects of O_3. In principle, the reverse is also true, *i.e.*, increased O_3, by reducing stomatal conductance, negating some of the benefits to growth that elevated CO_2 confers. However, the O_3 exposure necessary to induce stomatal closure is generally sufficient to cause visible leaf injury, and this disbenefit is likely to be far more damaging than reduced CO_2 uptake through stomata. More field-based studies are necessary to assess fully the effects of the two gases in combination.

Table 18.20 : Direct and interactive effects of CO_2 and O_3 treatments on growth, yield components and grain quality characteristics of soybean growing in open-top field chambers.

Chamber treatments		Shoot biomass [1]	Leaf area [1]	SLW [1]	Pods/plant	Seeds/plant	Weight of 100 seeds	Grain yield	Grain oil	Grain protein
CO_2	O_3	(g)	(dm²)	(mg/cm²)			(g)	(g)	(%)	(%)
CO_2 treatment means										
Ambient	-	55.3	55.7	2.88	51.7	128	16.9	436	20.4	40.4
+50 m mol/mol	-	61.1	58.1	3.07	57.8	140	17.7	497	22.3	39.3
+150 m mol/mol	-	59.2	58.0	2.99	60.9	144	17.6	509	22.1	39.2
LSD(0.05)		3.2	NS	NS	7.4	11	0.5	41	0.5	1.1
O_3 treatment means										
-	CF	58.9	60.1	2.73	63.7	153	18.2	558	21.9	39.4
-	NF	62.5	59.7	3.04	52.1	133	17.8	472	22.1	39.5
-	NF+O_3	54.3	52.0	3.17	54.6	127	16.2	412	21.7	40.1
LSD(0.05)	3.2	7.5	0.36	7.4	11	0.5	0.5	NS	NS	
CO_2 and O_3 treatment										
Ambient	CF	57.6	61.4	2.31	56.2	147	17.5	513	21.4	40.1
	NF	59.2	57.5	3.07	50.9	122	17.6	432	21.7	40.4
	NF+O_3	49.2	48.2	3.26	48.2	117	15.5	363	21.2	40.8
+50 m mol/mol	CF	60.4	61.8	2.99	66.6	155	19.1	594	22.1	39.6
	NF	66.7	58.0	3.12	50.6	135	17.7	474	22.9	38.6
	NF+O_3	56.4	54.4	3.11	55.8	130	16.3	424	22.1	39.7
+ 150 m mol/mol	CF	58.6	57.1	2.88	68.4	157	18.1	568	22.4	38.6
	NF	61.7	63.6	2.94	54.7	142	18.0	510	21.9	39.4
	NF+O_3	57.4	53.3	3.14	59.6	134	16.8	448	22.0	39.8
	Av.	58.6	57,3	2.98	56.8	137	17.4	480	21.9	39.6
LSD(0.05)		5.7	NS	0.36	NS	NS	0.9	53	NS	NS
CV (%)		5.3	10.7	9.3	9.9	6.1	2.4	7.0	1.6	2.2

[1] Vegetative samples were collected on 15 August, 1989.

NS = not significant : LSD = Least significant difference; CV = Coefficient of variance.

CO_2 and UV-B Radiation

Many studies of effects of UV-B radiation on crops are seriously flawed because UV exposures were either inadequately specified and/or were unrealistically large. Tevini, in a thorough review of effects of UV-B radiation on plants, pointed out that, since the spectrum of UV from artificial light sources differs from that in the solar spectrum, and since many photo-biological processes are strongly wavelength dependent, it is important to use weighting functions based on the action spectrum for specific responses before experiments in different exposure systems can be compared. Unfortunately, action spectra for many responses remain unknown. In addition, many plant responses to UV-B are larger when plants are growing at low light levels typical of growth chambers and some glasshouses than they are in full sunshine, probably because natural protective pigments are inadequately synthesized and repair processes are restricted in low light. For this reason, field-based studies of crops are most relevant for estimating responses of productivity and yield to UV-B. Even so, most field studies have used continuous UV-B exposures corresponding to 15 to 25% reductions in stratospheric O_3 *i.e.* roughly 30 to 50% increases in UV-B above present values. Such increases are much larger than are likely to occur as long-term means in agricultural regions under any likely depletion of stratospheric ozone in the next 30 years.

Effects of UV-B on yield of crops growing in the field have produced very variable results. Work in the late 1970s and early 1980s by Biggs and colleagues in Florida, using very high UV exposures, found yield reductions in only about half the crops studied. Work in Germany, also at high UV exposures, found no UV-B effects on cabbage, lettuce and rape. More recently, a large field-based study in Beltsville, Maryland, by Teramura, has revealed a high degree of intra-specific variability among soybean cultivars. The two most sensitive cultivars were grown for five seasons in the field in two UV treatments, corresponding to 16 and 25% ozone depletion simulations. In one cultivar (Essex) yield was reduced by 20 to 25% at the high UV treatment; in the other cultivar, yield generally increased by 10 to 22% in this treatment. Neither cultivar showed a consistent significant yield change at the lower UV exposure. The conclusion from these studies must be that, under any realistic scenario of increasing UV-B radiation, crop yields are unlikely to be altered to a significant extent. We therefore turn to the question of whether UV-B exposure alters the sensitivity of crops to other atmospheric changes.

Effects of the combination of elevated CO_2 and UV-B radiation on crop growth and yield have been addressed in very few studies. Teramura and his colleagues grew wheat, rice and soybean in a glasshouse experiment. Treatments in the factorial design were : ambient CO_2 (350 (m mol/mol), ambient UV, elevated CO_2 (650 μ mol/mol) and elevated UV-B (corresponding to a 10% decrease in stratospheric O_3 at the equator). Compared to the control, seed yield and total biomass increased significantly in elevated CO_2

for all three species. However, when UV-B and CO_2 were increased simultaneously, no increase in either seed yield (wheat and rice) or total biomass (rice) was observed compared to the control. In contrast, in soybean, the increases in seed yield and biomass induced by CO_2 alone were maintained in the high CO_2/high UV-B environment. Studies of leaf gas exchange indicated that UV-B reduced the apparent carboxylation efficiency in wheat and rice, but not in soybean.

Results of a CO_2/UV study on seedlings of sunflower and maize reported by Tevini are confounded by temperature changes in some treatments. A 25% increase in UV-B at 340 μ mol/mol CO_2 changed sunflower and maize dry weights (W) by -14 and -24%, respectively (compared to ambient controls). The same UV-B increment combined with a +2°C temperature change altered W by +5 % in sunflower and +31 % in maize. Adding a doubled CO_2 concentration to the elevated UV and temperature regime altered W by +19 and +32% for sunflower and maize, respectively. The incomplete experimental design limits interpretation of this study, but the results suggest that temperature and CO_2 are much more important influences on growth than any likely realistic increases in UV-B.

Ozone

There are a number of recent reviews of the mechanism of action of O_3 on plants, and of crop responses to O_3; consequently, we will only briefly summarize the main features. The response of plants to O_3 may be viewed as the culmination of a sequence of physical, bio-chemical and physiological events. The O_3 diffuses from the air into the leaf through stomata, and these exert control on O_3 uptake. Plants are able to detoxify O_3 or its metabolites and can repair or compensate for O_3 impacts, so visible signs of ozone injury may not occur if the rate of O_3 uptake is sufficiently small. When resources have to be devoted to repair or compensation mechanisms, or when O_3 uptake is too large for full repair to be achieved, growth and yield may be reduced in the absence of visible injury by O_3. The principal modes of action of O_3 on plants are through injury to proteins and membranes, reduction in photo-synthesis (*i.e.*, carbon gained), changes in allocation of carbohydrate, and acceleration of senescence. The main result of O_3 exposure is therefore a reduction in the capacity of the plant to accumulate photo-synthate through loss in photosynthetic capacity and increased foliage senescence.

Methods for exposing crops to ozone in experiments designed to develop yield/exposure responses are well established. The most commonly used technique is the open-top field chamber, which enables crop stands to be grown in either charcoal-filtered air or in air to which known amounts of O_3 have been added. Co-ordinated programmes in the United States and Europe have investigated crop responses to O_3 in open-top chambers, and relationships between yield and seasonal mean O_3 concentration are available for many of the major world crops. The seasonal mean index, however, has been shown to be inadequate in relating ozone exposure to effects, primarily because of its inability to consider duration of exposure and its consideration of all concentrations as equal in their effect. Both duration and concentration are important in the effect of ozone (Hogsett *et. al.*, 1988). In particular, studies have demonstrated that ozone's effect is cumulative and that higher concentrations are more important than lower concentrations in causing an effect. Consequently, the cumulative, peak-weighted indices are more appropriate because of their biological relevancy. Examples of such indices include SUM06 (in units of nmol/mol-h) which cumulates all concentrations equal to or greater than 0.06 nmol/mol, or SIGMOID which cumulates all concentrations during the season and weights all concentrations according to a sigmoid function. The US Environmental Protection Agency (EPA) Draft Ozone Criteria Document reviewed and tabulated the latest information concerning effects of ozone on crop yields using both the seasonal mean, and two of the cumulative, peak-weighted indices. The data for crop yield response are primarily from the National Crop Loss Assessment Network (NCLAN). Yield reductions in the NCLAN studies are calculated relative to a typical background O_3 concentration of SUM06 of O_3 or a 7-h seasonal mean of 25 nmol/mol. Yield losses of 10% or less in 50% of the crops (12 crop species, 54 studies including well-watered and drought studies) would occur with SUM06 exposure concentrations of 26.4 nmol/mol-h in three months, or 49 nmol/mol 7-h seasonal mean. Some species/cultivars were particularly sensitive; 11% would be expected to have a yield reduction of 10% at a 7-hr seasonal mean of less than 35 nmol/mol. Similarly, 18% of the crops/cultivars are expected to have yield reductions of 10% or less at 3-month SUM06 concentrations of less than 10 nmol/mol-h. Both of these exposure values are quite low compared to the decade-average (1982-1991) exposure (3 months) from all monitoring sites across the United States (SUM06 = 29.5 nmol/mol-h; 7-h mean = 54 nmol/mol). In general, grain crops are less sensitive than others, but within-species variability and sensitivity may be greater than between species. Krupa and Kickert reviewed various published lists of the sensitivity of crop yields to O_3 onion, spinach, potato, alfalfa and cotton appear to be particularly sensitive crops, although it is possible to find resistant cultivars of all these species. Legumes range between sensitive and intermediate, and cereal and grass crops tend to be relatively resistant.

There is also a very large literature describing biotic and abiotic factors that modify plant response to 0. We will concentrate in this paper on only a few interacting factors that are particularly relevant in terms of future changes in the atmospheric environment.

Ozone and Temperature

When interactions between ozone exposure and temperature were reviewed by the US EPA it was concluded that there were variable and conflicting results. More recent analysis suggests that many of these studies in controlled environments confounded effects of temperature with changes in vapour pressure deficit (VPD) because relative humidity was kept constant while temperature was increased. Since VPD can have profound effects on stomatal responses, evaporation rates and leaf expansion, it is not surprising that these early results have led to confusion. Recently, Todd *et. al.* designed an experiment with tomato seedlings in which differences in VPD at different temperatures were minimized during O_3 exposure. This study showed that, out of 11 growth variables measured, the only significant modifications by temperature of the effects of O_3 were on stem fresh weight and specific leaf area (leaf area/leaf dry weight). The authors suggested that VPD probably plays a more important role in determining sensitivity to O_3 than temperature. It is important to have more studies to clarify this issue, as many field-based studies of ozone responses use open-top chambers in which temperatures are increased by a few degrees above ambient. If either the changed temperature or the changed VPD in such chambers alters the sensitivity of the crops to O_3 there would be serious consequences for the application of much of our knowledge of the O_3 sensitivity of crops that has been gained from field-chamber experiments.

A second important O_3/temperature interaction, particularly affecting perennial species, including trees, is winter hardiness. Several studies have shown that exposures to O_3 reduce the frost-hardiness; Davison *et. al.* reviewed the subject. Whilst most of this research has been with woody species, Barnes *et. al.* showed that daily exposures to about 80 nmol/mol O_3 for 7 days significantly reduced the survival of pea seedlings after exposure to night time temperatures of about -4°C. It seems likely that, as with sulphur dioxide, exposure to ozone could alter the frost hardiness of a number of agricultural crops. This response would be most likely to be significant in the event of an early autumn frost on early-sown crops such as winter cereals. There are suggestions that exposure of trees in summer to ozone alters the hardiness the following winter. Eamus and Murray pointed out that, even in severe winters, there are brief periods of mild temperatures that induce partial dehardening of woody species. They speculated that O_3 increases tree predisposition to dehardening in such conditions and consequently puts trees at greater risk from low temperatures. There are likely to be significant differences between species in this phenomenon. For example, in Florida, Eissenstat *et. al.* found that, although O_3 reduced frost hardiness of citrus and avocado, the effects were small and the likelihood of significantly changing frost resistance is slight.

Ozone and Water Availability

The availability of soil water is one of the strongest influences on crop growth and productivity. When available, irrigation is commonly used to alleviate water stress, but in conditions of large evaporative demand, even irrigated crops may experience water stress. The most immediate effects of water stress are on plant-water potential, and may lead to reductions in leaf expansion rates and altered partitioning of dry matter, with increased assimilate partitioning to roots. When stomatal closure occurs, there is a resulting reduction in transpiration and CO_2 uptake and an increase in leaf temperature.

Water Stress and O_3 Responses

There has long been a general belief that water stress reduces the magnitude of any adverse effects of O_3 *i.e.*, leaf injury, growth and yield reductions. This belief was based on observations that stomatal closure, induced by drought, reduced the rate and quantity of O_3 absorbed by leaves. However, more recent results show that the O_3/water availability interaction is complex, and probably depends on the magnitude and timing of the water stress (which has seldom been adequately defined in air pollution experiments). One difficulty in quantifying the interaction is the very large effect that water stress alone may have on yield, so that any influence of O_3 is sometimes masked.

Temple *et. al.* studied the effects of O_3 and water stress on cotton growing in California. In this typically hot, dry season, the water-stressed plants wilted frequently, and yielded much less than the well-watered controls; nevertheless, when compared to yields of plants in a background O_3 concentration of 25 nmol/mol, and expressed as a percentage, the estimated yield loss at a seasonal mean 0, concentration of 50 nmol/mol was 7% in the well-watered plants and 2% in the water-stressed crop. Clearly the absolute yield loss from drought was much larger than the yield loss attributed to O_3 in either treatment.

Heagle *et. al.* reviewed six studies of soybean responses to O_3 and water stress, and concluded that in only three were there significant interactions, *i.e.*, the clear negative relationships between yield and O_3 exposure observed with well-watered plants were much reduced with water stressed plants. More recently, Heggestad and Lesser analysed three years of data for four soybean cultivars. They concluded that, in most cases, the relationships between yield and O_3 concentration had similar slopes for the water-stressed and well-watered treatments. If this conclusion applies more generally, the extra yield losses which would result from O_3 exposure could be estimated if a yield response to drought alone is known. Clearly, there is still uncertainty over the influence of water stress on the form of the yield response to O_3. This uncertainty is unlikely to be resolved until better experimental designs with more precise specifications of the degree of water stress are developed.

O_3 and Water-Use Efficiency

For irrigated crops where water supplies may be limited or expensive, it is important to know whether O_3 would influence the water-use efficiency (WUE) and consequently the amount of water required in a season. Reich *et. al.* exposed well-watered soybeans to 130 nmol/mol O_3 for 7 hours daily, and found a 25% decrease in WUE compared with controls in 10 nmol/mol O_3. Similar results were found for alfalfa, probably indicating that carbon dioxide uptake was reduced more by O_3 than water loss. Tingey *et. al.* also reported a significant decrease in WUE in soybeans exposed to episodic exposure regimes of 20-30 nmol/mol-h (3-month SUM06). The study found the O_3 did not close stomata, but rather affected CO_2 assimilation. A significant increase in leaf construction cost was reported, indicating that both carbon and water are used inefficiently by plants exposed to ozone. In contrast, Greitner and Winner found that O_3 exposure *increased* the WUE of radish and soybeans. This disagreement probably reflects variation with exposure dynamics, stage of development, and genetic variation between and within species.

Recently Barnes and Pfirrman investigated effects of O_3 and elevated CO_2 singly and in combination on the WUE of radish. Both gases reduced stomatal conductance, and the combination of gases reduced conductance still further. In the combination of O_3 and CO_2, there was initially no significant effect of O_3 (*i.e.*, the response of WUE was similar to that observed in elevated CO_2), but as growth progressed, the WUE was significantly reduced in the combination treatment compared to in elevated CO_2 alone. This may be a response to accelerated senescence induced by O_3 exposure.

In summary, relatively severe water stress may reduce the yield losses attributable to O_3 in some crop species but, in general, the yield loss resulting from the water stress outweighs the benefits of the O_3 protection. For well-watered crops, O_3 exposure may change the water-use efficiency compared with crops in clean air, but the direction and magnitude of change probably depend on exposure and genetic factors. The influence of O_3 and CO_2 in combination on WUE is potentially important and merits further study.

Ozone and UV-B Radiation

Krupa and Kickert found no reports of O_3/UV-B interactions in their review. They assessed potential risks on a geographical basis, using distributions of major crops, tropospheric ozone concentrations and UV-B irradiance. They suggested that interactions might involve episodic exposure to the two stresses, so that peaks of O_3 would coincide with lower UV-B irradiance and *vice versa*. Runeckles and Krupa have developed this concept further, arguing that, as tropospheric O_3 increases, UV-B irradiance at the surface is reduced, because the O_3 absorbs some UV. In theory, this effect could be important, in spite of the relatively small contribution of tropospheric O_3 to the total atmospheric O_3 column, because scattering by aerosols and molecules in the troposphere increases the radiation path length. In an attempt to quantify this effect, Albar compared the measured solar spectral irradiance I (1) at the ground near Nottingham, England, on two days when the tropospheric O_3 concentration was 51 and 84 nmol/mol, respectively, and stratospheric O_3 was constant. However I (1) at longer wavelengths was also greater (by about 20 to 25%), and this increase is most probably a consequence of less aerosol (dust) being present on the low O_3 day. Since this aerosol effect would also apply in the UV-B waveband, it seems likely that the UV-B increase *directly* attributable to reduced tropospheric O_3 was no more than 10% to 15%. Variations in aerosol from day to day seem likely to be more important than variations in tropospheric O_3 in modulating the intensity of UV-B at the ground, but further observations would be valuable.

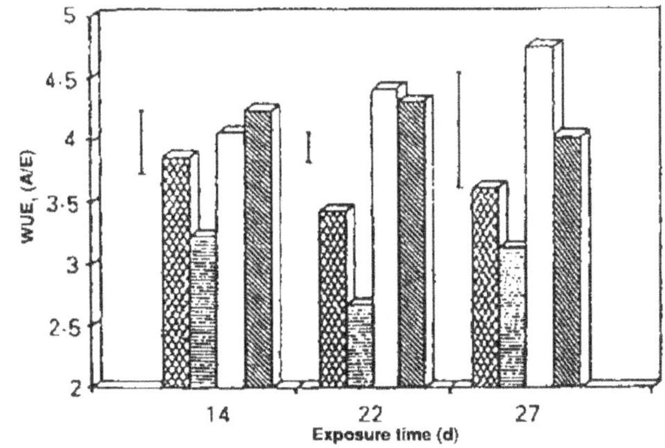

Figure 18.29 : Instantaneous water-use efficiency WUE (μ mol CO_2 assimilated (A) per mol HO transpired (E) for radish (*Raphanus sativus* L. cv. Cherry Belle) plants exposed in climatic chambers to CO_2 and O_3 Treatments were 350 μ mol/mol CO_2 + 20 nmol/mol O_3 (first columns) or + 80 nmol/mol O_3 (second columns); 750 μ mol/mol CO_2 + 20 nmol/mol O_3 (third columns) or + 80 nmol/mol O_3 (fourth columns). Error bars show lsd at $p=0.01$ for each of the sampling times.

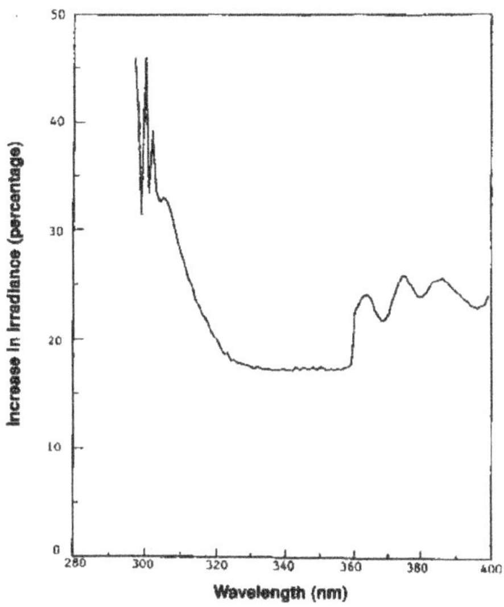

Figure 18.30 : Variation of percentage increase in solar spectral irradiance ((I(λ, 51)-I (λ, 84))/I (λ, 84)) with wavelength λ between days when the tropospheric ozone concentration was 84 nmol/mol O_3 and 51 nmol/mol O_3.

As mentioned earlier, many investigations of plant responses to UV-B are seriously flawed by inadequate specification and/or excessive UV exposure. An improved field-based system at Raleigh, North Carolina, using open-top chambers for studying crop responses to UV and O_3 has been used for three seasons of research on soybeans. The system exposed soybean crops from emergence to maturity to UV treatments (ranging from ambient to about twice ambient biologically effective UV-B) and to O_3 treatments giving seasonal mean 12 h/d O_3 concentrations from 14 to 83 μ mol/mol.

The ozone treatments resulted in reductions of photo-synthesis, accelerated senescence, and reduced yield, in agreement with many other published studies. In contrast, the UV treatments, even at this relatively large UV irradiance, did not induce any significant changes in photo-synthesis or yield, and there were no UV/O_3 interactions.

Fiscus *et. al.* attempted to reconcile the lack of response to UV reported in the Raleigh experiments with other reports which indicate that increased UV-B causes physiological dysfunction and reductions in crop yields. They concluded that three factors make the conclusions of other field and glasshouse studies, at best, hard to interpret and possibly misleading : a frequent failure to monitor UV-B adequately; a tendency to underestimate UV exposures when relying on model calculations which do not allow sufficiently for effects of dust in the atmosphere; and no adjustment of UV exposure for seasonal and daily weather changes (and hence a tendency towards unrealistically large exposures).

Although it would be useful to see further well-designed field studies of UV/O_3 interactions using a wide range of crop and natural species, on the evidence of the careful work of Booker, Miller, Fiscus and their colleagues, it seems unlikely that UV/O_3 interactions are of any importance for crop productivity, and it seems clear that O_3 poses a much greater threat to yields of many crops than any likely increases in UV-B radiation.

Concluding Remarks

In our introduction we emphasized the need for knowledge of how interacting environmental changes will affect crop productivity. Many examples in this paper show that misleading conclusions can be drawn by considering crop responses to single factors. However, the experimental demands of studying multiple factors are great, and it is necessary to define priorities. In our view, the following interactions should have high priority :

- Interactions of CO_2 and temperature. If the benefits of increased CO_2 in terms of increased potential carbon assimilation by C_3 species are not achievable at low temperatures, there are important implications for agriculture in cool climates. Equally, the merits of C_3 *versus* C_4 crops may need reassessing in warmer climates in view of the direct benefits of CO_2 to C_3 species.

- Canopy scale studies of effects of CO_2 on water-use efficiency, and on the interacting effects of O_3. Although it is clear that increased CO_2 enhances WUE at the leaf or single plant level, there are very few studies relevant to field-grown crops. Equally, the very limited studies of effects of O_3 and CO_2 on WUE suggest that responses are non-additive, and this is an important issue to resolve for agriculture in Europe, the eastern United States and other regions where tropospheric O_3 concentrations influence crop productivity.

It is disappointing that, in spite of considerable interest in the effects of UV-B radiation on plants, much of the research in the past decade has been seriously flawed by employing very unrealistic UV-B exposures and/or by using highly artificial growth conditions. On the evidence of the few well-designed studies of crop responses to UV-B that we have reviewed here, we do not believe that UV-B poses serious threats to crops, and have not identified any important interactions between UV-B and other stressors that would make further research of this type a particularly high priority.

Chapter 19

The Impact of Salinity Stress

INTRODUCTION

According to the FAO Land and Plant Nutrition Management Service, over 6% of the world's land is affected by either salinity or sodicity. The term *salt-affected* refers to soils that are saline or sodic, and these cover over 400 million hectares, which is over 6% of the world land area. Much of the world's land is not cultivated, but a significant proportion of cultivated land is salt-affected. Of the current 230 million ha of irrigated land, 45 million ha are salt-affected (19.5 per cent) and of the 1,500 million ha under dryland agriculture, 32 million are salt-affected to varying degrees (2.1 per cent).

Table 19.1 : Regional distribution of salt-affected soils, in million hectares.

Regions	Total area	Saline soils		Sodic soils	
	Mha	Mha	%	Mha	%
Africa	1,899	39	2.0	34	1.8
Asia, the Pacific and Australia	3,107	195	6.3	249	8.0
Europe	2,011	7	0.3	73	3.6
Latin America	2,039	61	3.0	51	2.5
Near East	1,802	92	5.1	14	0.8
North America	1,924	5	0.2	15	0.8
Total	12,781	397	3.1%	434	3.4%

Source : FAO Land and Plant Nutrition Management Service.

Salinity occurs through natural or human-induced processes that result in the accumulation of dissolved salts in the soil water to an extent that inhibits plant growth. Sodicity is a secondary result of salinity in clay soils, where leaching through either natural or human-induced processes has washed soluble salts into the sub-soil, and left sodium bound to the negative charges of the clay.

A *saline soil* is defined as having a high concentration of soluble salts, high enough to affect plant growth. Salt concentration in a soil is measured in terms of its electrical conductivity, as described in the section below on measurements. The USDA Salinity Laboratory defines a saline soil as having an EC_e of 4 dS/m or more. EC_e is the electrical conductivity of the 'saturated paste extract', that is, of the solution extracted from a soil sample after being mixed with sufficient water to produce a saturated paste. However, may crops are affected by soil with an EC_e less than 4 dS/m. The moisture content of a drained soil at field capacity may be much lower than the water content of its saturated paste. Further, under dryland agriculture, the soil water content might drop to half of field capacity during the life of the crop. The actual salinity of a rain-fed field whose soil had an EC_e of 4 dS/m could be 8-12 dS/m. As described below, this would severely limit yield of most crops.

TYPES AND CAUSES OF SALINITY

Natural or Primary Salinity

Primary salinity results from the accumulation of salts over long periods of time, through natural processes, in the soil or groundwater. It is caused by two natural processes.

The first is the weathering of parent materials containing soluble salts. Weathering processes break down rocks and release soluble salts of various types, mainly chlorides of sodium, calcium and magnesium, and to a lesser extent, sulphates and carbonates. Sodium chloride is the most soluble salt.

The second is the deposition of oceanic salt carried in wind and rain. 'Cyclic salts' are ocean salts carried inland by wind and deposited by rainfall, and are mainly sodium chloride. Rainwater contains from 6 to 50 mg/kg of salt, the concentration of salts decreasing with distance from the coast. If the concentration is 10 mg/kg, this would add 10 kg/ha of salt for each 100 mm of rainfall per year. Accumulation of this salt in the soil would be considerable over millennia. The amount of salt stored in the soil varies with the soil type, being low for sandy soils and high for soils contain a high percentage of clay minerals. It also varies inversely with average annual rainfall. For example, in Western Australia, the salt content of a 40 μ profile ranges from 170 to 950 tonne/ha for rainfall averaging from 1000 mm to 600 mm per year.

Table 19.2 : Concentration of salts in rain and seawater. The composition of rainwater varies greatly depending on prevailing winds and distance from the coast, and the table gives the composition of rainwater from a northern hemisphere source. It is measured as mg/kg or ppm (parts per million). The composition of seawater is uniform around the globe, and is expressed as g/kg or ppt (parts per thousand). The electrical conductivity of rainwater is about 0.01 dS/m, and of seawater is 55 dS/m.

Ion	rainwater (local)		seawater (global)	
	mg/kg (ppm)	(μmol/L) μM	g/kg (ppt)	(mmol/L) mM
Sodium (Na^+)	2.0	86	10.8	470
Chloride (Cl^-)	3.8	107	19.4	547
Sulfate (SO_4^{2-})	0.6	6	2.7	28
Magnesium (Mg^{2+})	0.3	11	1.3	53
Calcium (Ca^{2+})	0.1	2	0.4	10
Potassium (K^+)	0.3	8	0.4	10
Total	7.0		35.0	

Secondary or Human-induced Salinity

Secondary salinisation results from human activities that change the hydrologic balance of the soil between water applied (irrigation or rainfall) and water used by crops (transpiration). The most common causes are (i) land clearing and the replacement of perennial vegetation with annual crops, and (ii) irrigation schemes using salt-rich irrigation water or having insufficient drainage. Prior to human activities, in arid or semi-arid climates, the water used by natural vegetation was in balance with the rainfall, with the deep roots of native vegetation ensuring that the water tables were well below the surface. Clearing and irrigation changed this balance, so that rainfall on the one hand, and irrigation water on the other, provided more water than the crops could use. The excess water raises water table and mobilises salts previously stored in the sub-soil and brings them up to the root zone. Plants use the water and leave the salt behind until the soil water becomes too salty for further water uptake by roots. The water table continues to rise, and when it comes close to the surface, water evaporates leaving salts behind on the surface and thus forming a 'salt scald'. The mobilised salt can also move laterally to water courses and increase their salinity.

Irrigated lands of the world in 1987 totalled 227 Mha. In many irrigated areas, the water table has risen due to excessive amounts of applied water coupled with poor drainage. In most of the irrigation projects located in semi-arid and arid areas, the problems of water-logging and soil salinity have reached serious proportions even before the full potential of the irrigation project could be realised. Most of the irrigation systems of the world have caused secondary salinity, sodicity or water-logging. The proportion of salt-affected irrigated land in various countries ranges from a minimum of 9% to a maximum of 34%, with a world average of 20%. Irrigated land is only 15% of total cultivated land, but as irrigated land has at least twice the productivity of rainfed land, it may produce one-third of the world's food.

Table 19.3 : Global estimate of secondary salinisation in the world's irrigated lands.

Country	Total land area cropped	Area irrigated		Area of irrigated land that is salt-affected	
	Mha	Mha	%	Mha	%
China	97	45	46	6.7	15
India	169	42	25	7.0	17
Soviet Union	233	21	9	3.7	18
United States	190	18	10	4.2	23
Pakistan	21	16	78	4.2	26
Iran	15	6	39	1.7	30
Thailand	20	4	20	0.4	10
Egypt	3	3	100	0.9	33
Australia	47	2	4	0.2	9
Argentina	36	2	5	0.6	34
South Africa	13	1	9	0.1	9
Sub-total	843	159	19	29.6	20
World	1,474	227	15	45.4	20

Source : Ghassemi et. al. compiled from FAO data for 1987.

Irrigation water adds appreciable amounts of salt, even with good quality irrigation water containing only 200-500 mg/kg of soluble salt. Irrigation water with a salt content of 500 mg/kg (*i.e.* 500 mg/L) contains 0.5 tonnes of salt per 1,000 m^3. Since crops require 6,000-10,000 m^3 of water per hectare each year, one hectare of land will receive 3-5 tonnes of salt. Because the amount of salt removed by crops is negligible, salt will accumulate in the root zone, and must be leached by supplying more water than is required by the crops. If drainage is not adequate, the excess water causes the water table to rise, mobilising salts which accumulate in the root zone. When the crop is unable to use all the applied water, water-logging occurs.

Land clearing also changes the hydrological balance. In its natural state, native deep-rooted and perennial vegetation use almost all the rainwater that falls on the land. In arid or semi-arid climates the growth rate of the natural vegetation is limited by the availability of fresh rainwater. Salts will be flushed down by rain, and accumulate at the bottom of the root zone to the limiting concentration for roots to extract water, at approximately 50 dS/m. Clearing the deep-rooted native vegetation, and replacing it with shallow-rooted annual species that do not use all the rainfall, allows rainwater to escape below the roots, and 'recharge' the groundwater. Clearing of native vegetation for dryland agriculture can increase the rate of drainage by 100 times. In the Mallee Region of southern Australia, measurements show that drainage under native mallee vegetation is only 0.1 mm per year, but under annual crops it is 10 mm per year. This additional rainwater enters the groundwater or aquifers, causing the watertable to rise. In Australia, 2 Mha of land have been damaged by rising watertables due to land clearing, and another 15 Mha are at risk of salinisation by rising watertables over the next 50 years.

Soil Sodicity and Sub-soil Salinity

Sodic soils have a low concentration of soluble salts, but a high per cent of exchangeable Na^+; that is, Na^+ forms a high per cent of all cations bound to the negative charges on the clay particles that make up the soil complex. Sodicity is defined in terms of the threshold ESP (exchangable sodium percentage) that causes degradation of soil structure. The negatively charged clay particles are held together by divalent cations. When monovalent cations such as Na^+ displace the divalent cations on the soil complex, and the concentration of free soluble salts is low, the complex swells and the clay particles separate (*'disperse'*). The USDA Salinity Laboratory defines a sodic soil as having an ESP greater than 15, but in Australia it is considered sodic when the ESP is greater than 6. This lower threshold is due to Australian soils having a low content of other soluble cations, particularly Ca^{2+}, which help to stabilise clay colloids during leaching.

If the concentration of soluble salts is sufficiently low, hydrolysis of the sodic clay will occur, creating a highly alkaline soil. *Alkaline soils* are a type of sodic soil with a high pH due to carbonate salts, and are defined as having an ESP of 15 or more with a pH of 8.5-10.

The process of sodicity is complex and occurs over a long period of time. Initially, salts that have accumulated within the soil profile, from either airborne deposition or mineral weathering, cause the clay fraction of the soil to become saturated with sodium. Subsequently, leaching of the profile, either by rainwater over prolonged periods, or by irrigation with fresh water, lowers the electrolyte concentration and the clay particles disperse. Further leaching washes the dispersed clay particles deeper into the profile where they block pores and hinder infiltration of water. The soil then is very slow to drain, and is readily water-logged.

In semi-arid environments, soil profiles are commonly *saline/sodic*, where the salt has accumulated due to the low permeability of the sodic sub-soil. In theory, if sufficient salts accumulate, the threshold electrolyte concentration for flocculation will be exceeded and the clay will flocculate and take on pseudo-structure. However, given that permeability and leaching will then increase, the subsequent dilution of salts will cause colloids to disperse. Consequently, a quasi-steady state between flocculation and dispersion processes is maintained.

Saline/sodic soils are widespread in arid and semi-arid lands of the world, with a large component in Australia (over 250 million ha). Water infiltration is slow, and salts derived from rainfall or weathering reactions accumulate in saturated zones in the sub-soil. The term '*transient salinity*' denotes the seasonal and spatial variation of salt accumulation in the root zone not influenced by groundwater processes and rising water table. This transient salinity fluctuates in depth, due mainly to seasonal rainfall patterns. Transient salinity is extensive in many landscapes dominated by sub-soil sodicity. Probably, two thirds of the agricultural area of Australia has a potential for transient salinity not associated with groundwater.

MEASURING SOIL SALINITY

Soil salinity is measured by its electrical conductivity. The SI unit of electrical conductivity (EC) is dS/m. Table 19.4 shows the relationship to other units of conductivity, and to NaCl concentration (10 mM NaCl has an EC close to 1 dS/m). Originally the conductivity was measured in a saturated paste extract (EC_e), but the method is tedious as first the saturated paste has to be made, second the water needs to be extracted by a powerful vacuum pump, and third a very sandy soil does not make a saturated paste. A more convenient and universal method is a '1 : 5 extract'. The soil is dried or compressed, shaken with 5 g deionised water per 1 g soil, the probe of a hand-held conductivity meter is placed in the suspension, and the EC measured. The EC of the soil that was sampled can be calculated if its water content is known at the time of sampling. Alternatively, the measurement can be related to the field capacity of the soil using conversion factors. As a rough guide, a sandy soil will have a field capacity of 0.2 g water per g soil, and a clay soil about 0.4 g/g.

Table 19.4 : Units for measuring salinity, and conversion factors.

Conversion factors relating total dissolved salts or pure NaCl to an electrical conductivity (EC) of 1 dS/m (1 deciSiemen/metre) are given, along with equivalent units of various types, old and new.

The conversion of EC of 1 dS/m to total dissovled salts (640 mg/L) assumes a composition of salts that is common in groundwater across the world. The exact factor varies from 530 (if the salt is predominantly NaCl) to 900 (if the salts are formed predominantly from divalent ions).

Measurement and units	Application	1 dS/m is equal to:	Equivalent units
Conductivity (dS/m)	soils	1	1 dS/m = 1 mS/cm = 1 mmho/cm
Conductivity (µS/cm)	irrigation and river water	1000 µS/cm	1 µS/cm = 1 µmho/cm
Total dissolved salts (mg/L)	irrigation and river water	640 mg/L (approx.)	1 mg/L = 1 mg/kg = 1 ppm
Molarity of NaCl (mM)	laboratory	10 mM	1 mM = 1 mmol/L

Table 19.5 : Electrical conductivity (EC) of pure solutions at 20°C (dS/m).

The solutions represent those of salts found in soils or in seawater. Data from the Handbook of Physics and Chemistry . (Note that 1 dS/m = 1 mmho/cm).

Solution	EC (dS/m)
10 mM NaCl	1.0
100 mM NaCl	9.8
500 mM NaCl	42.2
10 mM KCl	1.2
10 mM CaCl2	1.8
10 mM MgCl2	1.6
50 mM MgCl2	8.1

The electrical conductivity of irrigation or river water is measured with the same hand-held conductivity meter as above, but is expressed in units 1000 times magnified, as channel or river water would normally have a very low concentration of salts. River water quality is often expressed as dS/cm

(1000 x dS/m). Irrigation water quality is often expressed as total soluble salts, an international convention being that 1 dS/m is equivalent to 640 mg/L of mixed salts.

As 10 mM NaCl has an EC of almost exactly 1.0 dS/m, and as it is the most common salt in solution in ground or river water, a 'rule of thumb' is to muliply the EC by ten to equate it to salt molarity.

Soil salinity on a large scale is mapped with an electromagnetic (EM) conductivity meter. This instrument estimates the bulk electrical conductivity of the soil, which depends on the salinity of the soil solution, its water content, and the type and amount of clay in the soil. The output needs to be calibrated by chemical measurements of cores taken from the field. The ground EM conductivity meters consist of a small transmitter coil, energised with an alternative current. This current generates a primary magnetic field in the ground, which induces small currents which generate their own secondary magnetic field, which is smaller and proportional to the soil conductivity. A receiver coil close by measures the primary and secondary magnetic fields. The EM38 meter (Geonics Limited) is designed for agricultural surveys and measures to 1.5 m depth in the vertical mode and 0.75 m in the horizontal mode. An EM38 survey can be used to map the extent of sub-soil salinity as well as discharge areas.

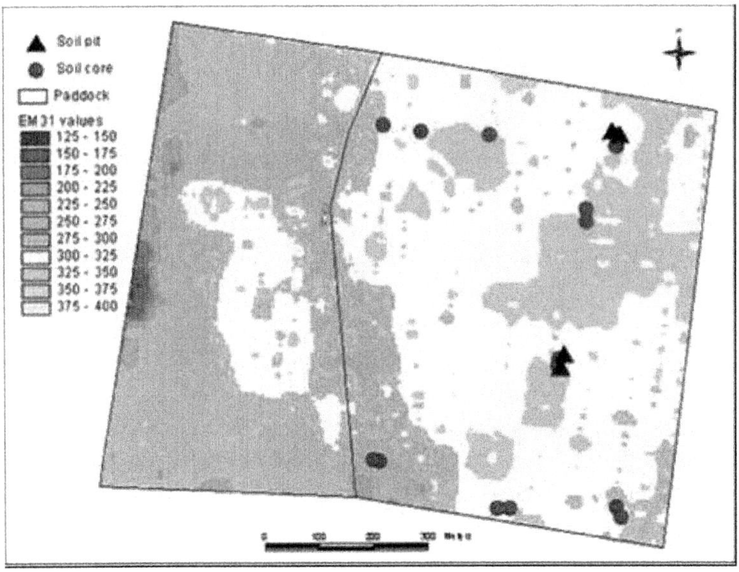

Figure 19.1 : Example of an EM survey taken on a paddock basis at ground level.

Airborne EM mapping is feasible now that global positioning systems have become available for accurate and rapid mapping. In the airborne method, the transmitter is slung below a plane flying at 150 m, and the receiver is towed behind. The plane generates a magnetic field, and the moving field passsing through the ground creates a secondary electric field whenever it hits something conductive. The receiver trailing behind the plane collects the data.

THE REPERCUSSIONS OF SALINITY

The Effect of Salinity on Plants

Salts in the soil water may inhibit plant growth for two reasons. First, the presence of salt in the soil solution reduces the ability of the plant to take up water, and this leads to reductions in the growth rate. This is referred to as the osmotic or water-deficit effect of salinity. Second, if excessive amounts of salt enter the plant in the transpiration stream there will be injury to cells in the transpiring leaves and this may cause further reductions in growth. This is called the salt-specific or ion-excess effect of salinity. The definition of salt tolerance is usually the per cent biomass production in saline soil relative to plants in non-saline soil, after growth for an extended period of time. For slow-growing, long-lived, or uncultivated species it is often difficult to assess the reduction in biomass production, so per cent survival is often used.

As salinity is often caused by rising water tables, it can be accompanied by water-logging. Water-logging itself inhibits plant growth and also reduces the ability of the roots to exclude salt, thus increasing the uptake rate of salt and its accumulation in shoots.

Variation in Salt Tolerance between Species

Wheat is one of the more salt-tolerant crop species, and many cultivars that have been selected for yield in water-limited conditions do not suffer a 50% reduction in biomass until salinities reach 15 dS/m (approximately 150 mM NaCl). Rice is more salt-sensitive, and many cultivars suffer a 50% reduction in growth at half this concentration of salts. Maize falls in between these two species in terms of salt sensitivity.

Another criterion of salt tolerance of crops is their yield in saline versus non-saline conditions. A survey of salt tolerance of crops, vegetables and fruit trees was made by the USDA Salinity Laboratory. This shows for each species a threshold salinity below which there is no reduction in yield, and then a regression for the reduction in yield with increasing salinity. The data in some cases are for a single cultivar of the species, or a limited number of cultivars at a single site, so they are not necessarily representative of the species. Further, the data are related to an EC_e value, which is not an appropriate reference point for a sandy soil, or for many current soil salinity estimates that based merely on a 1 : 5 extract. However, the data are useful in that they show the wide range of tolerance across species, and also show that yield has a different pattern of response than does vegetative biomass Yield always shows a threshold in response to a range of salinities, but with young plants a threshold is rarely seen.

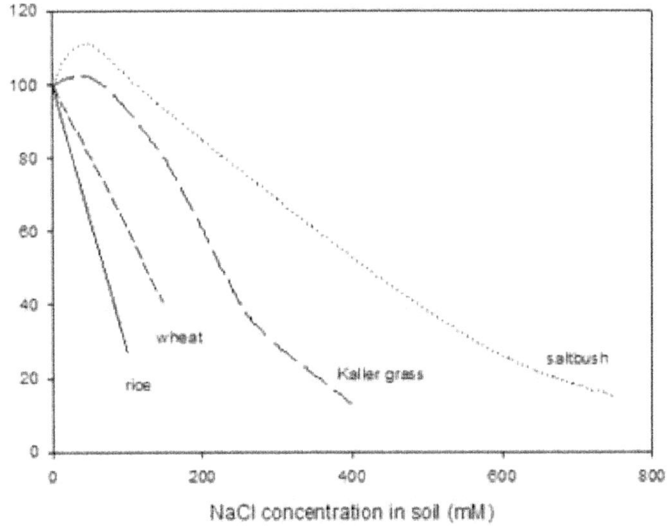

Figure 19.2 : Biomass production of four diverse and important plant species in a range of salinities. Wheat is one of the more salt-tolerant crops, and rice is one of the more salt-sensitive crops. Two halophytes : a saltbush species *Atriplex amnicola* and a grass *Diplachne* (syn. *Leptochloa*) *fusca* or Kallar grass. Both halophytes show outstanding salt tolerance with high growth rates and are being used in Australia and Asia for grazing on saline land.

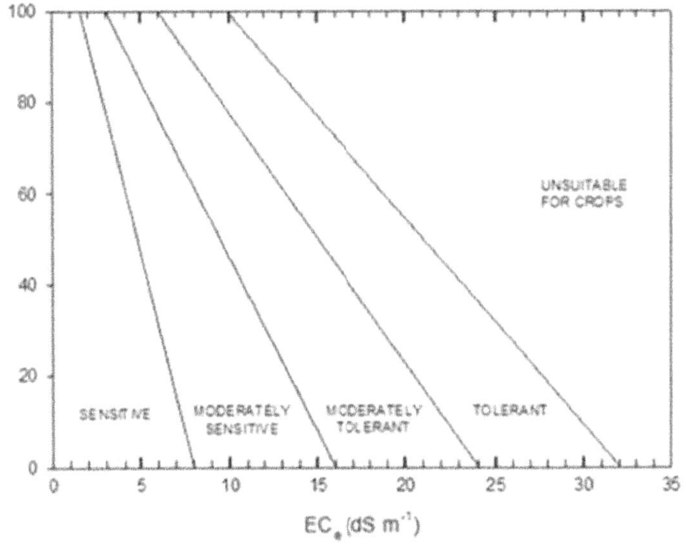

Figure 19.3 : Categories for classifying crop tolerance to salinity according to the USDA Salinity Lab. Note that the ECe is more applicable to an irrigated than a rainfed field, in the latter the soil moisture content might be 2-4 time less than in a saturated paste.

Causes of the Growth Reduction under Saline Conditions

The effects of a saline soil are two-fold : there are effects of the salt outside the roots, and there are effects of the salt taken up by plants.

The salt in the soil solution (the "osmotic stress") reduces leaf growth and to a lesser extent root growth, and decreases stomatal conductance and thereby photo-synthesis. The cellular and metabolic processes involved are in common to drought-affected plants. The rate at which new leaves are produced depends largely on the water potential of the soil solution, in the same way as for a drought-stressed plant. Salts themselves do not build up in the growing tissues at concentrations that inhibit growth : meristematic tissues are fed largely by the phloem from which salt is effectively excluded, and rapidly elongating cells can accommodate the salt that arrives in the xylem within their expanding vacuoles. So, the salt taken up by the plant does not directly inhibit the growth of new leaves.

The salt within the plant enhances the senescence of old leaves. Continued transport of salt into transpiring leaves over a long period of time eventually results in very high Na^+ and Cl^- concentrations, and they die. The rate of leaf death is crucial for the survival of the plant. If new leaves are continually produced at a rate greater than that at which old leaves die, then there might be enough photo-synthesising leaves for the plant to produce some flowers and seeds. However, if the rate of leaf death exceeds the rate at which new leaves are produced, then the plant may not survive to produce seed. For an annual plant there is a race against time to initiate flowers and form seeds, while the leaf area is still adequate to supply the necessary photosynthate. For perennial species, there is an opportunity to enter a state of dormancy, and thus survive the stress.

The two responses occur sequentially, giving rise to a two-phase growth response to salinity. The first phase of growth reduction is quickly apparent, and is due to the salt outside the roots. It is essentially a water stress or osmotic phase, for which there is surprisingly little genotypic difference. Then there is a second phase of growth reduction, which takes time to develop, and results from internal injury. The experiment was conducted with two genotypes with contrasting rates of Na^+ uptake, and known differences in salt tolerance; previous experiments had shown that the genotype with the low Na^+ uptake rate had a higher survival of high salinity. This is called the 'Phase 1' response, and is due to the osmotic effect of the salt. Then after 4 weeks, the genotypes separated; the one with the low Na^+ uptake rate continued to grow, although still at a reduced rate compared to the controls in non-saline solution, but the one with the high Na^+ uptake rate produced little biomass and many individuals died. This is the 'Phase 2' response, and is due to genotypic differences in coping with the Na^+ or Cl^- ions in the soil, as distinct from the osmotic stress.

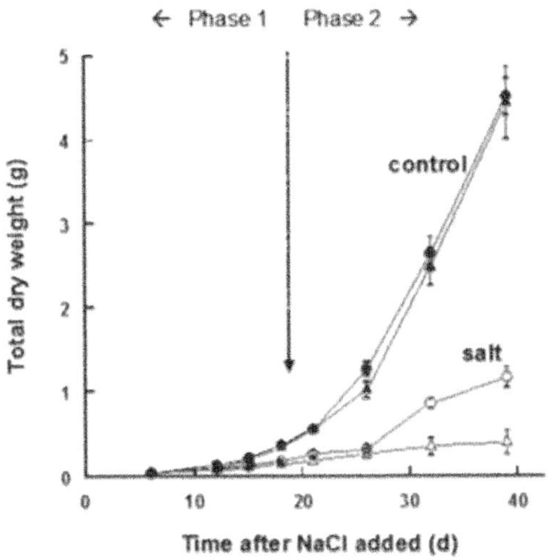

Figure 19.4 : Two accessions of the diploid wheat progenitor *Ae. tauschii* grown in supported hydroponics in control solution (closed symbols) and in 150 mM NaCl (open symbols). Circles denote the tolerant accession, triangles the sensitive one. The arrow marks the time at which symptoms of salt injury could be seen on the sensitive accession; at that time the proportion of dead leaves was 10% for the sensitive and 1% for the tolerant accession.

These results illustrate the principle that the initial growth reduction is due to the osmotic effect of the salt outside the roots, and that what distinguishes a salt-sensitive plant from a more tolerant one is the inability to prevent salt from reaching toxic levels in the transpiring leaves, which takes time.

To grow in saline conditions, plants must maintain a high water status in the face of soil water deficits and potential ion toxicity. A plant can only grow or survive in a saline soil if it can both continue to take up water *and* exclude a large proportion of the salt in the soil solution.

The Required Extent of Salt Exclusion

Roots must exclude most of the Na^+ and Cl^- dissolved in the soil solution or the salt will gradually build up with time in the shoot and become so high that it kills it. To prevent salt building up with time in the shoot, roots should exclude 98% of the salt in the soil solution, allowing only 2% to be transported in the xylem to the shoots. This value of 2% can be calculated from the following equation :

The concentration at which NaCl accumulates in the shoot depends on the salt concentration in the soil solution, the percentage of salt taken up by roots, and the percentage of water retained in the leaves :

$$[NaCl]_{shoot} = [NaCl]_{soil} \times \frac{\%\ \text{salt taken up}}{\%\ \text{water retained}} \qquad \text{(Eqn. 1)}$$

Plants retain only about 2% of the water they transpire, *i.e.* they take up about 50 time more water from the soil than they retain in their shoot tissues. The percentage of transpired water that is retained in the shoot can be calculated from the product of the water use efficiency (*wue*; mass of shoot produced per mass of H_2O transpired) and the shoot water content (*wc*; shoot H_2O per shoot mass) :

Water use efficiency (WUE) of plants growing at moderate evaporation demand are usually in the range of 3-6 mg g^{-1}, the variation due to extremes of evaporative demand, rather than a peculiarity of the species. For a water use efficiency of 4 mg g^{-1} and a shoot H_2O : DW ratio of 5 : 1, about 20 mg of water is retained in the shoot for every g of water transpired (Eqn. 2). That is, the shoot retains only 2% of the water transpired. In order to prevent the salt concentration in the shoot increasing above that in the soil, then only 2% of the salt should be allowed into the shoot, *i.e.* 98% should be excluded.

$$\% \text{ water retained} = wue \times wc \times 100 \qquad \text{(Eqn. 2)}$$

A soil salinity of 100 mM NaCl or 10 dS m^{-1} is about as high as most crops will tolerate without a significant reduction in growth or yield, and a concentration of 100 mM NaCl on a whole shoot basis is about as high as is desirable because it will include some old leaves with much higher salt concentrations, as well as younger leaves or other tissues with lower concentrations. So for plants to grow for extended periods of time in soils with salinity of this order of magnitude, roots should ensure that no more than 2% gets to the shoots. Roots themselves do not accumulate excessively high concentrations of salt. The Na^+ and Cl^- concentration in roots is rarely higher than in the external solution, and often is lower.

The Relationship between Transpiration and Salt Uptake

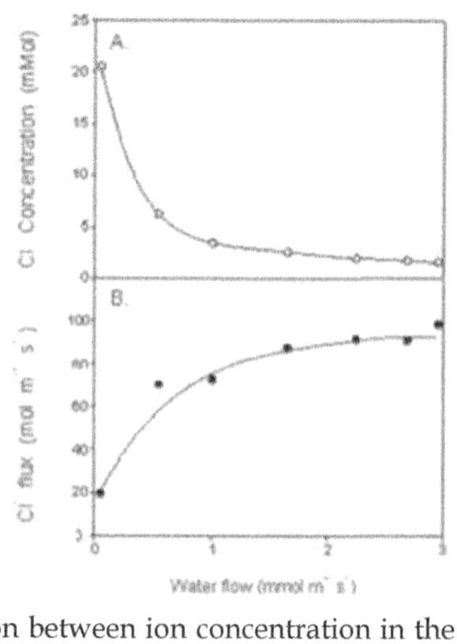

Figure 19.5 : The relation between ion concentration in the xylem (A), ion flux to the shoot (B), and transpiration (water flow).

The fundamental processes governing the relationship between water and ion flow through roots are complex and not well understood. NaCl does not move passively with the transpiration stream, neither is its movement entirely independent of it, at least in some species, or over certain ranges of transpiration. As water flow increased from a very low to a moderate rate, there was an increase in Cl^- flux, showing that the movement of the ion through the root was enhanced as water flow started to increase. However, when the water flow increased from moderate to high rates, there was little or no further increase in Cl^- flux, showing that the movement of the ion was independent of further increases in water flow. This relationship also holds for Na^+ and K^+.

Measurements of ion concentrations in leaves of most plants grown at different humidities are consistent with this pattern, that salt transport to leaves is substantially affected only if transpiration is greatly affected. An effect might be seen more with species that are very poor excluders, such as lupin as mentioned above, and rice, which carry much more salt in an apoplastic or transpirational "bypass" pathway than other species. In rice, the percentage of water moving through a bypass pathway from roots to shoots was estimated as 5.5% of the total water transpired, and could account for all the Na^+ transported to the shoots, whereas in wheat only 0.4% of the water moved along a bypass pathway and could not account for most of the Na^+ transported.

Mechanisms of Control of Salt Transport

Here we look at differences between species in the ability to tolerate the salt-specific component of salinity. Differences in tolerating the osmotic stress itself are in common with drought tolerance.

The rate at which old leaves die depends on the rate at which salts accumulate to toxic levels. Thus, control of the rate at which salt arrives in leaves is essential, as are mechanisms that reduce the toxicity of the salt. For species lacking the ability to compartmentalise salts in the vacuoles to high concentrations, continued transport of salt to the leaves will eventually result in either excessive build-up of salts in the cell walls or in the cytoplasm. The former will cause death through dehydration, and the latter will cause death through poisoning of metabolic systems such as photo-synthesis or respiration.

Control at the Whole Plant Level

Control of salt transport into and through the plant takes place at five sites in the plant. Control occurs in the root cortex, at the loading of the xylem, at the retrieval from the xylem in upper parts of the roots. These three processes serve to reduce the transport to the leaves. Control in the shoot occurs by the exclusion of salt from the phloem sap flowing to meristematic regions of the shoot. An additional mechanism occurs in most halophytes :

specialised cells to excrete salt from leaves. However, halophytes also rely on the first four mechanisms to reduce the flux of salt to the leaves - excretion is an additional backup for plants growing in very saline site, and for perennial species.

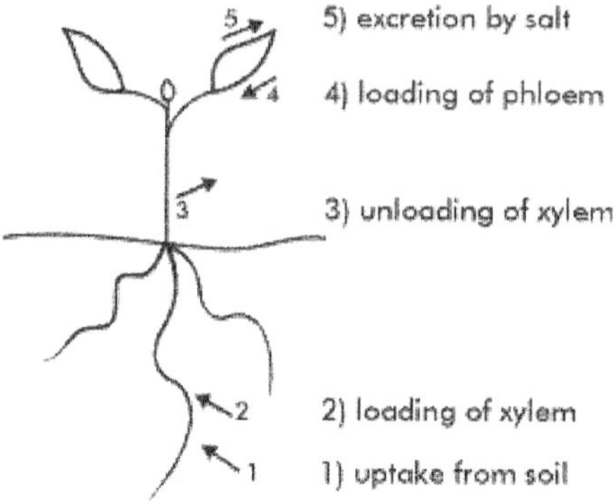

Figure 19.6 : Control points at which salt transport is regulated. These are : 1. selectivity of uptake from the soil solution, 2. loading of the xylem, 3. removal of salt from the xylem in the upper part of the plant, 4. loading of the phloem and 5. excretion through salt glands or bladders. For a salt tolerant plant growing for some time in a soil solution of 100 mM NaCl, the root concentrations of Na^+ and Cl^- are typically about 50 mM, the xylem concentration about 5 mM, and the concentration in the oldest leaf as high as 500 mM.

Exclusion is particularly important for perennial species whose leaves may live for a year or more. For these species there is greater need to regulate the incoming salt load than for annual species whose leaves may live for only one month.

There are contributory features that function to maintain low rates of salt accumulation in leaves. High shoot/root ratios and high intrinsic growth rates, and absence of an apoplastic pathway in roots all will serve to reduce the rate at which salt enters the transpiration stream and accumulates in the shoot.

$$\text{Ion concentration in shoot (mol g}^{-1}\text{)} = \frac{\text{Ion uptake rate (mol g}^{-1}\text{d}^{-1}\text{)}}{\text{Relative growth rate (g g}^{-1}\text{ d}^{-1}\text{)}} \qquad \text{(Eqn. 3)}$$

Two principles emerge from this equation. First, a fast-growing plant will have a lower concentration in the shoot than a slow growing plant, for the same uptake rate. Note that it is the relative growth rate (RGR) not the size itself that influences the ion concentration in the shoot. The notion of 'vigour' as affecting ion concentration therefore needs to be defined carefully, as RGR is independent of plant size. Second, any increase in shoot ion concentration with increasing salinity may not be due to increase ion uptake rate—it could be due to decreased RGR due to the osmotic effect of the salt. This also means that an additional stress such as water-logging that depress RGR will also cause an increase in concentration in the shoot for the same uptake rate.

Control at the Cellular Level : Ion Compartmentation

There is no evidence of adaptations in enzymes to the presence of salt, so mechanisms for salt tolerance at the cellular level involve keeping the salt out of the cytoplasm, and sequestering it in the vacuole. That this occurs in most species is indicated by the high concentrations found in leaves that are still functioning normally, concentrations well over 200 mM, which are known to completely repress enzyme activity *in vitro* Generally, Na^+ starts to inhibit most enzymes at a concentration above 100 mM. The concentration at which Cl^-becomes toxic is even less well defined, but is probably in the same range as that for Na^+. If Na^+ and Cl^- are sequestered in the vacuole of the cell, K^+ and organic solutes should accumulate in the cytoplasm and organelles to balance the osmotic pressure of the ions in the vacuole. The organic solutes that accumulate most commonly under salinity are proline and glycinebetaine, although other molecules can accumulate to lesser degrees.

Salt tolerant species have transport systems on the tonoplast that can sequester Na^+ and Cl^- at high concentrations within the vacuoles, while maintaining much lower concentrations in the cytoplasmic compartments. The ion channels and transporters that regulate the net movement of salt across cell membranes were described in several recent reviews.

In summary, roots do most of the work in protecting the plant from excessive uptake of salts, and filter out most of the salt in the soil while taking up water. Even so, there are mechanisms for coping with the continuous delivery of relatively small amounts of salt that arrive in the leaves, the most important being the cellular compartmentalisation of salts in the vacuoles of the mesophyll cells. This strategy allows plants to minimize or delay the toxic effects of high concentrations of ions on important and sensitive cytoplasmic processes. The rate at which leaves die is the rate at which salts accumulate to toxic levels, so genotypes that have poor control of the rate at which salt arrives in leaves, or a poor ability to sequester that salt in cell vacuoles, have a greater rate of leaf death.

Chapter 20

The Mitigation of Drought Stress

MITIGATION OF DROUGHT STRESS BY CROP MANAGEMENT

Managing Drought Stress by Supplemental Irrigation

Irrigation, where available, is the major means for combating drought conditions. It is a prime approach to the intensification of agriculture and the generation of stable income. The development of irrigation depends on various environmental, economical and social factors on both the macro- and micro-scales.

There are hazards in irrigation if practiced indiscriminately, such as soil erosion, soil salination, soil leaching and soil disease infection. Irrigation as such is not an important topic in this site. Links to irrigation sites throughout this section should provide additional information.

General Crop Irrigation Guidelines

The key to planning irrigation system and scheduling is knowledge of the crop, the soil properties and the potential evapotranspiration (PET) of the specific crop at the site. This information can also be used to estimate dryland crop water use and deficit at any given time during the crop cycle, which is actually an index of crop drought stress.

The Penman-Monteith potential evapotranspiration equation is recommended by the FAO as the standard method for estimating reference and crop evapotranspiration. The new method has been proved to have a global validity as a standardized reference for grass evapotranspiration and it has been recognized by both the International Commission for Irrigation and Drainage and by the World Meteorological Organization.

The (FAO) Penman-Monteith method was developed by defining the reference crop as a hypothetical crop with an assumed height of 0.12 μ having a surface resistance of 70 s m^{-1} and an albedo of 0.23, closely resembling the evaporation of an extensive surface of green grass of uniform height, actively growing and adequately watered.

Irrigation to Control Drought in Various Crops

Deficit (or supplemental) irrigation is the more common irrigation practice for crops not designated *a priori* for fully irrigated conditions and maximized yield. Supplementary irrigation is a practice dictated by constraints, which can be derived from the limited availability of water, irrigation equipment, the cost of water, or other economic and technical constraints. With supplemental irrigation the amount of water applied to the crop in irrigation is well below the full requirement of the crop. The water-use efficiency of supplemental irrigation is generally high if applied logically. It can be applied to save the crop in case of un-expected drought or as a planned practice to supplement the expected total seasonal rainfall. The practice may vary extensively with crop and region. In many environments, and especially the Mediterranean region, if only a single supplementary irrigation is given it is usually more effective if applied pre-planting. As such the crop enters the season with a stored supply, which can insure growth despite unexpected transient rainfall fluctuations.

Managing the Dryland Crop Environment

Modern dryland farming is a system of low inputs combined with soil and water conservation practices and risk reducing strategies. The system can be sustainable if practiced properly. Water shortage is the main limiting factor, but successful dryland systems also maintain reasonable practices to eliminate other limiting factors (poor nutrient status, weeds, biotic stresses, etc.'), which can reduce the effectiveness by which the crop uses the limited moisture. However, as water shortage *a priori* dictates a limit on yield, all other inputs must be carefully adjusted downwards to fit the expected low economic return.

The most advanced systems have been developed in the Great Plains of the USA, Canada and Australia, while traditional systems employed in Asia and the Middle East also offer important insights. In the USA, the lesson learned during the "dustbowl" years in the early 1930's prompted extensive legistration and investments in developing sustainable dryland farming systems. These systems and the associated technological progress such as plant breeding, brought about an increase in mean winter wheat yield from 0.5 ton ha^{-2} in 1930 to about 2 ton ha^{-2} in 1980. In Southern Australia the "ley farmimg" system was developed in the 1920's and adopted widely in the late 1940's. The system involves a rotation between a self-seeding legume grown for several years and wheat. The farmer grows wheat and raises sheep while the legume serves to sustain soil fertility (mainly nitrogen). This system has become less popular in recent years with the increase in economic pressures and other considerations.

The lesson learned from the American and Australian experience is that the development of a sustainable dryland farming system involves the following principles, not necessarily in their order of importance :

1. Improved soil and water conservation practices and the associated reduced tillage systems.
2. Optimization of the fit between crop growth cycle and the available moisture.
3. Weed control.
4. Soil fertility management.
5. Optimized plant population density and spatial arrangement of plants with respect to the expected soil moisture regime.
6. Control of soil biotic stress factors that reduce root development.
7. Improved forage/livestock/grains integration and rotation.
8. Avoidance of mono-cropping and enhancement of crop diversification.
9. The increase of precipitation by cloud seeding, as an ongoing experiment.

SOIL AND WATER CONSERVATION

Fallow and Conservation Tillage

The fallow system is designed to conserve soil moisture from one season to another or from one year to the other, depending on climate and crop. Increasing storage of soil moisture by the fallow system with or without conservation tillage is standard agricultural practice in dryland farming. The benefit of fallow and conservation tillage in terms of increasing available soil moisture to the crop depends on soil water-holding capacity, climate, topography and management practices. Fallow efficiency, in terms of per cent increase in soil moisture availability to the crop measured at planting date normally ranges from about 5% to 30%. While these amounts are not impressive they can make a difference between crop failure and success. The fallow carries additional benefits such as improved soil nutrients availability and the eradication of certain soil-born pests, such as nematodes.

Conservation tillage is not a novel concept or practice, which has recently gained wider and sometimes an enthusiastic acceptance. It involves the principle of minimized tillage operations to conserve soil structure and to maintain ground cover by mulch, such as stubble. These practices reduce water runoff and increase soil infiltration. Conservation tillage has become the cornerstone of dryland systems in certain regions of the USA, Canada, Australia and other regions. While the benefits of conservation tillage are well-documented it has also been noted that crop residues under this system may promote certain crop diseases. To obtain some real impressions on the subject spend an hour in a farmers' meeting in California on the subject.

Deep tillage is a system to overcome hardpan, very high bulk density and compacted soils. It can be performed by deep plowing or deep ripping. Deep plowing involves actual plowing to depth which is an expensive operation. It is uncommon in dryland farming. Deep ripping is less expensive and often used in crop production. The important consideration in deep ripping is to operate at the correct depth in order to break the hardpan, no less and no more.

Furrow Dikes and Soil Pitting

These techniques constitute a field surface tillage manipulation to minimize runoff away from the field.

Furrow dikes are furrows, which are divided into short basins by small dikes. This is achieved by special equipment. The system is very amenable to row crops such as cotton, corn and sorghum and it can be integrated with or without furrow irrigation. It is generally considered effective for increasing rainfall capture and raising dryland yield where annual rainfall ranges between 500 and 800 mm.

Soil pitting (left side photograph) involves the formation of small depressions at close proximity to reduce runoff from rainstorms. The crop is planted over this modified surface. Experiments performed with wheat in nine farmer demonstration plots in Southern Israel during 1988 showed that pitting increased yield by an average of 7.5% at a mean yield of 3.25 ton/ha. Unlike furrow dikes these system is not limited to row crops.

Water Harvesting/Spreading

Figure 20.1.

This is a broad term to describe various methods to collect runoff from large contributing areas and concentrate it for use in smaller crop area. This is an ancient practice already adopted by Nabatian desert settlements in the Middle East several centuries A.D. The photo on left represents a view from the ancient city of Avdant in the Negev region of Israel. In the front there are several ancient water spreading plots while at the back is a modern experimental farm (set-up by Prof. Evenari) seeking to evaluate the effectiveness of these systems in sustaining agriculture with around 100 mm of annual rainfall.

Presently, the basic water harvesting systems involve an external contributing area to induce runoff. This area is physically or chemically treated for maximizing runoff. The water is diverted into a receiving area comprising of cultivated plots, individual trees or small terraces. The contributing area may lie in the agricultural field (a system sometimes referred to as "conservation bench terrace") or outside the field in the natural watershed system. In the Avdat photo the small valley is a water-shed system experiencing flash flood once or twice a year. The size ratio between the contributing and the receiving areas is determined by the expected rainfall events, crop water requirements, soil characteristics and topography. The resulting yield increase in the receiving (crop) area is proportional to the amount of water gained.

Diversification of Farming

Diversification of farming is an ancient but an effective approach to reduce the risk associated with farming in unpredictable environments. Reduced diversification to the extent of mono-cropping is possible only with a high level of control over the crop environmental conditions. Such control method (irrigation, chemical pest control, etc.') are among the main reasons for the more recent environmental quality problems found to be associated with mono-cropping. Diversification of cropping to reduce risk is especially important under dryland conditions. It is achieved on several levels, as described by Pandey *et. al.* (#4194) for the case of traditional rain-fed rice in Eastern India.

1. **Spatial diversification of fields.** The farmer's land is divided into several fields or plots which may differ in their topography, soil and hydraulic properties. Some fields may be prone to flooding while others do not hold water. Certain fields may be on a warmer slope while others on a cooler one. The different field conditions allow to achieve a better fit between the crop and the environment and to reduce the general probability of stress affecting the farmer.
2. **Crop diversification is an important feature of traditional farming.** It takes an advantage of the generally low correlation between crops in performance when grown in a single stress environment. Crops differ in their response to a given environment and this difference is used to reduce the risk associated with growing one crop. "Mixed cropping" or "intercropping" is an example of a traditional and a successful approach to crop diversification on a single parcel of land, where two or more crops are grown together in various possible configurations. If for some

reason only one crop is grown, a certain (though lower) level of risk reduction can be achieved by varietal diversification. Planting of several crop varieties offer a better probability for reducing loss due to environmental stress, as compared with growing one variety only. For environmental stress conditions varietal diversification is based mainly on differential phenology, primarily flowering date. A typical example is a transient frost or heat wave that is likely to occur around flowering time of the specific crop. Damage reduction can be achieved when the crop is sown to several varieties of different flowering dates.

3. **Temporal diversification may achieve the same result as varietal diversification, when phenology is concerned.** The purpose of setting a distinct planting date is to optimize crop development with respect to seasonal climate, mainly rainfall in rain-fed agriculture. Ideally the crop is planted at the beginning of the rainy season, rainfall peaks when crop evapotranspiration peaks and it terminates just before harvest time. When such conditions are reasonable predictable, planting date can be set to optimize production. Where the timing of rainfall is very unpredictable, adopting more than one planting date for the given crop can reduce the risk involved with untimely rainfall and a given planting date.

Cloud Seeding

Cloud seeding is a form of weather modification attempt. The process of cloud seeding involves deposition of cloud condensation nuclei (CCN) into a specific region of the cloud. Seeding may be achieved from above or through the clouds by aircraft, and from below where CCN are carried into the cloud by updrafts. With either method, the CCN must reach the super cooled cloud region, where water molecules remain unfrozen at temperatures below 0C. Experiments in cloud seeding have been performed for the last 60 years. The results and benefits of this practice are still under debate.

MITIGATION OF DROUGHT STRESS BY CROP PLANT BREEDING

The Nature of Drought Resistance

Drought Resistance and Crop Yield

Crop plant breeding for drought resistance has long been part of the breeding process in most crops that are grown under dryland conditions. During the period of the pre-scientific agriculture the genetic improvement of plant adaptation to dry conditions was simply attained by repeatedly selecting plants that appeared to do well when drought stress occurred. As a result of many generations of selection by generations of farmers we now encounter such materials, which are defined as "landraces" of the crop. Such landraces were shown to possess distinct drought resistance traits. Later, as scientific agriculture developed and following the emergence of Mendelian genetics, elaborate biometrical and statistical methods for quantitative genetics analysis were developed to enable selection for yield and yield stability more effectively and efficiently. An important factor of yield stability is coping with drought and other abiotic plant stresses. As crop physiology emerged and developed, yield-based selection programs were augmented by observing plants under carefully managed stress environments, followed by the development of physiological selection criteria for stress resistance. More recently, molecular methods, such as marker-assisted selection are being adopted to facilitate more efficient selection for distinct components of abiotic stress resistance. Finally, bio-technology is experimenting with genetic transformation, open the way for additional genetic solutions to breeding for drought resistance.

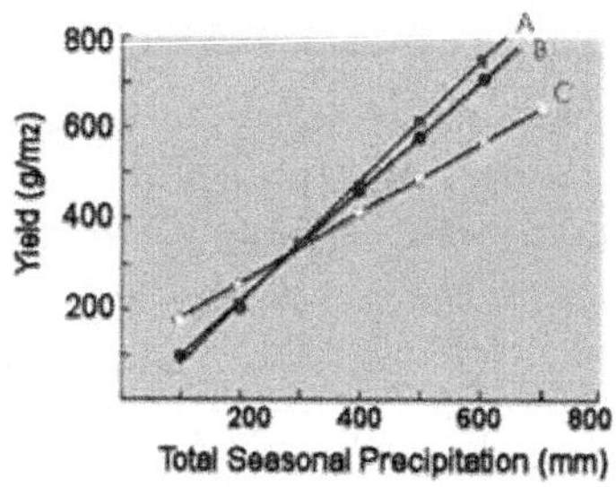

Figure 20.2 : The schematic association between yield and total seasonal precipitation for 3 different wheat cultivars.

Looking at crop drought resistance from a botanical perspective it must be realized at the onset that there is a vast difference between drought resistance in natural vegetation and in crop plants. Natural vegetation has evolved to conserve the species. Henceforth, plant survival and the capacity to produce at least one seed per life cycle despite stress is the most powerful component of natural selection. On the other hand, drought resistance in modern agriculture requires sustaining economically viable plant production despite stress. Crop survival is of a lesser consequence to economical farming. On the other hand, plant survival can be a critical factor in subsistence agriculture, where the ability of a crop to survive drought and produce some yield at all may translate into a difference between famine and livelihood. Breeding for drought resistance is therefore very tightly linked to the target environment of the crop, not only with respect to its physical and chemical features but also its social grounds.

For a variety of reasons there is a general trade-off between a genetically high yield potential and drought resistance. At the same time there is a yield advantage under drought stress brought about by a high yield potential, to a limit.

Wheat cultivar C is different from A and B in that it has a lower yield potential (yield at high moisture conditions) but as moisture becoming deficient C turns out to be superior to A and B. In terms of yield, C may be defined as drought resistant while cultivars A and B are of high yield potential but are relatively drought susceptible. The "cross-over" where the advantage of C over A and B under stress begins to be expressed is at about 300 mm or at a yield level of about 300 g m^{-2}. Hence, drought resistance of C is expressed only when stress is sever (<300 mm). Still, it is extremely important to realize that the high yielding cultivars A and B are superior to the drought resistant C when drought stress is moderate (*e.g.* at 400 to 500 mm). A high yield potential therefore ascribes an advantage under moderate stress conditions. Cases where drought resistance has been improved together with yield potential exist but they are very rare and exceptional and cannot be used to indicate a general rule. Real drought resistance cannot be field-tested or evaluated if yield level is above around 300-400-g m^{-2}, schematically. Consider the principle not the actual numbers. The actual numbers were obtained for wheat and barley by several studies.

The Components of Drought Resistance

Drought resistance in crop plants is conditioned by two major pathways : Dehydration avoidance and dehydration tolerance. Dehydration avoidance is the capacity to avoid plant tissues and cells dehydration under drought stress. Dehydration tolerance is the capacity to sustain function when the plant is dehydrated. Plant survival can be conditioned by either avoidance or tolerance.

As discussed above under the impact of stress, moisture stress signals the expression of certain stress responsive genes, which are responsible for a chain of events and gene "networking", expressed at various levels of plant organization. It has been assumed almost axiomatically that stress responsive genes are involved in adaptation; henceforth that they are 'stress adaptive'. It was later realized that not every stress responsive gene is necessarily adaptive in terms of drought resistance or survival or crop productivity under moisture stress.

Irrespective of the role and function of stress adaptive genes in plant drought resistance, it should be recognized that not only certain stress adaptive genes might determine plant performance under drought stress. Genes that are expressed irrespective of the environment also condition plant function and performance under stress. These genes are expressed constitutively and determine various plant traits irrespective of any stress. An example for a constitutive plant trait that may control drought resistance is potential root depth (maximum root length). Stress and soil conditions can affect root depth in several ways but potentially a deep rooted genotype will maintain its advantage over a potentially shallow rooted genotype under conditions of deep soil moisture. For this difference to be expressed plants do not have to be subjected to drought stress conditions. Therefore, drought resistance and plant production under drought stress is determined by constitutive and adaptive plant traits.

Dehydration Avoidance

Plant Development and Size

Plant size as expressed mainly in terms of single plant leaf area or leaf area index (LAI) has a major control over water-use. At the same time small plants and reduced leaf area are generally conducive to low potential productivity. Botanists have long recognized small plants bearing small leaves as typical ecotypes of xeric environments. While such plants withstand drought very well their growth rate and biomass are relatively low.

In the domain of plant breeding, cultivars developed for dryland conditions by selecting mainly for yield under such conditions often resulted in plants of moderate size and water-use. For example this can be seen in dryland temperate cereals as well as upland rice, which tend to have moderate tillering. On the other hand it has also been shown that early plant are important traits for dry conditions. The reason is in the rapid ground cover achieved and the subsequent decrease in water loss by direct soil evaporation at this stage. However, other benefits for seedling vigor were also noted, such as the nitrogen status of the plant. Early flowering which determines 'drought escape' generally involves a reduction in adult plant size and leaf area leading to reduced water-use. Thus, small plant size and small leaf area are very often linked to improved dehydration avoidance and lower potential yield, a tradeoff discussed elsewhere.

The Root

The most important control of plant water status is with the root, whereas the roots is the main engine for meeting transpirational demand. Two major dimensions describe the root : root depth (or maximum length) and root-length density. The more practically and commonly important dimension for most breeding scenarios is root depth, which facilitate deep soil moisture extraction where such moisture is available. It is a primary component of drought resistance. The development of lateral roots at very shallow soil depth may have a role in capturing small amount of intermittent rainfall.

Root depth in the cereals is generally associated with a small number of main thick axes. Such fibrous root system is typically seen in upland rice, which has a deeper root system, in contrast to lowland rice with the shallower roots. The control of root growth is not only in the root. In the cereals, tillering is associated with production of new crown roots from each developing tiller. Such profuse rooting can be at the expense of the growth of existing roots into deep soil. Hence, limited tillering in cereals and grasses has been repeatedly observed to be associated with relatively deeper root extension.

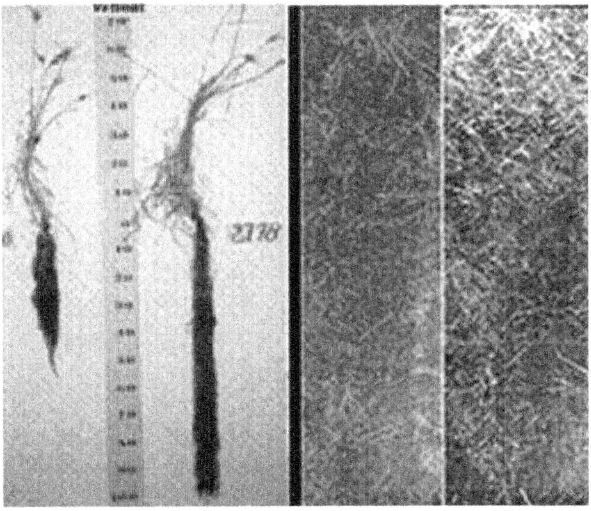

Figure 20.3 : Left panel : hydroponically grown roots of two wheat cultivars differing in root length. Right panel : roots of two sorghum cultivars in soil in a root observation box (1.8 m tall), differing in root-length density on the given date.

In certain soils a hardpan can limit deep root growth and the capacity for hardpan penetration by roots becomes a critical factor in drought resistance. The factors, which may support axial root force and hardpan penetration, are not known and most research in this area has been performed mostly with seedlings. In mature plants the penetration of hardpan by roots seems to be better in plants that constitutively develop fibrous and thick roots.

Many drought environments present a situation where rainfall is low and soil depth that contains moisture is permanently shallow. For example, in many of the drier Mediterranean wheat-growing regions the wetted soil depth of around 60-80 cm is shallower than the normal maximum root depth of wheat (=>100 cm). Under such conditions a deep root is not an issue. Other plant factors may then become far more important in the control and use of the limited soil moisture, such as shoot developmental characteristics (*e.g.* leaf area development or growth duration), osmotic adjustment, leaf surface properties, etc.' Greater root length density within this limited soil horizon might allow extracting more moisture from a given soil volume which in certain cases should provide several more days before wilting.

Another scenario of seasonal soil moisture status is where the crop is grown on stored soil moisture and there is little effective rainfall during the growing season. Under such conditions the main consideration is to manage seasonal soil moisture use such that sufficient moisture will remain for carrying the crop to maturity. It is to be expected that with the available moisture the crop might grow luxuriously leading to a large leaf area and an even greater water requirement towards the latter part of the season. Hence, short growth duration, small leaf area and perhaps a higher root hydraulic resistance can achieve the control of seasonal water use. The last option has been researched at the CSIRO Australia and an increase wheat root hydraulic resistance was effectively attained by selection for smaller root xylem element diameter. It is not known whether this approach has found its way into actual application in wheat breeding or whether such a cultivar was released.

Whatever may be the constitutive form and function of roots, the environment can modify the root in a pronounced way. Off course, soil conditions in terms of topsoil moisture and deep soil hardness alter root growth and depth. Drought stress generally inhibit total root mass (while it can modify its distribution). Root-length density may locally increase in wet regions in the soil while it might decrease in the drying parts. As soil moisture deficit develop throughout the profile, the proportion of dry to wet soil increase so that the proportion of dead to live roots increase. There is hardly evidence to show that total root mass increase with drought stress. The shoot/root mass ratios consistently decrease under drought stress, which is a universal expression of adaptation. The ratio changes mainly due to the reduction in shoot mass.

The root system is highly dynamic and as long as it is not senesced or diseased it is capable of regrowth from meristems in the root axes and meristems in the root crown (in cereals and grasses). The renewal of root branching into wet soil immediately after rainfall is considered as an important factor in plant recovery from drought stress. Root hairs are considered an important component of root length density and the capacity for soil moisture extraction via improved contact with the soil.

Roots are a major target and a candidate for marker assisted selection (MAS) for the apparent reason that phenotypic selection for root traits is a slow and impractical in large populations. Still, practical results from MAS for root traits are limited.

Plant Surface

Plant surface structure, form and composition carry a major impact on the plant interaction with the environment. Plant surface absorbs solar energy part of which is used for photo-synthesis and most of which must be dissipated. Energy is dissipated by reflection, emission and the dissipation of latent energy by transpiration. Plant surface structure determines the reflective properties of the leaves and their resistance to transpiration. Leaf resistance to transpiration is off course largely determined by stomatal activity. However, plant surface structure determines the hydraulics of leaf surface, which affect the rate of water removal from the leaf surface, upon transpiration. Therefore, plant surface help to avoid dehydration by two channels : improved reflectance of incoming g radiation (*i.e.* decreasing net radiation) and by improved cuticular hydraulic resistance.

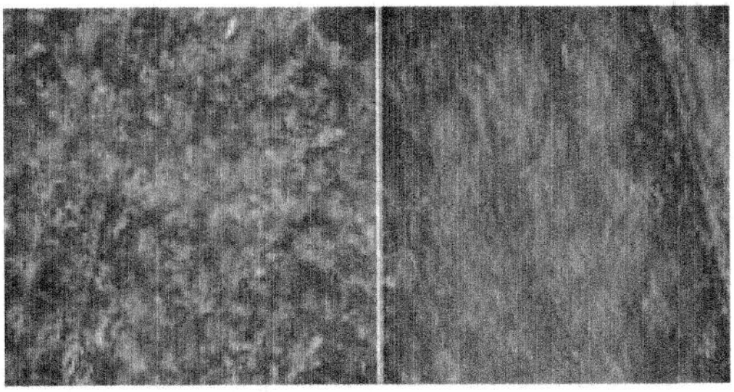

Figure 20.4 : Sorghum leaf epicuticular wax by the scanning electron microscope; left normal (Bm genotype); right low wax (bm genotype).

Figure 20.5 : Leaf pubescence in the wild plant Solanum elaeagnifolium (Silverleaf) (right) as compared with cotton (left).

After the stomata, the secondary site for water loss by transpiration is the cuticle. The hydraulic permeability of the cuticle is determined by the wax embedded in the cuticle matrix as well as by the wax deposited over the cuticle. High cuticular permeability not only affects non-stomatal transpiration pathway but it may also directly affect water loss from guard cells and therefore their water status and stomatal aperture. The lower wax (*bm*) genotype had far greater total leaf transpiration than the *Bm* genotype.

Epicuticular wax is deposited in different forms and structures, mostly as a function of its composition. The environment also affects the density of epicuticular wax. Conditions of water stress, high temperature, and high radiation increase its density. The full genetic potential for wax deposition is therefore best evaluated in plants subjected to stress.

In practical terms, the quantitative effect of wax on transpiration is finite, and for a given plant, the increase in epicuticular wax load beyond a given threshold would not reduce transpiration. Sorghum typically represents relatively high potential epicuticular wax deposition while rice represents species that lack in this respect, as estimated by quantifying epicuticular wax and by rate of cuticular transpiration. Hence, there is a potential for improving drought resistance in rice by genetically increasing epicuticular wax load.

The shape and angles of the cuticular wax deposits of may affect the spectral properties of the leaf. Thus, for example, the glaucous appearance of some wheat genotypes is determined by the structural properties of the wax deposits. Increased glaucousness was found to result in an increase in leaf reflectance of wheat and sorghum within the spectrum range of at least 400 to 700 nm and possibly also at the UV-B. This increase in reflectance may result in a reduction in net radiation and leaf temperatures in glaucous genotypes.

Leaf pubescence is a common feature in xerophytic plants as well as in some crop plants, such as soybean. Generally it increases reflectance from the leaf within the range of 400 to 700 nm and sometimes up to 900 nm, resulting in lower leaf temperatures under high irradiance. It is sometimes argued that the increased reflectance in the photo-synthetically active spectrum would reduce photo-synthesis under non-stress conditions. Under conditions of stress, there is a trade-off between the effect of pubescence towards the reduced stress load and its possible effect on photo-synthesis. Increased leaf pubescence may increase the leaf boundary-layer resistance by up to 50%. However, it has been argued that this should carry a relatively small effect on water and CO_2 exchange, as compared with the effect of pubescence on the radiative properties of the leaf.

Leaf colour can affect the thermal properties of the leaf. In both wheat and barley there are 'yellow leaf' cultivars, which have about a third less chlorophyll than the 'normal' ones. The 'yellow' cultivars tend to perform relatively better under drought stress as compared with the normal green. Yellow leaves are more reflective and their temperature is relatively lower than that of green ones. Beyond this difference in reflective properties, the low chlorophyll lines seem to sustain lower injury to the photo-system under conditions of high irradiance and water deficit.

Osmotic Adjustment (O.A.)

Figure 20.6 : Differential response to sever drought stress of a high O.A. line (left) and a low O.A. line (right) of wheat.

When water deficit develop various solutes accumulate in cells and subsequently tissue osmotic potential is reduced. O.A. is derived from the net increase in cellular osmolality caused by the accumulation of solutes such as various ions (mainly potassium), sugars, poly-sugars (*e.g.* fructan), amino acids (*e.g.* proline), glycinebetaine, etc.' O.A. occurs when cellular water deficit exceeds a certain threshold, which is not universally determined. Nor has the exact signaling for O.A. been resolved. O.A. is a slow process requiring time, and very rapid desiccation in experiments or even in the field may not allow for O.A.. Ideally the rate of plant dehydration should not be faster than about 0.1 MPa day^{-1}. Practically, it should take about 2 weeks from fully hydrated state to wilting on order for the full capacity and impact of O.A. to be expressed in whole plant, depending on species and the growth history of the specific plant. The rate of O.A. varies greatly among species and cultivars. A minimal rate of O.A., which can be considered as effective, is about 0.3 MPa and rates of up to 1.5 to 2.0 MPa were observed in certain cereal cultivars. Some crop plants generally tend to be better at O.A. than others with cowpea, japonica rice and maize generally having lower rates while indica rice, sorghum and wheat tend to express higher rates.

O.A. is probably one of the most crucial components of dehydration avoidance and drought resistance in general. It help maintain cellular turgor at a given leaf water potential and thus delay wilting. O.A. enables to sustain growth and productivity at lower plant water status. Irrespective of the effect on turgor maintenance, the accumulated solutes can protect cellular proteins, various enzymes, cellular organelles, and cellular membranes against desiccation injury. Hence, cell and tissues may continue to function despite the progressing desiccation. This is why the accumulated osmotic solutes are sometimes defined as "protectants". One consequence of O.A. at the whole plant level is the continued growth of roots and the extraction of deeper soil moisture. Finally, O.A. is crucial for the conservation of meristem viability under desiccation towards the recovery of function upon dehydration. O.A. in different cultivars of wheat, sorghum, various pulses and brassicas has been shown to be positively associated with biomass and/or yield under drought stress.

Upon rehydration the various solutes are recycled and metabolized to the extent that the accumulated sugars, for example, are considered as an important energy resource for recovery growth.

Extensive genetic engineering efforts are being made to use the phenomenon of O.A. in the design of stress resistant plants. Most experiments involve transgenic model plants that were modified to constitutively express the accumulation of osmolytes. Such transgenics that accumulate glycinebetaine, D-ononitol, mannitol, and trehalose gave positive or inconclusive results with respect to stress resistance, and work in this area is developing rapidly.

Non-senescence (Delayed Senescence or 'Staygreen' -SG)

Figure 20.7 : "Stay green" (left) and "normal" (right) cultivars of sorghum under post-flowering drought stress.

Plant senescence is a genetically programmed process, accelerated by environmental stress such as drought, heat, and nitrogen deficiency. The primary expression of leaf senescence is the breakdown of chlorophyll and the subsequent collapse of photo-synthesis. Leaf greenness as measured by chlorophyll content or by leaf reflectance properties (using the Minolta chlorophyll meter for example is becoming an acceptable estimate of senescence (and leaf nitrogen status). In various crops certain genotypes were identified as expressing delayed senescence or non-senscent or stay-green phenotype. These genotypes generally sustain leaf greenness and photo-synthesis for a longer time and consequently tend to yield more. Since drought stress accelerates senescence, SG genotypes are important in sustaining green leaf area under stress.

SG does not present a uniform expression across different crop plants. In sorghum for example SG can be associated with high stem soluble carbohydrate content and greater resistance to lodging caused by stem 'charcoal rot'. In sorghum and millet at least, SG genotypes sustain higher RWC under stress as compared with normal ones. This is why SG is discussed under 'dehydration avoidance'. Maintenance of RWC is not necessarily an expected result of delayed chlorophyll loss or delayed leaf protein breakdown. Furthermore, certain SG genotypes of sorghum are expressed better when exposed to drought stress. Hence, the phenotypic selection of SG in sorghum (and perhaps other crops) is more effective under post-flowering drought stress.

SG is at least partly regulated by endogenous plant hormones, whereas in certain cases an increase in kinetin in leaves promoted SG. In other cases

SG was associated with decrease in plant ethylene content. Such hormonal regulation can involve both nitrogen and water status of leaves.

The expression of SG and plant senescence in general can be markedly influenced by intra-plant interactions which involve assimilate partitioning and endogenous hormonal balance. A simple exercise to obtain a SG phenotype in grain producing crops is by detaching the inflorescence at flowering. Grain set and grain growth generally enhance leaf senescence by enhancing carbohydrate and nitrogen export from leaves into the grain.

Very low yielding or partially sterile plants may present some delay in senescence when subjected to drought stress during grain filling. There are ongoing attempts to achieve genetic transformation of SG trait by either promotion of endogenous kinetin or by antisense suppression of ethylene. QTLs for SG are being identified in several crops and marker assisted selection for the trait is becoming possible in sorghum and probably other crops in the future.

Dehydration Tolerance

Effect of stress kinetics on differential gene expression of immature ears of maize. Plants were grown in buckets where drought stress reached the point of null photosynthesis (Pn) in 5 days. Plants were grown in the field where the same state of stress was reached after >5 weeks.

Stress Phenotyping	Stress kinetics	% genes responding
Stress in a large pot	Rapid (5 days)	27
Stress in the field	Slow (4 weeks)	2

Cellular and molecular adaptive processes in response to water deficit do not occur until a certain level of water deficit has been reached. Cellular and molecular adaptive responses serve one or more of the following major functions : (a) reduce whole plant growth in order to reduce plant water-use; (b) reduce the rate of cellular water loss and retain cellular hydration; and (c) protect various cellular structures and functions as cells desiccate.

With modern genomic tools it becomes fairly straightforward to reveal hundreds of genes that are up regulated or down regulated in response to plant tissue water loss. However, research into the function of most of these genes is not as developed. Subsequently the exact function at the whole plant level of the found gene responses to cellular water deficit is not well understood to the extent that they can be used in plant breeding. In terms of application to plant breeding dehydration tolerance is the capacity to function in a dehydrated state which often (but not always) means the involvement of stress responsive and adaptive genes. Most of the information that is relevant for application to breeding is derived from whole plant physiological studies while some rudimental information comes from genomics.

Plant physiology always cautioned that the evaluation of plant response to drought stress and the evaluation of plant adaptation require sufficient time under stress. Adaptive plant responses to drought stress do not only depend on the level of tissue desiccation but also on its rate. It was well established that fast or slow desiccation may have totally different impact on results in terms of adaptation. Very rapid desiccation often exercised in laboratory experiments is totally irrelevant even though statistically significant results can be obtained. Tissue desiccation under natural conditions is slow. Confirmation of this axiom is now received from a gene expression study in maize as presented in Table, which speaks for itself.

Stem Reserve Utilization

The current source of carbon for grain filling is assimilation by the light intercepting viable green leaf area. This source is normally diminishing due to natural senescence and the effect of various stresses. At the same time the demand by the growing kernel is increasing, in addition to the demand posed by maintenance respiration of the live plant biomass. When the demand by the grain is not fully supplied by the source of current assimilation, then plant reserves can provide the balance. Small grains and cereal stems as well as several other crops store carbohydrates in the form of glucose, fructose, sucrose, fructans or starch. Type of storage depends on the species. Total storage in cereal plant roots or leaves is relatively small to that in the stem (including leaf sheaths). This storage is commonly analyzed as total non-structural carbohydrates (TNC) or water-soluble carbohydrates (WSC) and it is available for translocation to the grain.

Usage of stem reserves depends on the available storage and the rate and duration of mobilization of storage to the grain. The size of the storage strongly depends on favourable growing conditions before anthesis and genotype. Developmentally, potential stem storage as a sink will also be determined by stem length and stem weight density (stem dry weight per unit stem length). Stem length, as affected by the height genes is important in affecting stem reserve storage, as demonstrated in sorghum. The Rht1 and Rht2 dwarfing genes of wheat were also found to limit the reserve storage by about one third as a consequence of a reduction in stem length. This may be one of the reasons for the general greaterer drought susceptibility of the dwarf high yielding wheat cultivars.

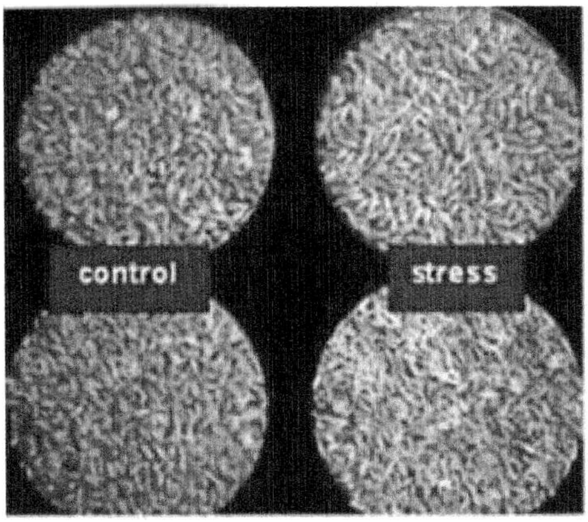

Figure 20.8 : Grain of two wheat cultivars subjected to sever drought stress during grain filling (right). Top : cultivar with superior capacity for stem reserve utilization; bottom : normal cultivar. Note the shriveled grain under stress in the latter.

Stem reserve mobilization or the percentage of stem reserves in total grain mass is affected by sink size, by the environment and by cultivar. It is not surprising therefore, those different estimates of the percentage of grain yield that is accounted by stem reserves range from 9 to 100%.

The demand by the grain yield sink is a primary factor in determining stem reserve mobilization. When manual de-graining reduced sink size, more reserves were stored in the stem, as compared with intact ears. The availability of storage at grain filling does not necessarily assure mobilization. There are cases on record where despite stress conditions the available storage was not utilized. This may be traced to problems in enzymatic conversion of storage to transportable constituents or sometimes inhibition of sink processes. For example, under heat stress starch synthesis in the wheat grain might be inhibited by a thermolabile enzyme (such as soluble starch synthase) and available stem reserves would not be in demand. Heat tolerant starch synthase is therefore also essential for grain filling under heat stress. Hormonal signaling might also be involved in reserve mobilization.

The reduction in current assimilation during grain filling, under different stresses, will induce greater stem reserve mobilization to the grain. What is important is the reduction in assimilation and not the nature of stress causing the reduction. Thus, stem reserve mobilization is a solid source of carbon for grain filling under any stress (such as heat stress also), which would inhibit current photo-synthesis, including biotic stresses such as late developed leaf diseases. Tolerance to Septoria leaf blotch in wheat was expressed in sustained grain filling under sever epiphytotic. It has been demonstrated that mobilized stem reserve is a major component of Septoria tolerance in wheat.

The full potential for stem reserve utilization of a cereal cultivar can be experimentally assessed by growing plants under favourable conditions and then detaching all leaf blades and shading the inflorescence at the onset of grain filling. Grain weights per inflorescence in such treated plants as compared with controls provide a reliable estimate. It appears that wheat genotypes differ in their capacity to store stem reserves. Cultivars that have this high capacity must also possess relatively long grain filling period in order to allow sufficient time for reserves to be mobilized into the grain.

A possible "penalty" for high stem reserve utilization capacity is accelerated shoot senescence, due to the export of C and N storage into the grain. Thus, it seems that the two factors may not be recombined and the breeder will probably have to opt for either stem reserve mobilization or delayed senescence trait as mechanisms supporting grain filling under stress.

Cellular Membranes Stability

The fluid mosaic model of the cellular membranes (CM) describes the membrane as a bi-layer of phospholipids and glycolipids studded and spanned by proteins partially or fully solvated by the lipid matrix. CM is central site for various cellular functions; especially those associated with membrane bound enzymes and transport of water and solutes. The function and role of the CM under extreme temperature stress is somewhat clearer than with drought stress.

The phospholipids in terms of their quantity and composition are generally considered as the more crucial components of cellular membrane stability under drought stress. The most notable factor in cellular membrane function under desiccation and heat stress is that plant previous exposure to moderate stress signal a hardening ("acclimation") effect that is expressed in increased membrane stability under stress. Cellular membrane stability under stress has been shown to be positively correlated with yield advantage under stress, more often for heat than for drought stress. The simplest way to assess membrane stability is by measuring the leakage of cellular electrolytes under stress.

Water passage through both the plasma membrane and the tonoplast is crucial to cell life and specific proteins inserted in the membrane largely regulate it. These 'water-channel' proteins, also termed aquaporins respond to various signals and "molecular switches". These pores are highly selective to water and they play an import role in cellular water relations in response to plant water deficit and osmotic stress. For example, maize root aquaporins were found to be stimulated by water deficit, resulting in improved water transport. The study of aquaporins and their function is now at the forefront of research on cell water relations.

Antioxidation

Oxidative Stress is a general term used to describe a state of damage caused by reactive oxygen species (ROS). ROS, such as free radicals and peroxides, represent a class of molecules that are derived from the metabolism of oxygen. There are many different sources of ROS that can cause oxidative damage to an organism. Most come from endogenous sources as by-products of normal and essential reactions, such as energy generation from mitochondria or the detoxification reactions. Free radicals are unstable because they have unpaired electrons in their molecular structure. This causes them to react almost instantly with any substance in their vicinity. Free radicals destroy cellular membranes, enzymes and DNA.

Antioxidants are active substances naturally occurring in all organisms which detoxify free radicals. These are for example superoxide dismutase (SOD), catalase, glutathione reductase or ascorbate peroxidase. SOD, for example, converts the $O2^{o}$ to H_2O_2 and Catalase converts H_2O_2 to molecular oxygen.

Drought, as well as other stresses cause oxidative stress in plants and antioxidant abundance and activity is important for the protection of metabolism under stress. When various studies are reviewed it is unclear whether the genetic enhancement of antioxidant production in plants beyond the natural level is indeed required to alleviate drought stress at the whole plant level and whether the naturally occurring active antioxidants are not sufficient to protect the plant. It has been shown that drought induced oxidative stress related genes and that this was associated with increased levels of various antioxidants in plants. The most important information in this respect is coming from the developing work with transgenic plants, which over-express antioxidant production. When these studies are taken as a whole, no clear conclusions can be made yet with respect to the possible importance of breeding for overproduction of antioxidants towards the improvement of plant production under drought stress.

Stress Proteins and Chaperons

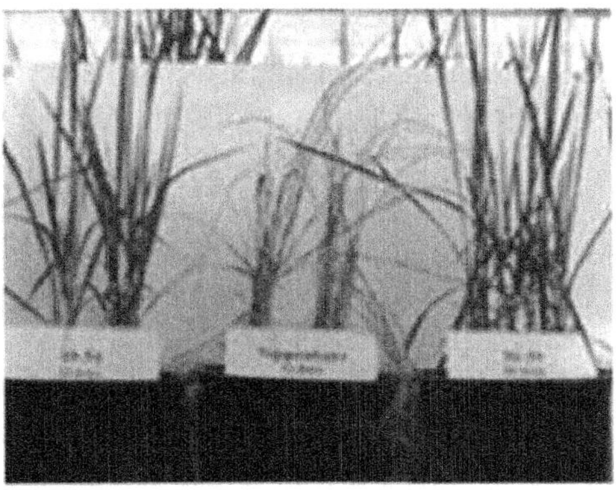

Figure 20.9 : Rice transgenic plants over expressing the HVA1 barley embryo LEA protein and subjected to drought stress. The middle pot is the 'wild type' (control) plant. These transgenics were developed by the late Prof. R. Wu and associates at Cornell University. The photograph was taken from a study by Dr. H.T. Nguyen at Texas Tech University.

Stress proteins is a large group of different proteins induced by different environmental and biotic stress in various organisms ranging from prokaryotes to man. A group of relatively small molecular weight proteins is developmentally regulated in growing seed such as that of barley. Their accumulation during embryo development has a role in protecting the embryo as the seed matures and desiccates during maturation (typically to about 10% water content). These are defined as 'late embryogenesis abundant' (LEA) proteins. Further research found that LEA proteins consist of a family, including several similar proteins such as dehydrins. These are not limited to seed embryo and they can be induced by drought stress in various plant tissues. Some are ABA responsive while others are not. Additional information especially on LEA protein and desiccation tolerance in seed is available on line.

Work with transgenic plants indicated that the LEA family of stress protein might have a role in drought and osmotic stress resistance. Their exact function is not clear but it may involve osmotic adjustment or protection of cellular membranes or organelles during desiccation. They may also act as molecular chaperons and in that respect they are very similar to low molecular weight heat shock proteins (HSP). In this role they may conserve protein structure during stress. The LEA family of proteins may carry an important potential for enhancing stress resistance.

Abscisic Acid (ABA) Accumulation and Its Consequences

ABA accumulates in various plant parts subjected to desiccation. ABA responsive genes are often assumed to be stress adaptive. The rational is that if plants under stress consistently respond by producing ABA in leaves and root, then ABA must be important for coping with stress. The most prominent effect of ABA accumulation in the plant is stomatal closure and reduced transpiration. Thus, when transgenic model plants such as Arabidopsis or tobacco are engineered to over-produce ABA they maintain turgor when grown in pots subjected to dehydration. Thus it is often concluded that ABA promotes drought resistance. Turgor maintenance by stomatal closure is important for survival but not for plant production under drought stress. Furthermore, a review of the literature indicates that ABA has numerous and critical negative effects on plants especially when crop productivity is considered.

Genotypes of wheat that were selected for a high ABA accumulation under drought stress were found to be no better or even worse than the normal ones in terms of function and yield under drought stress. Selection for low leaf ABA content in maize was correlated with reduced yield under conditions of limited water supply.

On the other hand ABA may have an important role in regulating an orderly shutdown of plant functions towards a state of dormancy, as the case is for the maturing seed. Dormancy is essential for surviving extreme plant desiccation. ABA mediated dormancy is crucial for attaining freezing tolerance, which involves cellular desiccation. The value of plant survival

Table 20.1 : Effects of ABA on plant processes involved with growth and reproduction.

Growth	
General growth	Inhibition
Cell division	Decrease
Cell expansion	Decrease
Leaf initiation	Inhibition
Germination	Decrease
Root growth	Increase
Tillering	Decrease
Dormancy	Improved
Reproduction	
Flowering (annuals)	Advance
Flower induction	Inhibition
Flower abscission	Increase
Pollen viability	Decrease
Seed set	Decrease

Figure 20.10 : Wheat seedlings grown in vermiculite and severely desiccated after which they were irrigated. The seedling on the right received 0.1 μmol of ABA in the irrigation water before the onset of stress. Control seedlings are on the left.

under sever desiccation depends on the agricultural ecosystem concerned. It can be important in subsistence agricultural where plant recovery from severe drought stress can provide some growth and production. Plant survival is of lesser consequence to commercial crop production as practiced in developed countries. Even the ability of seedling survival and recovery after a prolonged drought in commercial wheat production does not carry great impact when re-seeding the crop is a viable option. A commercial crop based on recovered seedlings is likely to be inferior to that grown from newly planted seed. The present knowledge on ABA and it role in plant adaptation to drought stress as well as in general plant production capacity does not allow yet to formulate a breeding strategy with respect to ABA.

Drought Escape and Plant Phenology

Figure 20.11 : Early flowering (left) and late flowering (right) sorghum cultivars under late-season drought stress. The late cultivar will not flower at all due to stress.

Short growth duration (generally defined by early flowering) constitutes an important attribute of 'drought escape', especially for conditions of a late-season drought stress. On the other hand, longer growth duration is often associated with high yield potential. Consequently, using drought escape as a solution may involve a cost in terms of reduced yield potential. This is serious, especially when the moisture environment is absolutely unpredictable and may vary to a large extent between years. The more predictable the environment is, the easier it is to optimize phenology. The unpredictability of the environment may reach a state where short growth duration is a drawback, especially in indeterminate plants that offer a potential for regrowth and productivity upon recovery. Longer growth duration in both determinate and indeterminate plants would improve the probability for regrowth upon recovery simply because, on the same calendar day, late-maturing genotypes are younger than early ones and younger plants recover better. The final decision on the optimum growth duration has, of course, to consider additional factors, such as late-season disease and insect pressure or periods of frost.

Early maturity leads to reduced, total seasonal evapotranspiration simply because of the shorter time in the field. However, as growth duration is genetically linked with leaf number, early genotypes tend to have a small transpiring leaf-area index. Thus, early genotypes show reduced evapotranspiration during most growth stages, up to the point where a full ground cover is achieved. At most growth stages, root-length density and total root length per plant is generally greater in a late than in an early cultivar. This should be reflected in an advantage for the late genotype under conditions where extensive rooting is required.

A phenological feature specific to maize is the timing of anthesis with respect to silking, defined as anthesis-to-silking interval (ASI). Evidently a short interval is desirable whereas a large interval results in poor pollination. The maize program at CIMMYT dedicated many years of work to research the trait and explore its significance in tropical maize breeding for stress environments. Maize germplasm can vary for ASI irrespective of the effect of stress; a short ASI is a universally important trait for maize production. However, stress, and especially drought during the reproductive stage may extend ASI and thus reduce yield. Maize genotypes may vary in ASI under drought stress from few days up to a month or more. The effect on yield of change in ASI between null and 10 days is exponential. Selection for short ASI under drought stress proved to be an effective approach to improve drought resistance of tropical maize. QTLs (quantitative trait loci) controlling ASI were located and marker-assisted selection is possible.

Different crop plants may advance (*e.g.* wheat) or delay (*e.g.* rice) their flowering when stress occurs before flowering. The rate of delay is a function of plant water deficit and probably also ABA signaling. The rate of change in flowering time under stress can be taken as an index of genotypic rate of stress in the field.

Water-use Efficiency (WUE)

WUE is not a component of drought resistance but the term implies greater production for a given amount of limited water. Namely "more crop per drop". This is not necessarily the case. High WUE result (in most cases) from "less drop per crop".

WUE was originally developed by agriculture engineers as a ratio between yield and irrigation water in order to assess returns for irrigation input and cost. WUE is an important yardstick to measure irrigation efficiency. The WUE term was later adopted by soil scientists and agronomists for a wider use in agronomy, including dryland-rainfed crop production. Physiologists found the term useful also at the leaf level in studies of gas exchange where WUE (*i.e.* "transpiration ratio") is defined as the ratio of carbon fixation to transpiration. WUE can therefore be used at various levels of the crop, from the single leaf to the field.

Studies of water use efficiency at the whole plant and field level were cumbersome due to the work load and costs involved in assessing whole plant or crop water use, especially when large plant populations in plant breeding were considered. The breakthrough came with the development of better understanding of stomatal dynamics, gas exchange and photo-system function, leading to the carbon isotope discrimination (delta) assay as a heritable marker for WUE at the whole plant level. In the majority of cases low carbon isotope discrimination (low delta) as measured in the grain or the leaves was found to be well correlated with high WUE across variable genetic materials and *vice versa,* with few exceptions where delta was not associated with WUE.

An important contribution of the carbon isotope discrimination method was that it enhanced research on WUE and provided extensive data on the subject especially in the context of breeding and genetic diversity. At the same time the large volume of published information on delta, WUE and their implications towards selection for water limited environments created some confusion in the plant breeding community. Confusion was largely created by the fact that the relations between delta (WUE) and yield were sometimes positive and sometimes negative, depending on the crop growing conditions. Plant breeders discussing carbon isotope discrimination and WUE expressed confusion on two primary questions : (1) under what environmental conditions selection for carbon isotope discrimination is expected to result in yield gain, and (2) which direction should selection be made, high (low delta) or low (high delta) WUE.

Beyond these questions the real issue is whether selection for high WUE is universally associated with drought resistance and improved plant production under drought stress. WUE is often equated in a simplistic manner with drought resistance without considering the fact that it is a ratio between two physiological (photo-synthesis and transpiration) or agronomic (yield and crop water use) variables. This ratio it is often susceptible to misinterpretation, especially when the dynamics of the nominator and the denominator are ignored. When all studies of carbon isotope discrimination in breeding population are taken together it can be seen that higher WUE is derived from a reduction in water use rather than from an increase in production. Reduced water-use under dryland conditions is contradictory to productivity. Thus genotypes of high WUE under drought stress tend to be less productive under stress - with few specific exceptions. A more detailed discussion of WUE in the context of plant breeding for plant production under water limited environments is presented by Blum. *This discussion concludes that the target of plant breeding for water limited environments is effective use of water (EUW) rather than WUE.*

Photo-synthetic Systems and Plant Production under Stress

Plant science is still seeking ways to genetically increase productivity for a given unit of water-use under drought stress. The key is in photo-synthesis. The C4 photo-synthetic metabolism as compared with the more widely common C3 type photo-synthetic metabolism is intimately associated with superior productivity at given water-use. The C4 pathway of photo-synthesis as found in maize, sorghum, pearl millet, and various forage grasses is essentially a pumping mechanism that moves CO_2 from the mesophyll cells and causes high CO_2 concentrations in the specific biochemically active vascular-bundle sheath cells. This mechanism goes hand in hand with certain anatomical and morphological features of the C4 plant ("Kranz leaf anatomy") that are inseparable from the system as a whole. The CO_2-concentrating mechanism results in a high utilization efficiency of low inter-cellular CO_2 concentrations. This is due to the PEP carboxylase enzyme in the C4 plant, which unlike RuBP carboxylase is insensitive to atmospheric O_2 concentrations. Atmospheric O_2 concentrations are strongly inhibitive to CO_2 uptake in C3 plants where CO_2 is fixed directly by RuBP carboxylase. In C4 plants CO_2 fixation is carried out in the bundle-sheath cells using CO_2 from decarboxylated C4 acids in the mesophyll cells. This sequence results in sufficiently high CO_2 concentration maintained at the bundle sheath cell. The efficiency of the CO_2 fixation pathway in the C4 plant bears significance toward its transpiration-ratio. For a given rate of transpiration, photo-synthesis is greater in C4 than in C3 plants. This advantage is also translated into a greater plant or crop WUE, hich is not always necessarily related to drought resistance and a relatively better yield under stress. For reasons other than the biochemistry of photo-synthesis (say, deep roots or O.A.) a certain C3 crop might produce better than a C4 one under drought stress. However, under well-watered conditions the greater WUE of the C4 plant is most likely translated into better economic returns on the cost of irrigation. The normal WUE (for grain yield) of supplemental irrigation in grain sorghum (C4) is about 20 kg mm-1 ha-1, as compared with 10 kg mm-1 ha-1 in wheat (C3).

Whereas WUE is often confused with drought resistance it is very important to take note of a comparative study of C4 and C3 Panicoid grasses (#10259). It concluded that declining C4 photo-synthesis with water deficit was mainly a consequence of metabolic limitations to CO_2 assimilation, whereas, in the C3 species, stomatal limitations had a prevailing role in the drought-induced decrease in photo-synthesis. The drought-sensitive metabolism of the C4 plants could explain the observed slower recovery of photo-synthesis upon re-watering, in comparison with C3 plants which recovered a greater proportion of photo-synthesis through increased stomatal conductance. Therefore, within the Panicoid grasses, the high WUE C4 species are metabolically more sensitive to drought than the lower WUE C3 species and recover more slowly from drought.

Chapter 21

HEAT STRESS AND ITS IMPACT

THE ENVIRONMENTAL AND PHYSIOLOGICAL NATURE OF HEAT STRESS

Heat stress often is defined as where temperatures are hot enough for sufficient time that they cause *irreversible* damage to plant function or development. In addition, high temperatures can increase the rate of reproductive development, which shortens the time for photo-synthesis to contribute to fruit or seed production. I also will consider this as a heat-stress effect even though it may not cause permanent (*irreversible*) damage to development because the acceleration does substantially reduce total fruit or grain yield.

The extent to which heat stress occurs in specific climatic zones is a complex issue. Plants can be damaged in different ways by either high day or high night temperatures and by either high air or high soil temperatures. Also, crop species and cultivars differ in their sensitivity to high temperatures. Cool-season annual species are more sensitive to hot weather than warm-season annuals. We did not include sunflower in the table because it is unusual in that during the vegetative stage it grows well in cool conditions and during the reproductive stage it grows well in hot conditions.

Table 21.1 : Annual crop species adapted to cool and warm seasons.

Cool-season annuals	Warm-season annuals
Barley, brassicas, canola, fava bean, flax, garbanzo bean, Irish potato, lentil, lettuce, lupine, mustard, oat, pea, radish, rye, spinach, triticale, turnip, vetch, wheat	common bean, cotton, cowpea, cucurbits , finger millet, grain amaranth, lima bean, maize, mung bean, pearl millet, pepper, pigeon pea, rice sesame, sorghum, soybean, sunflower, sweet potato, tobacco, tomato

High day temperatures can have direct damaging effects associated with hot tissue temperatures or indirect effects associated with the plant-water-deficits that can arise due to high evaporative demands. Evaporative demand exhibits near exponential increases with increases in day-time temperatures and can result in high transpiration rates and low plant water potentials.

The extent of heat stress that can occur in a specific climatic zone depends on the probability of high temperatures occurring and their duration during the day or night. Where global climate change is occurring these probabilities may not be predicted well based only on historical records for specific locations. Heat stress is a complex function of intensity (temperature degrees), duration and rate of increase in temperature. The magnitude of heat stress rapidly increases as temperature increases above a threshold level and complex acclimation effects can occur that depend on temperature and other environmental factors.

High soil temperatures can reduce plant emergence. The maximum threshold temperatures for germination and emergence are higher for warm-season than for cool-season annuals. For example, the threshold maximum seed zone temperature for emergence of cowpea is about 37^{0}C compared with 25 to 33^{0}C for lettuce.

During the vegetative stage, high day temperatures can cause damage to components of leaf photo-synthesis, reducing carbon dioxide assimilation rates compared with environments having more optimal temperatures. Sensitivity of photo-synthesis to heat mainly may be due to damage to components of photo-system II located in the thylakoid membranes of the chloroplast and membrane properties. Membrane thermostability has been evaluated by measuring electrolyte leakage from leaf disks subjected to extreme temperatures. More stable membranes exhibit slower electrolyte leakage. Studies comparing responses to heat of contrasting species indicated that photo-system II of the cool season species, wheat, is more sensitive to heat than photo-system II of rice and pearl millet, which are warm season species adapted to much higher temperatures.

Extreme temperatures can cause premature death of plants. Among the cool-season annuals, pea is very sensitive to high day temperatures with death of the plant occurring when air temperatures exceed about 35^{0}C for sufficient duration, whereas barley is very heat tolerant, especially during grain filling. For warm season annuals, cowpea can produce substantial biomass when growing in one of the hottest crop production environments on earth (maximum day-time air temperatures in a weather station shelter of about 50^{0}C), although its vegetative development may exhibit abnormalities such as leaf fasciations. For monocotyledons, including both cool-season and warm-season annuals, high daytime temperatures can cause leaf firing which involves necrosis of the leaf tips and this symptom also can be caused by drought.

Reproductive development of many crop species is damaged by heat such that they produce no flowers or if they produce flowers they may set no fruit or seeds. The reviews of Hall discuss the detrimental effects of heat stress on reproductive development that has been reported for cowpea, common bean, tomato, cotton, rice, wheat, maize and sorghum. I will examine the detrimental effects of heat stress on cowpea because of the comprehensive information available for this species and the likelihood that many of the same phenomena occur with other warm-season crop species.

Controlled-environment studies in which cowpea plants were subjected to separately controlled root and shoot and day and night temperatures demonstrated that pod set (the proportion of flowers producing pods) was damaged by moderately high night temperature of the shoot. It was surprising that night temperature would have this effect since much hotter day temperatures did not damage pod set of cowpea. Reciprocal artificial pollinations between plants grown under high and optimal night temperatures indicated the low pod set was caused by male sterility and that the pistils did not appear to be damaged by high night temperature. The detrimental effects of high night temperature on pod set also were shown to occur in field conditions. In these experiments a unique experimental approach was used in which plots of cowpea plants were subjected to different increments of higher night temperatures during early stages of flowering using enclosure systems placed over the plots only during the night-time.

Possible mechanisms for the sensitivity of pod set to high night temperatures have been proposed. Mutters and Hall demonstrated that there is a distinct period during the 24-hours cycle when pollen development in cowpea is sensitive to high night temperatures. Plants subjected to high temperature during the last six hours of the night exhibited substantially decreased pollen viability and pod set, whereas plants subjected to high temperature during the first six hours of a twelve-hours night exhibited no damage. Mutters and Hall hypothesized that these results could be explained if a heat-sensitive process in pollen development is under circadian control and only occurs in the late night period. Note that if a heat-sensitive process is under circadian control and if genetic variation exists for the time in the 24-hours cycle when this process occurs, evolution in hot environments would favour plants in which the heat-sensitive process occurs at the coolest time which is just prior to dawn. The damaging effect of high night temperature on pod set was greater in long days than in short days, and red and far red light treatments indicated it is a phytochrome-mediated response.

Studies were conducted in which cowpeas were transferred between growth chambers having high or optimal night temperatures. They demonstrated that the stage of floral development most sensitive to high night temperature occurs 9 to 7 days prior to anthesis, which is after meiosis and coincides with release of pollen microspores from the tetrads. Damage due to high night temperature was associated with premature degeneration of the tapetal layer that provides nutrients to developing pollen, infertile pollen and in some genotypes anthers did not dehiscence. The transfer of proline from the tapetal layer to pollen was inhibited.

Comparisons of heat-sensitive and heat-tolerant cowpeas showed a genotypic association between sensitivity to heat during pod set and rapid leakage of electrolytes from leaf discs subjected to heat stress. Possibly, the damage to pollen development by high night temperatures may be in some way associated with a heat-induced malfunction in membrane properties.

Floral bud development also can be damaged by heat such that plants do not produce flowers. For cowpea, two weeks or more of consecutive or interrupted hot nights during the first month after germination caused complete suppression of floral bud development. In extreme cases the floral buds become necrotic and die. In field conditions, the damage occurs under long days but not under short days. However, responses to red and far red light indicated the effect was only partially consistent with the system being mediated by phytochrome. The damaging effect of high night temperature and long days on floral bud development also depended on light quality whereas the damaging effect on pod set did not depend on light quality. When growth chambers were used with relatively large amounts of fluorescent light and little incandescent light, such that the red/far red ratio was 1.9, floral buds were not suppressed in long-day high night temperature conditions, but pod set was very low. This artificial light system provides a useful experimental method for studying the effects of heat stress on pod set without complications due to heat stress effects on floral bud development. When growth chambers were used with lighting systems providing a red/far red ratio of 1.3 to 1.6, floral bud suppression was observed that was similar to what is obtained under long-day high night temperature conditions in the field where sunlight has a red/far red ratio of about 1.2.

There are two important conclusions from these studies. First, the use of growth chambers with lighting systems that mainly depend on fluorescent lights can result in either serious artifacts or methodological advantages when studying plant reproductive responses to heat stress. Second that intense shading of floral buds could reduce the red/far red ratio below 1.2 in field conditions and intensify the floral bud suppression effect. In densely sown fields of cowpea, individual plants that are suffering from competition and are tall and spindly can exhibit floral bud suppression even though night temperatures are not too hot.

Pods of different cowpea genotypes produce 9 to 20 ovules with many cultivars having 15 ovules, but pods rarely produce these many seeds per pod. Under optimal conditions two-thirds of the ovules may produce seed, whereas with high day or high night temperature and other stresses, such as drought, fewer seeds are produced per pod. For most cultivars and stresses it is the ovules at the blossom end of the pod that suffer embryo abortion and do not produce seed, resulting in the production of "pinched" pods.

Cowpea seeds produced under high day temperatures can have asymmetrical twisted cotyledons. Germination of the seed is not influenced and this effect of heat stress may not be a major problem for commercial production. In contrast, heat-induced brown discolouration of cowpea seed coats can occur with some cultivars and be a major problem causing consumers to reject grain. Higher night temperatures resulted in progressively larger numbers of seed with larger areas of brown discolouration on seed coats.

The extent to which high-temperature damage to photo-synthesis or reproductive development affect fruit or grain yield probably depends on the extent to which the photo-synthetic source and the reproductive sink are limiting fruit or grain yield, and this may vary among species and cultivars.

Surface and internal tissues of tomato and citrus fruit can be damaged by the combination of high temperatures and intense solar radiation. High tissue temperatures also can damage cambium layers in exposed trunks and branches.

REPERCUSSIONS OF HEAT STRESS

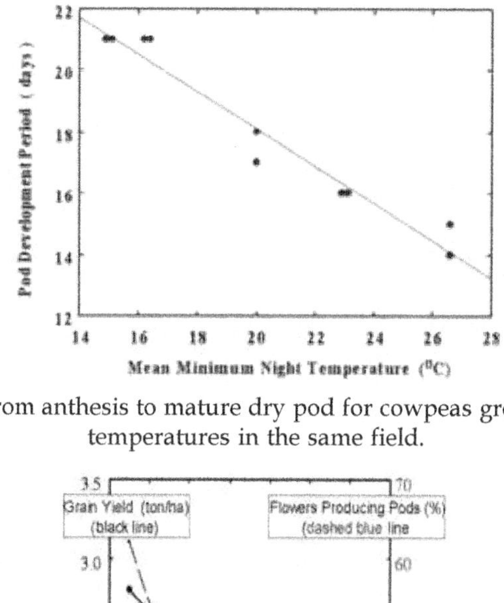

Figure 21.1 : Days from anthesis to mature dry pod for cowpeas grown under nighttime temperatures in the same field.

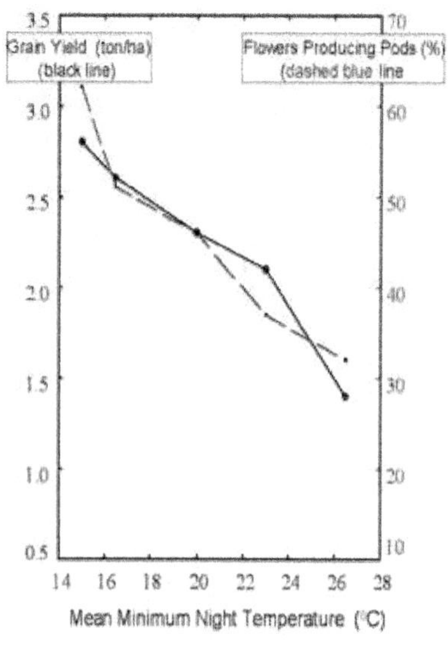

Figure 21.2 : Grain yield and % flowers producing pods for cowpeas grown under different night-time temperatures in the same field.

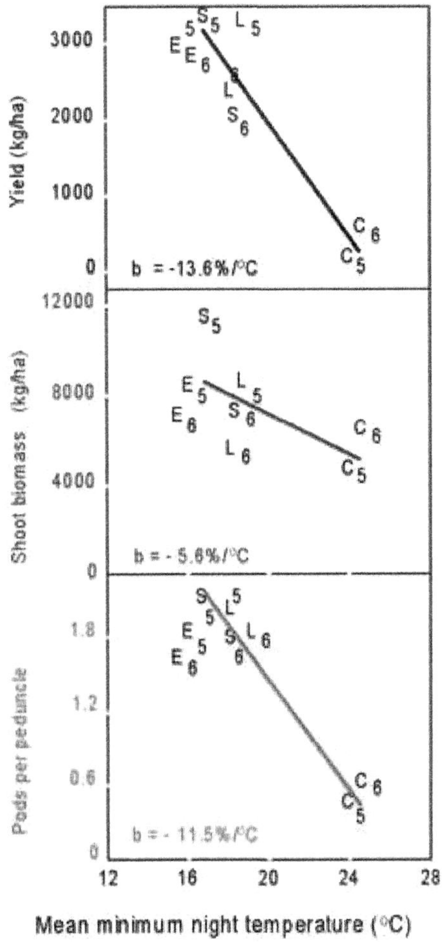

Figure 21.3 : Plant production of heat-susceptible cowpea lines grown in different fields with contrasting thermal regimes.

The reduction in maximal emergence of annual crops due to hot soils can be so pronounced that yield of the economic product is reduced substantially. This can be a major problem for crops, such as lettuce, whose canopies are not very plastic and cannot compensate for the low emergence. Heat stress at emergence is a major problem for cool-season crops that are sown in the late summer in hot sub-tropical zones with the objective of producing a crop during the cooler weather in the fall. Warm-season crops also can experience this problem in those tropical zones and seasons where soil temperature can be extremely hot at sowing.

The acceleration of reproductive development by high temperatures may partially explain why the potential grain yields of warm season crops, such as rice and cowpea, usually are higher in the sub-tropics than the tropics. The extent of the acceleration of development of cowpea has been determined by subjecting plants to different night temperatures in field conditions using temporary enclosures imposed only at night with regulated heating systems. Under the cool night temperatures of sub-tropical California (minimum night temperatures of 16°C), individual cowpea pods took 21 days to proceed from anthesis to mature dry pod. In contrast, the pod development period of the same cowpea cultivar was only 14 days with the higher night temperatures that occur under tropical conditions (minimum night temperatures of 26°C).

More rapid pod development may increase the extent of embryo abortion, and individual seed usually are smaller in tropical compared with sub-tropical conditions for the same cowpea cultivar. Cowpeas subjected to elevated night-temperature treatments produced smaller seed. The more rapid development of individual fruits also results in the overall reproductive period of the plant being shorter. Grain yields of cowpea cultivars grown with optimal management are much less in tropical zones than in sub-tropical zones mainly due to the shorter overall reproductive period caused by the hot nights of tropical zones. Acceleration of reproductive development also is a problem for cool season crops, such as wheat, growing in environments that are hot during reproductive development.

Direct evidence for heat-stress effects on grain yield was provided by the studies of Nielsen and Hall in which cowpea was subjected to different increments of elevated night temperature under field conditions in a sub-tropical location in California. For minimum night temperatures greater than 15°C there were linear reductions in both grain yield and the proportion of flowers that set pods with 50% reductions occurring at minimum night temperatures of about 26°C.

Evaluating the same set of cultivars over a range of environments that mainly differ in temperature provides a more indirect approach for evaluating heat stress effects on the performance of crops that is of interest to farmers. Six heat-susceptible cowpea genotypes, including a California cultivar, were grown in eight field environments with contrasting temperatures but similar high levels of solar radiation and optimal management. Grain yield was negatively correlated with mean minimum night temperatures during the three-week period beginning one week before first flowering. For minimum night temperatures greater than 16.5°C grain yield decreased 13.6 % per °C associated with a similar decrease (11.5 % per °C) in number of pods per peduncle but only a small decrease (5.6 % per °C) in shoot biomass production.

Chapter 22

Impact of Mineral Deficiency Stress

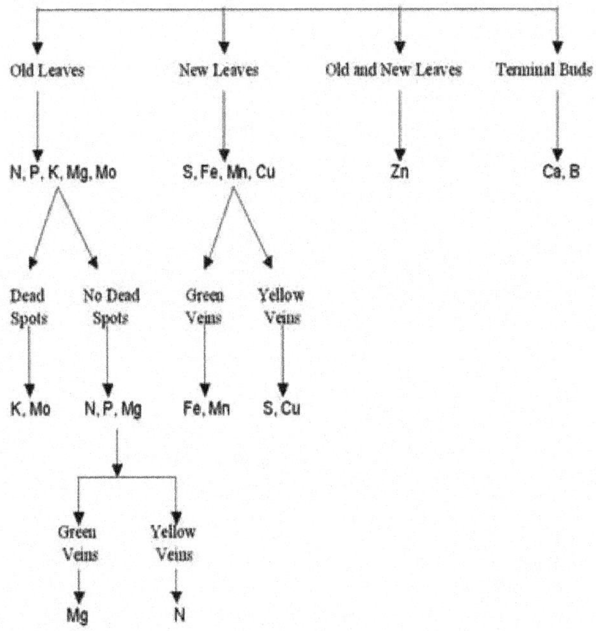

Figure 22.1 : Flowchart for the identification of deficiency symptoms according to Reddy and Reddi.

A mineral element is considered as essential, when plants cannot complete reproductive stage of life cycle due to its deficiency. Deficiency must be corrected only by supplying the element in question and when the element is directly involved in the metabolism of the plant. Based on these criteria, sixteen elements so far were identified as essential. These are : carbon, hydrogen, oxygen, nitrogen, phosphorus, potassium, calcium, magnesium, sulfur, iron, manganese, zinc, copper, boron, molybdenum and chlorine. Most of the carbon as carbon dioxide enters the plant from the air; hydrogen and oxygen are taken up as water. The rest of the elements are taken up from the soil solution as mineral nutrients. Among these nutrients N, P, K, Ca, Mg, and S are considered major or macro-nutrients, because they are required in large quantities that range between 1 to 150 g per kg of plant dry matter. Fe, Zn, Mn, Cu, B, Mo and Cl are minor or micro-nutrients that are required at rates of 0.1 to 100 mg per kg of plant dry matter. Chloride is essential in micro quantities but can accumulate in the plant in large quantities when present in high concentrations in the soil solution.

Table 22.1 : Soil conditions inducing nutrient deficiencies in crop plants (from Fageria *et. al.* (1997a).

Nutrient	Conditions inducing deficiency
N	Excess leaching with heavy rainfall, low organic matter content of soils, burning of crop residue, denitrification on (losses of N_2 and N_2O to the atmosphere)
P	Acidic (high Al contend, organic, leached and calcareous soils, high rate of liming
K	Sandy, organic, leached and eroded soils, intensive cropping system without fertilizer addition
Ca	Acidic, alkali, or sodic soils
Mg	Low clay content sodic low Mg soils
S	Low organic matter content of soils; continued use of N and P fertilizers containing no sulfur, burning of crop residue
Fe	Calcareous soils, soils high in P, Mn, Cu or Zn, high rate of liming
Zn	Highly leached acidic soils, calcareous soils, high levels of Ca, Mg and P in the soils
Mn	Calcareous silt and clays, high organic matter, calcareous soils
B	Sandy soils, naturally acidic leached soils, alkaline soils with free lime
Mo	Highly podzolized soils, well drained calcareous soils.

All the essential nutrients are required by plants in balanced proportions. Deviation from this may result in nutritional disorders. Early detecting of nutritional deficiency stress is important. Stress might extend to the entire plant with loss of yield if relief of stress is not employed. Continuous shortage of a nutrient or nutrients might cause plant death. When two or more elements are deficient simultaneously, the composite picture of symptoms may resemble no single known deficiency. Mineral deficiency symptoms are sometimes confused with other complex field events such as damage caused by insect-pest, disease, salt stress, water stress, pollution, light and temperature injury and herbicide damage. Toxicity of Mo or Se is similar to P deficiency, Fe deficiency in Mango is similar to Chloride toxicity. Therefore, it is necessary to critically observe and define these deficiency symptoms. The deficiency symptoms might be distinguished based on the plant part that shows deficiency symptoms, presence or absence of dead spots and entire leaf or interveinal chlorosis.

Generally, nutrient deficiency in the plant occurs when a nutrient is insufficient in the growth medium and/ or cannot be absorbed and assimilated by the plants due to unfavourable environmental conditions. Nutrient disorders limit crop production in all types of soil around the world.

VISIBLE SYMPTOMS OF STRESS

Nitrogen (N)

The characteristic deficiency symptom of nitrogen is the appearance of uniform yellowing of leaves including the veins, this being more pronounced on older leaves as expressed in rabbit-eye and blueberries; Fescue; *Ailanthus triphysa*; chili and sugarcane. The leaves become stiff and erect. In dicotyledonous crops the leaves detach easily under extreme deficiency condition. Cereal crops show characteristics 'V' shaped yellowing at the tip of lower leaves. O'Sullivan *et. al.*, observed relatively small and pale green leaves with dull appearance in sweet potato. If such condition of nitrogen stress do persist, the result is a decreased foliage growth and shoot growth. See for example : black pepper; douglas-fir and sapota.

Phosphorus (P)

In phosphorus deficiency, leaves remain small, erect, unusually dark green with greenish red in sweet potato, bluish green in chili, brown in birdsfoot trefoil or purplish tinge in sugar maple; blueberry and sugarcane. The under side develops bronzy appearance. The root growth is also restricted under phosphorus stress in black pepper. Anthocyanin pigment increases in leaves of barley and *Arabidopsis thaliana* under phosphorus stress,

Potassium (K)

Under potassium stress condition, yellowing of leaves starts from the tips or margins of leaves extending towards the center of leaf base. The yellowing is interveinal and irregular in the leaves of tomato and blueberry. These yellow parts become necrotic (dead spots) with leaf curling in tobacco; sugar maple; sapota and sugarcane. There is a sharp difference between green, yellow and necrotic parts.

Calcium (Ca)

Calcium stress in plants results in chlorosis of young leaves along the veins of birdsfoot trefoil and blueberry, if deficiency persist longer, bleaching of upper half leaf followed by leaf tip curling do occur in black pepper and sugarcane. The growing bud leaf becomes chlorotic white with base remaining green, the distortion of the tips of shoots *i.e.* dieback was observed by Edwards and Hortan, in peach seedlings. Similarly, Spehar and Galway, found brown spots on leaves, reduced expansion and premature leaf senescence under Ca stress in soybean crop. Stress during fruiting in tomato increases susceptibility to blossom end rot. Calcium stress is also responsible for other disorders such as bitter pit in apple; leaf tip burn in cabbage and lettuce; black heart of celery; cavity spot of carrots; vitrescence in melons.

Magnesium (Mg)

Magnesium deficiency causes yellowing, but differs from that of nitrogen. The yellowing takes place in between veins of older leaves of *Picea abies* and veins remain green, this is followed by necrosis of tissues in birdsfoot trefoil, melons. black pepper and blueberry. Mg deficiency my be induced in tomatoes by high levels of ammonium in the nutrient solution.

Sulfur (S)

Sulfur deficiency cause leaves to become yellowish in black pepper; potato and *Brassica oleracea* and it appears similar to nitrogen deficiency, but the symptoms are first visible on younger leaves. The affected leaves are narrow and the veins are paler and chlorotic than interveinal portion, especially towards the base with marginal necrosis in sugarcane.

Iron (Fe)

The principal veins remain conspicuously green and surrounding portion of the younger leaves turn yellow tending towards whiteness in chickpea; groundnut; radish, cauliflower, cabbage and sorghum; lentil and soybean. Under sever deficiency, most part of the leaf becomes white.

Zinc (Zn)

The leaves become narrow and small in chili, the lamina becomes chlorotic in sweet potato, sour orange seedlings and chickpea, while veins remain green. Subsequently, dead spots develop all over the leaf including veins, tips and margins under sever deficiency, shoot growth is reduced. Khaira disease in rice results due to zinc deficiency. Shoot elongation is reduced and a tuft or rosette of distinctly narrow leaves is produced at the shoot terminal in apple and pear. The symptoms are termed 'little leaf' or 'rosette'.

Boron (B)

Boron deficiency causes yellowing or chlorosis of youngest leaves and stems which starts from the base to the tip. Rosetting of terminal shoots of potato. Leaf tip burn, elongate and become whitish brown in rice. Death of terminal bud occurs in extreme cases. Boron deficiency causes brown heart in radish and crown choking in coconut.

Manganese (Mn)

The principal veins as well as smaller veins are green, the interveinal portion become chlorotic in *Ailanthus triphysa* followed by necrosis and browning of interveinal tissue in melons. The affected young leaves remain small and abscise before older leaves in birdsfoot trefoil.

Molybdenum (Mo)

The common symptoms of Mo deficiency in plants include a general yellowing, marginal and interveinal chlorosis, marginal necrosis, rolling, scorching and downward curling of margins in poinsettia cultivars and in various field, horticulture and forage crops. The deficiency of molybdenum in cauliflower causes the disorder described as 'Whiptail'.

Copper (Cu)

In copper deficiency, visible foliar symptoms appear on young leaves as chlorosis changing to necrosis; rolling, wilting and twisting of leaves in wheat. The later affected leaves appear papery and twisted in rice.

Chlorine (Cl)

The symptoms of chlorine deficiency develop first on the older leaves. Discrete patches of pale green chlorotic tissue appear between the main vein near the tip of the leaf, downward cupping of some of the older leaves of Kiwifruit was observed by Smith *et. al.,*. The leaflets of youngest leaves shrivel completely, older leaflets develop a brown necrosis which start near the tip and extend backwards particularly at the margins of red clover.

Nickel (Ni)

Plant growth is reduced and older leaves turn chlorotic giving plants a nitrogen deficient phenotype, when grown on urea-based nutrient solutions not supplemented with Ni in tomato and soybean. Similar results were obtained in oilseed-rape, zucchini and soybean by Gerendás and Sattelmacher.

Chapter 23

Oxidative Stress

ACTIVATION OF OXYGEN

One of the paradoxes of life on this planet is that the molecule that sustains aerobic life, oxygen, is not only fundamentally essential for energy metabolism and respiration, but it has been implicated in many diseases and degenerative conditions. A common element in such diverse human disorders as ageing, arthritis, cancer, Lou Gehrig's disease and many others is the involvement of partially reduced forms of oxygen. Our realisation of the significance of oxygen in disorders and stress-induced dysfunctions in cultivated plants is recent due in no small part to the difficulty in detecting and tracing oxygen molecules, to the multitude of forms and intermediates that oxygen can assume, and to the extreme reactivity and rate of the chemical reactions involved. As a consequence we often in our experiments can only look for the "footprints" of oxygen reactions in our attempts to determine cause-effect relationships in stress responses. The following chapter describes our current understanding of the general principles of activated oxygen.

Atmospheric oxygen in its ground-state is distinctive among the gaseous elements because it is a biradical, or in other words it has two unpaired electrons. This feature makes oxygen paramagnetic; it also makes oxygen very unlikely to participate in reactions with organic molecules unless it is "activated". The requirement for activation occurs because the two unpaired electrons in oxygen have parallel spins. According to Pauli's exclusion principle, this precludes reactions with a divalent reductant, unless this reductant also has two unpaired electrons with parallel spin opposite to that of the oxygen, which is a very rare occurrence. Hence, oxygen is usually non-reactive to organic molecules which have paired electrons with opposite spins. This spin restriction means that the most common mechanisms of oxygen reduction in bio-chemical reactions are those involving transfer of only a single electron (monovalent reduction).

Activation of oxygen may occur by two different mechanisms : absorption of sufficient energy to reverse the spin on one of the unpaired electrons, or monovalent reduction. The biradical form of oxygen is in a triplet ground state because the electrons have parallel spins. If triplet oxygen absorbs sufficient energy to reverse the spin of one of its unpaired electrons, it will form the singlet state, in which the two electrons have opposite spins. This activation overcomes the spin restriction and singlet oxygen can consequently participate in reactions involving the simultaneous transfer of two electrons (divalent reduction). Since paired electrons are common in organic molecules, singlet oxygen is much more reactive towards organic molecules than its triplet counterpart.

Triplet Oxygen (ground state)	· O-O ·
Singlet Oxygen	O-O :
Superoxide	· O-O :
Perhydroxyl Radical	· O-O : H
Hydrogen Peroxide	H : O-O : H
Hydroxyl Radical	H : O ·
Hydroxyl Ion	H : O :
Water	H : O : H

Figure 23.1 : Nomenclature of the various forms of oxygen.

The first step in the reduction of oxygen forming superoxide is endothermic but subsequent reductions are exothermic.

Superoxide can act as either an oxidant or a reductant; it can oxidise sulphur, ascorbic acid or NADPH; it can reduce cytochrome C and metal ions. A dismutation reaction leading to the formation of hydrogen peroxide and oxygen can occur spontaneously or is catalysed by the enzyme superoxide dismutase. In its protonated form (pKa = 4.8) superoxide forms the perhydroxyl radical (OOH) which is a powerful oxidant, but its biological relevance is probably minor because of its low concentration at physiological pH.

The univalent reduction of superoxide produces hydrogen peroxide which is not a free radical because all of its electrons are paired. Very often the reduction products of oxygen are referred to by biologists as oxygen free radicals which is a misnomer because in chemistry a free radical is defined as an atom or molecule with an unpaired electron. It is more appropriate to refer to the intermediate reduction products of oxygen as activated not as free radicals because triplet oxygen (ground state) is a radical and hydrogen peroxide is not.

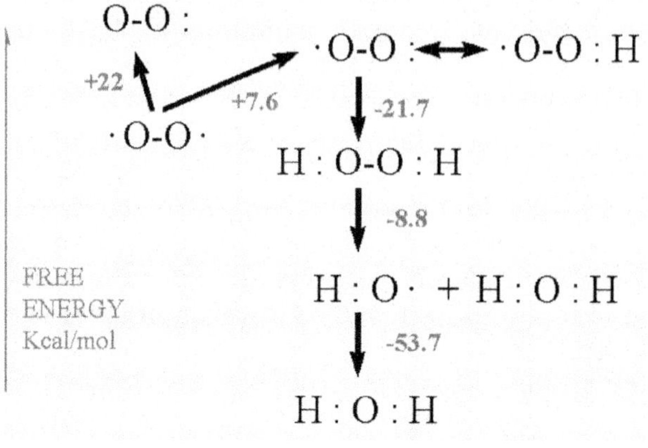

Figure 23.2 : The activation states of oxygen. Non-activated oxygen is a biradical. From this triplet state it can be activated by either reversing the spin on one of the unpaired electrons to form the singlet state or by reduction. The first reduction reaction is endothermic forming superoxide. Subsequent reductions form hydrogen peroxide, hydroxyl radical and water. The electronic state for each activation step is shown with the energy of the reaction in Kcal/mole.

Hydrogen peroxide is noteworthy because it readily permeates membranes and it is therefore not compartmentalised in the cell. Numerous enzymes (peroxidases) use hydrogen peroxide as a substrate in oxidation reactions involving the synthesis of complex organic molecules. The well-known reactivity of hydrogen peroxide is not due to its reactivity *per se*, but requires the presence of a metal reductant to form the highly reactive hydroxyl radical which is the strongest oxidizing agent known and reacts with organic molecules at diffusion-limited rates.

Fenton described in the late nineteenth century the oxidising potential of hydrogen peroxide mixed with ferrous salts. Forty years later, Haber and Weiss identified the hydroxyl radical as the oxidising species in these reactions :

(1) $Fe^{2+} + H_2O_2 \rightarrow Fe^{3+} + \bullet OH + OH^-$

In biological systems the availability of ferrous ions limits the rate of reaction, but the recycling of iron from the ferric to the ferrous form by a reducing agent can maintain an ongoing Fenton reaction leading to the generation of hydroxyl radicals. One suitable reducing agent is superoxide which participates in the overall reaction 2 as two half reactions shown in reactions 3 and 4 :

(2) $\bullet O_2^- + H_2O_2 \rightarrow O_2 + \bullet OH + OH^-$

(3) $Fe^{2+} + H_2O_2 \rightarrow Fe^{3+} + \bullet OH + OH^-$

(4) $\bullet O_2^- + Fe^{3+} \rightarrow O_2 + Fe^{2+}$

Therefore, in the presence of trace amounts of iron, the reaction of superoxide and hydrogen peroxide will form the destructive hydroxyl radical and initiate the oxidation of organic substrates. Metals other than iron may also participate in these electron transfer reactions by cycling between oxidised and reduced states.

The oxidation of organic substances may proceed by two possible reactions Ä addition of OH to the organic molecule, or abstraction of a hydrogen atom from it. In the addition reaction, the hydroxyl radical adds to an organic substrate forming a hydroxylated product that is further oxidised by ferrous ions, oxygen or other agents to a stable, oxidised product. The hydroxylated products can also dismutate to form cross-linked products.

(5) $\bullet OH + R \rightarrow \bullet ROH$

(6) $\bullet ROH + Fe^{3+} \rightarrow ROH + Fe^{2+} + H^{+}$

(7) $\bullet ROH + O_2 \rightarrow ROH + O_2^{-} + H^{+}$

(8) $\bullet ROH + \bullet ROH \rightarrow R - R + 2H_2O$

In the abstraction reaction, the hydroxyl radical oxidises the organic substrate forming water and an organic radical. The latter product has a single unpaired electron and thus can react with oxygen in the triplet ground-state. The addition of triplet oxygen to the carbon radical can lead to the formation of a peroxyl radical which can readily abstract hydrogen from another organic molecule leading to the formation of a second carbon radical. This chain reaction is why oxygen free radicals cause damage far in excess of their initial concentration.

(9) $\bullet OH + RH \rightarrow R\bullet + H_2O$

(10) $\bullet R + O_2 \rightarrow ROO\bullet$

(11) $ROO\bullet + RH \rightarrow R\bullet + ROOH$

BIOLOGICAL REACTIONS OF OXYGEN RADICALS

The reactions of activated oxygen with organic substrates are complex even *in vitro* with homogenous solutions, but in biological systems there are even more complications due to the surface properties of membranes, electrical charges, binding properties of macro-molecules, and compartmentalisation of enzymes, substrates and catalysts. Thus, various sites even within a single cell differ in the nature and extent of reactions with oxygen.

The nature of the oxidative injury that causes cell death is not always obvious. The mechanisms by which oxygen radicals damage membrane lipids are well accepted, and consequently oxidative damage is often exclusively associated with these peroxidation reactions in membrane lipids. What is sometimes overlooked in our research on environmental stress in plants is that activated forms of oxygen also degrade proteins and nucleic acids, reactions which can also be very lethal. In this section some of the major reactions of activated oxygen with lipids, protein, and nucleic acids are reviewed.

OXIDATIVE DAMAGE TO LIPIDS

Classical Peroxidation Reactions

The reactions of oxygen free radicals with polyunsaturated lipids have been extensively researched because of their involvement in rancidity and the development of undesirable odours and flavours in foods. Historically these reactions are the most frequently cited consequence of oxygen radical production in plant cells. Perhaps the mechanisms were so well established by oil chemists long before the recognition of their importance in biology that plant biologists applied these mechanisms directly to their experimental systems, rarely questioning their validity or transposability. This has delayed recognition of the presence of free radical reactions in plant membranes. The complexity of the biological membrane is well established and the reader is referred elsewhere for more detailed considerations of its structure. The lipid bilayer membrane is composed of a mixture of phospholipids and glycolipids that have fatty acid chains attached to carbon 1 and 2 of the glycerol backbone by an ester linkage. The peroxidation reactions differ among these fatty acids depending on the number and position of the double bonds on the acyl chain and the reader is referred to Frankel for a detailed review. The following is a simplified summary of these reactions for a general lipid, 'R', and for a specific fatty acid, linoleate, which is common in plant cell membranes.

The peroxidation of lipids involves three distinct steps : initiation, propagation and termination. The initiation reaction between an unsaturated fatty acid (*e.g.* linoleate) and the hydroxyl radical involves the abstraction of an H atom from the methylvinyl group on the fatty acid; in the case of linoleate this occurs at carbon-11. The remaining carbon centred radical, forms a resonance structure sharing this unpaired electron among carbons 9 to 13. In the propagation reactions, this resonance structure reacts with triplet oxygen, which remember is a biradical having two unpaired electrons and therefore reacts readily with other radicals. This reaction forms a peroxy radical. In the case of linoleate, addition occurs at either carbon-9 or -13. The peroxy radical then abstracts an H atom from a second fatty acid forming a lipid hydroperoxide and leaving another carbon centred free radical that can participate in a second H abstraction. Therefore, once one hydroxyl radical initiates the peroxidation reaction by abstracting a single H atom, it creates a carbon radical product (R) that is capable of reacting with ground state oxygen in a chain reaction. The role of the hydroxyl radical is analogous to a "spark" that starts a fire. The basis for the hydroxyl radical's extreme reactivity in lipid systems is that at very low concentrations it initiates a chain reaction involving triplet oxygen, the most abundant form of oxygen in the cell.

The lipid hydroperoxide is unstable in the presence of Fe or other metal catalysts because ROOH will participate in a Fenton reaction leading to the formation of reactive alkoxy radicals :

(12) $ROOH + Fe^{2+} \rightarrow OH^{-} + RO\bullet + Fe^{3+}$

Therefore, in the presence of Fe, the chain reactions are not only propagated but amplified. Note that two radicals are produced by the summation of reactions 9 to 12. Among the degradation products of ROOH are aldehydes, such as malondialdehyde, and hydrocarbons, such as ethane and ethylene, that are commonly measured end products of lipid peroxidation.

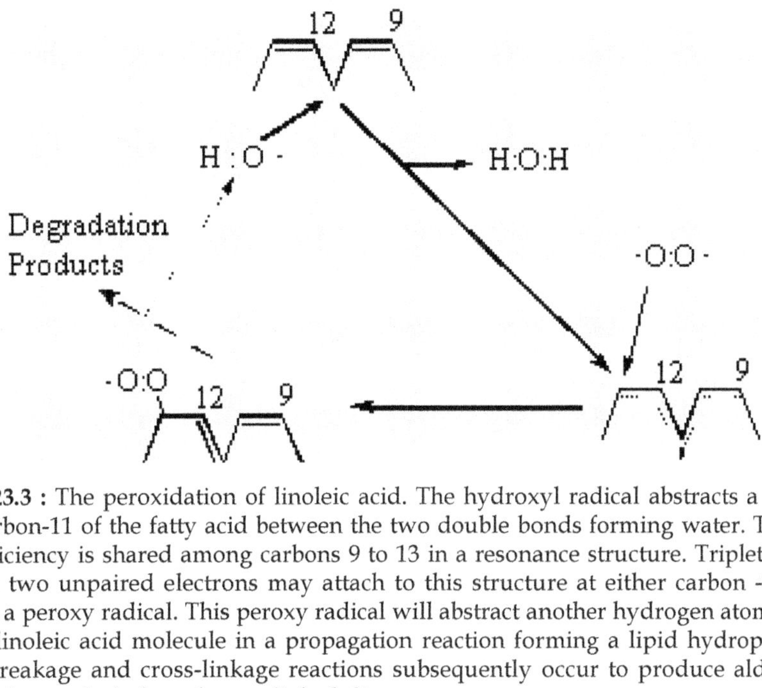

Figure 23.3 : The peroxidation of linoleic acid. The hydroxyl radical abstracts a H atom from carbon-11 of the fatty acid between the two double bonds forming water. The electron deficiency is shared among carbons 9 to 13 in a resonance structure. Triplet oxygen that has two unpaired electrons may attach to this structure at either carbon -9 or -13 forming a peroxy radical. This peroxy radical will abstract another hydrogen atom from a second linoleic acid molecule in a propagation reaction forming a lipid hydroperoxide. Chain breakage and cross-linkage reactions subsequently occur to produce aldehydes, hydrocarbons, alcohols and cross-linked dimers.

The peroxidation reactions in membrane lipids are terminated when the carbon or peroxy radicals cross-link to form conjugated products that are not radicals, such as those shown in reactions 13 to 15 :

(13) $R\bullet + R\bullet \rightarrow R - R$

(14) $R\bullet + ROO\bullet \rightarrow ROOR$

(15) $ROO\bullet + ROO\bullet \rightarrow ROOR + O_2$

Typically high molecular weight, cross-linked fatty acids and phospholipids accumulate in peroxidised membrane lipid samples.

Singlet oxygen can react readily with unsaturated fatty acids producing a complex mixture of hydroperoxides. Again, the chemistry of these reactions is based on foods. Oxidation of unsaturated fatty acids by singlet oxygen produces distinctly different products than the hydroxyl radical. Once formed the lipid hydroperoxides will decompose into a variety of products, some of which can produce oxygen free radicals in the presence of metal catalysts.

Unique Reactions in Plant Membranes

The above mechanisms predict that oxygen free radical or lipid peroxidation reactions in plant membranes would selectively degrade unsaturated fatty acids and accumulate aldehydes, hydrocarbons, and cross-linked products. When examining the effects of environmental stresses on plant membranes, many studies have measured the products of lipid peroxidation, such as malondialdehyde and/or ethane and concluded that oxygen free radicals are involved in these stress responses. When the substrates of these reactions, the membrane fatty acids, have been examined, it has been very often observed that the unsaturated fatty acids are not selectively degraded, and therefore these reports have concluded that oxygen free radicals are not involved in these stress responses. This controversy has caused many to rule out the involvement of oxygen free radicals in processes such as seed ageing. However, *in vitro* experiments that have treated plant membranes with Fenton reaction products have shown that degradation of plant membrane lipids by oxygen free radicals does not involve selective loss of unsaturated fatty acids. In both samples, there was destruction of fatty acids and their recovery from solution was lower after the free radical treatment. In the liposome sample there was selective degradation of unsaturated fatty acids; the proportion of linoleic and linolenic acids relative to the other fatty acids declined. In contrast, treatment of the wheat microsomal membranes caused degradation to the phospholipids but no change in the proportion of the fatty acids. Inother words, there was not selective degradation of the unstaurated fatty acids. Clearly other reactions to those described above were occurring in these plant membranes.

An alternative to the classical mechanism of lipid peroxidation was proposed by Niehaus based on his observation that most esters react with superoxide by cleaving the C-O bond. Since the fatty acid chains are attached to the glycerol backbone of the phospholipid molecule by an ester bond, superoxide attack on a phospholipid bilayer would produce free fatty acids by de-esterification reactions. This was experimentally observed by Senaratna *et. al.* in microsomal membranes from soybean seed axes treated *in vitro* with superoxide from xanthine oxidase.

Table 23.1 : Degradation of phospholipid and esterified fatty acid in two membrane systems by Fenton reaction products.

Data for each fatty acid are expressed as its proportion (%) of all fatty acids recovered (*i.e.*each sample totals 100%). Phospholipid (PL) recovery is expressed as a % of the original PL before treatment.

Adapted from McKersie *et. al.*

Fatty Acid	Liposomes		Wheat Microsomes	
	Before	After	Before	After
16:0	21	34	29	30
18:0	3	5	1	1
18:1	7	10	6	7
18:2	61	47	28	29
18:3	8	4	37	35
PL Recovery	100	51	100	65

Kinetic analysis indicates that superoxide attack on esters occurs by a nucleophilic addition mechanism.

The peroxy radical RC(O)OO would abstract an H atom to form a hydroperoxide that would decompose into an acid $RCOO^{\text{Ä}}$. In the case of a phospholipid, $RCOO^{\text{Ä}}$ would be a free fatty acid.

These reactions would not be selective for unsaturated fatty acids and therefore degradation of fatty acids attached to phospholipid molecules would be random. Experimentally if these reactions occurred in plant membranes, we would observe loss of phospholipids but no change in fatty acid unsaturation. Although this is commonly observed in stress conditions in plants, most researchers usually attribute these products to the action of an enzyme, such as phospholipase or a non-specific lipase.

Enzymatic and chemical reactions can clearly give the same products and the distinction between these mechanisms is clouded even further by the realisation that the formation of superoxide is the result of a dysfunctioning enzyme. Therefore, the debate is somewhat philosophical. Are there enzymes whose primary function is to degrade (turnover) phospholipid molecules during periods of stress? or are there enzymes that primarily function in redox transfer of electrons; but that dysfunction to form superoxide during stress?

It is not at all clear why some plant membranes such as the microsomal membranes from wheat crowns, exhibit de-esterification reactions instead of peroxidation reactions. Presumably, it is due to differences in composition, possibly the presence of specific lipid-soluble, membrane antioxidants (phenols, flavonoids, quinones), whose effect on free radical reactions are poorly understood.

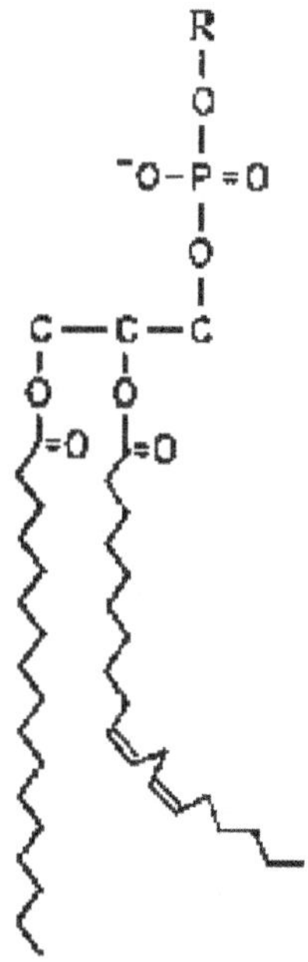

Figure 23.4 : The structure of a typical phospholipid molecule. Phospholipids form the essential structural component of plant membranes. Their amphipathic nature dicates that the phospholipid headgroup is oriented toward the external aqueous phase and the fatty acid tails are oriented towards the interior hydrophobic phase. The fatty acids are attached by ester linkages to the sn-1 (palmitoyl; 16 : 0) and sn-2 (linoleoyl; 18-2) positions of glycerol. The phosphate headgroup (R) is attached to the Sn-3 position. There are two common sites of oxygen free radical attack on the phospholipid molecule - the unsaturated double binds of the fatty acid and the ester linkage between glycerol and the fatty acid.

Oxidative Damage to Proteins

Oxidative attack on proteins results in site-specific amino acid modifications, fragmentation of the peptide chain, aggregation of cross-linked reaction products, altered electrical charge and increased susceptibility to proteolysis. The amino acids in a peptide differ in their susceptibility to attack, and the various forms of activated oxygen differ in their potential reactivity. Primary, secondary, and tertiary protein structures alter the relative susceptibility of certain amino acids. In spite of this complexity, generalisations can be made. Sulphur containing amino acids, and thiol groups specifically, are very susceptible sites. Activated oxygen can abstract an H atom from cysteine residues to form a thiyl radical that will cross-link to a second thiyl radical to form disulphide bridges. Alternatively, oxygen can add to a methionine residue to form methionine sulphoxide derivatives. Reduction of both of these may be accomplished in microbial systems by thioredoxin and thioredoxin reductase. A protein-methionine-S-oxide reductase has been measured in pea chloroplasts. This enzyme reduces the methionyl sulfoxide back to methionyl residues in the presence of thioredoxin. In some instances this enzyme has restored the biological activity of a protein, but this function in plants has not been described.

Other forms of free radical attack on proteins are not reversible. For example, the oxidation of iron-sulphur centres by superoxide destroys enzymatic function. Many amino acids undergo specific irreversible modifications when a protein is oxidised. For example, tryptophan is readily cross-linked to form bityrosine products. Histidine, lysine, proline, arginine, and serine form carbonyl groups on oxidation. The oxidative degradation of protein is enhanced in the presence of metal cofactors that are capable of redox cycling, such as Fe. In these cases, the metal binds to a divalent cation binding site on the protein. The metal then reacts with hydrogen peroxide in a Fenton reaction to form a hydroxyl radical that rapidly oxidises an amino acid residue at or near the cation binding site of the protein. This site-specific alteration of an amino acid usually inactivates the enzyme by destruction of the cation binding site.

Oxidative modification of specific amino acids is one mechanism of marking a protein for proteolysis. In *E. coli* there are specific proteases that degrade oxidised proteins and similar specificity is expected in plants. It is well documented that the various peptide components of photo-system II turnover at different frequencies; the D1 protein specifically is noted for its high rate of turnover, and it is assumed that this is a consequence of oxidative attack at specific sites on the protein.

Oxidative Damage to DNA

Activated oxygen and agents that generate oxygen free radicals, such as ionising radiation, induce numerous lesions in DNA that cause deletions, mutations and other lethal genetic effects. Characterisation of this damage to DNA has indicated that both the sugar and the base moieties are susceptible to oxidation, causing base degradation, single strand breakage, and cross-linking to protein. Degradation of the base will produce numerous products, including 8-hydroxyguanine, hydroxymethyl urea, urea, thymine glycol, thymine and adenine ring-opened and -saturated products.

The principle cause of single strand breaks is oxidation of the sugar moiety by the hydroxyl radical. *In vitro* neither hydrogen peroxide alone nor superoxide cause strand breaks under physiological conditions, and therefore, their toxicity *in vivo* is most likely the result of Fenton reactions with a metal catalyst. At least in *E. coli* these Fenton reactions can be driven by NADH. For example, the ndh mutant in *E. coli* accumulates NADH as a result of the mutant's inability to donate electrons from NADH to respiratory pathways; as a result, the mutant is hypersensitive to hydrogen peroxide. Studies of other *E. coli* mutants have lead to the conclusion that a Fenton active metal is bound to DNA, probably chelated to phosphodiester linkage. If the bound metal is reduced by a small diffusible molecule, such as NAD(P)H or superoxide, it will react with hydrogen peroxide to form the hydroxyl radical. The short-lived hydroxyl radical then oxidises an adjacent sugar or base causing breakage of the DNA chain.

Cross-linking of DNA to protein is another consequence of hydroxyl radical attack on either DNA or its associated proteins. Treatment with ionising radiation or other hydroxyl radical generating agents causes covalent leakages such as thymine-cysteine addicts, between DNA and protein. When these cross-linkages exist, separation of protein from DNA by various extraction methods is ineffective. Although DNA-protein cross-links are about an order of magnitude less abundant than single strand breaks, they are not as readily repaired, and may be lethal if replication or transcription precedes repair.

DNA is an obvious weak link in a cell's ability to tolerate oxygen free radical attack. First, it seems that DNA is effective in binding metals that are involved in Fenton reactions, and secondly less damage can be tolerated in DNA than other macro-molecules. As a consequence, the cell has a number of DNA repair enzymes. One reason why eukaryotic organisms have compartmentalised DNA in the nucleus, away from sites of redox cycling that are high in NAD(P)H and other reductants, may be to avoid oxidative damage.

SITES OF ACTIVATED OXYGEN PRODUCTION

As indicated above, there are two forms of activated oxygen that are formed by distinctly different mechanisms. The reduction of oxygen to form superoxide, hydrogen peroxide and hydroxyl radicals is the principle mechanism of oxygen activation in most biological systems. However in photosynthetic plants, the formation of singlet oxygen by the photosystems has importance. Activated oxygen is often formed as a component of metabolism to enable "complex" chemical reactions, such as the oxidation of xenobiotics or the polymerisation of lignin, but in other instances activated oxygen is formed by the dysfunctioning of enzymes or electron transport systems, as a result of perturbations in metabolism caused by chemical or environmental stress.

Chloroplasts

As described by Elstner, there are at least four sites within the chloroplast that can activate oxygen.

(1) PSI can reduce oxygen by the Mehler reaction which is an important mechanism of oxygen activation in the chloroplast. The reducing side of PSI is thought to contribute significantly to the monovalent reduction of oxygen under conditions where NADP is limiting. This would occur, for example, if the Calvin cycle did not oxidise NADPH as rapidly as PSI supplied electrons.

(2) Photoactivated chlorophyll normally transfers its excitation energy to the PS reaction centres, but under conditions that prevent the captured light energy from being utilised in the electron transport systems, this energy can excite oxygen from the triplet to singlet form. These conditions include stomatal closure caused by drought, damage to the membrane transport systems, lack of specific nutrients, or the presence of xenobiotic chemicals such as pollutants or herbicides.

(3) The oxidising side of PSII facilitates four single electron transfers from water to the PSII reaction centre releasing triplet or ground-state oxygen. Leaks of electrons from this site to molecular oxygen, or release of partially reduced oxygen products are thought to make a relatively minor contribution to activated oxygen production, but nonetheless it has been demonstrated that certain alcohols can be reduced by PSII.

1. Photorespiration is the most obvious oxygenation pathway in the chloroplast. Rubisco catalyses the addition of oxygen to carbon 2 of RuBP forming phosphoglycolate and phosphoglycerate. Although this does not generate activated oxygen in the chloroplast, the subsequent metabolism of glycolate in the peroxisomes does.

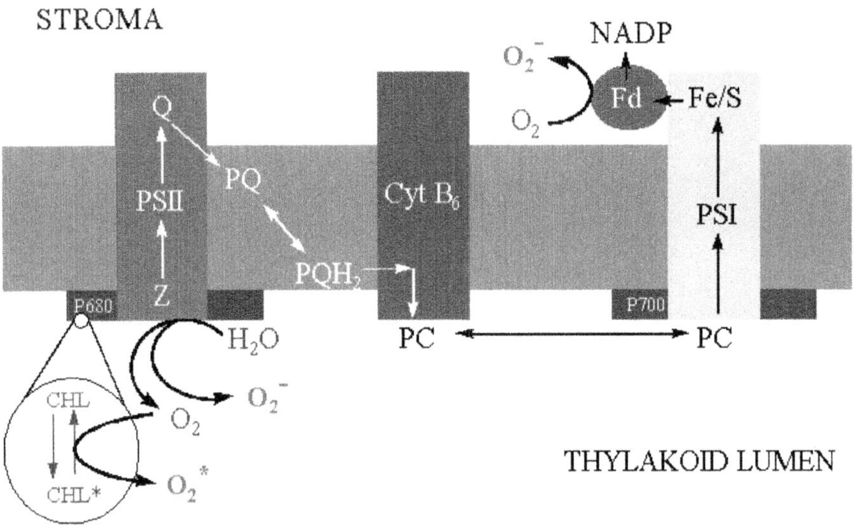

Figure 23.5 : Schematic representation of the electron transport system in the thylakoid membrane showing three possible sites of activated oxygen production.

a) Singlet oxygen may be produced from triplet chlorophyll in the light harvesting complex.

b) Superoxide and hydrogen peroxide may "leak" from the oxidizing (water-splitting) side of PSII.

c) Triplet oxygen may be reduced to superoxide by ferredoxin on the reducing side of PSI.

Mitochondria

Most oxygen is consumed by the cytochrome oxidase enzyme in the mitochondrial electron transport system, and involves the sequential transfer of four electrons to oxygen, releasing water. Plant mitochondria have an additional site of oxygen reduction at the alternative oxidase, distinguished from cytochrome oxidase by its resistance to cyanide. However, neither of these sites produce significant quantities of superoxide. However, isolated mitochondria produce H_2O_2 and O_2 in the presence of NADH. Antimycin A, which blocks electron flow after ubiquinone enhances oxygen reduction. Presumably other conditions which also increase the reduction of ubiquinone favour reduction of oxygen in the ubiquinone Ä cytochrome b region of the chain. The various Fe-S proteins and NADH dehydrogenase have also been implicated as possible sites of superoxide and hydrogen peroxide formation.

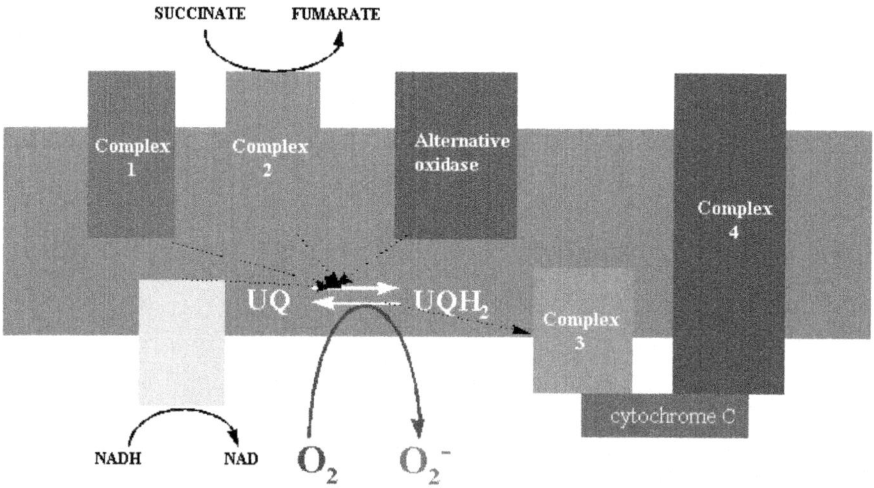

Figure 23.6 : Schematic representation of the electron transport system in the mitochondrial membrane showing a possible site of superoxide production by reduced ubiquinones.

Endoplasmic Reticulum

Various oxidative processes, including oxidation, hydroxylations, dealkylations, deaminations, dehalogenation and desaturation, occur on the smooth endoplasmic reticulum. Mixed function oxygenases that contain a heme moiety add an oxygen atom into an organic substrate using NAD(P)H as the electron donor. The generalised reaction catalysed by cytochrome P_{450} is :

(17) $RH + NADPH + H^{+} + O_2 \rightarrow ROH + NADP^{+} + H_3O$

The best characterised cytochrome P_{450} in plants is cinnamate-4-hydroxylase which functions in flavonoid and lignin bio-synthesis, but other mixed function oxidases function in other bio-chemical pathways including gibberellin and sterol bio-synthesis. Activation of oxygen by these systems is an essential prerequisite to oxygen addition reactions in the synthesis of these "complex" metabolites. Superoxide is produced by microsomal NAD(P)H dependent electron transport involving cytochrome P_{450}. After the univalent reduction of the substrate (RH) and the addition of triplet oxygen to form the complex P_{450} - RHOO the complex may decompose to P_{450}-RH and release superoxide.

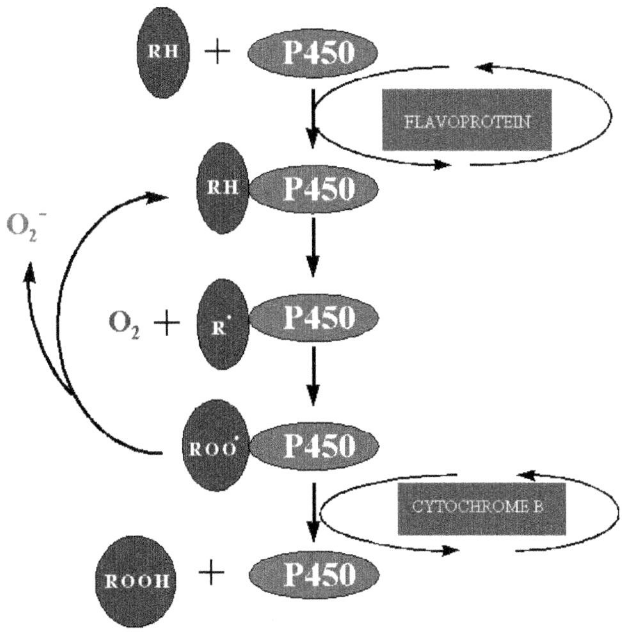

Figure 2 : Schematic representation of the cytochrome P_{450} electron transport system on the endoplasmic reticulum showing one possible site of superoxide production.

Cytochrome P_{450} reacts first with its organic substrate, RH. The complex is reduced by a flavoprotein to form a radical intermediate that can readily react with triplet oxygen because each has one unpaired electron. This oxygenated complex may be reduced by cytochrome b or occasionally the comples may decompose releasing superoxide.

Microbodies

Peroxisomes and glyoxysomes are organelles with a single membrane that compartmentalises enzymes involved in the ß-oxidation of fatty acids, and the glyoxylic acid cycle including glycolate oxidase, catalase and various peroxidases. Glycolate oxidase produces H_2O_2 in a two electron transfer from glycolate to oxygen. Xanthine oxidase, urate oxidase and NADH oxidase generate superoxide as a consequence of the oxidation of their substrates. The xanthine oxidase reaction is often used *in vitro* as a source of superoxide producing one mole of superoxide during the conversion of xanthine to uric acid.

Plasma Membranes

A superoxide-generating NAD(P)H oxidase activity has been clearly identified in plasmalemma enriched fractions. These flavoproteins may produce superoxide by the redox cycling of certain quinones or nitrogenous compounds. In the root, NAD(P)H oxidase reduces Fe^{3+} to Fe^{2+} converting it to a form that can be transported. Dysfunction of this root enzyme will produce superoxide. An auxin-activated NADH oxidase has been associated with acidification of the cell wall and auxin-stimulated cell elongation.

The plant NAD(P)H oxidase may have an analogous function to the animal enzyme. Leucocytes contain an NADH oxidase on the outer membrane surface which is activated in response to a foreign agent, generating superoxide that initiates oxidative reactions that destroy the potential pathogen. In plants, fungal elicitors cause a similar formation of superoxide that has been linked to the hypersensitive response to some pathogenic fungi. Wounding, heat shock and xenobiotics transiently activate this superoxide generating reaction, and consequently, it has been proposed that these superoxide generating reactions may serve as a signal in plant cells to elicit responses to biological, physical or chemical stress.

Cell Walls

Although it is not immediately obvious, cell walls are active sites of metabolism, and also oxygen activation. Some of these reactions may be involved in the defense reactions against pathogens as described above. Others may involve the degradation or compartmentation of xenobiotic chemicals. However, the most common reactions are bio-synthetic. For example, the phenylpropanoid precursors of lignin are crosslinked by H_2O_2 dependent reactions, that randomly link the sub-units to form lignin. NADH is generated by a cell wall malate dehydrogenase, and then used to form H_2O_2, possibly by the NADH oxidase on the plasmalemma. Diamine oxidases are also involved in production of activated oxygen in the cell wall using diamines or polyamines (putrescine, spermidine, cadaverine, etc.) to reduce a quinone that will autoxidize, forming peroxides.

DEFENCE MECHANISMS

Superoxide Dismutase

Superoxide dismutase (SOD) was first isolated by Mann and Keilis and thought to be a copper storage protein. Subsequently, the enzyme was identified by a number of names, erythrocuprein, indophenol oxidase, and tetrazolium oxidase until its catalytic function was discovered by McCord and Fridovitch. SOD is now known to catalyse the dismutation of superoxide to hydrogen peroxide and oxygen :

(18) Superoxide Dismutase

$$\cdot O_2^{\cdot} + \cdot O_2^{\cdot} + 2H^+ \longrightarrow H_2O_2 + O_2$$

Mn-SOD : mitochondria

Fe-SOD : chloroplast

CuZn-SOD : chloroplast

CuZn-SOD : cytosol

Therefore, the activity of this enzyme determines the relative proportions of the two constituents of the Haber-Weiss reaction that generates hydroxyl radicals. Since SOD is present in all aerobic organisms and most (if not all) sub-cellular compartments that generate activated oxygen, it has been assumed that SOD has a central role in the defence against oxidative stress. There are three distinct types of SOD classified on the basis of the metal cofactor : the copper/zinc (Cu/Zn - SOD), the manganese (Mn-SOD) and the iron (Fe-SOD) isozymes. These isozymes can be separated by native polyacrylamide gel electrophoresis, their activity detected by negative staining and identified on the basis of their sensitivity to KCN and H_2O_2. The Mn-SOD is resistant to both inhibitors, whereas the Cu/Zn-SOD is sensitive to both inhibitors; Fe-SOD is resistant to KCN, and sensitive to H_2O_2. The sub-cellular distribution of these isozymes is also distinctive. The Mn-SOD is found in the mitochondria of eukaryotic cells; some Cu/Zn-SOD isozymes are found in the cytosol, others in the chloroplasts of higher plants. The Fe-SOD isozymes are often not detected in plants, but when detected, Fe-SOD is usually associated with the chloroplast compartment. The prokaryotic Mn-SOD and Fe-SOD, and the eukaryotic Cu/Zn-SOD enzymes are dimers, whereas the Mn-SOD of mitochondria are tetramers. Peroxisomes and glyoxysomes of watermelons (*Citrillus vulgaris*) have been shown to contain both Cu/Zn- and Mn-SOD activity, but there are no reports of extra-cellular SOD enzymes in plants. All forms of the SOD are nuclear-encoded and are targeted to their respective sub-cellular compartments by an amino terminal targeting sequence. Several forms of SOD have been cloned from a variety of plants.

Prokaryotic cells, and many eukaryotic algae contain only the Mn-SOD and Fe-SOD isozymes which are believed to be more ancient forms. In the bacteria *E. coli*, SOD activity is transcriptionally regulated by the SOX RS operon but investigations into the regulatory mechanism of SOD expression in plants are only beginning. To date it has been shown that SOD activity is increased in cells in response to diverse environmental and xenobiotic stresses including paraquat, high light, water-logging and drought. Apparently, each of the SOD isozymes are independently regulated according to the degree of oxidative stress experienced in the respective sub-cellular compartments, but how this is communicated at the molecular level is unknown. Bowler *et. al.* have suggested that this role may be served by unique lipid peroxidation products from each organelle that diffuse from the site of oxidative damage to the nucleus where they would enhance transcription of specific SOD genes.

Several reviews on superoxide dismutase have recently been published which describe the characteristics of the enzymes, the cloned cDNA sequences and genes, and the effects of over-expression in transgenic plants (Bowler, *et. al.*, 1994, Doke, *et. al.*, 1994, Foyer, *et. al.*, 1994, Gressel and Galun, 1994, Scandalios, 1993, Van Camp, *et. al.*, 1994)

Catalase

Catalase is a heme-containing enzyme that catalyses the dismutation of hydrogen peroxide into water and oxygen. The enzyme is found in all aerobic eukaryotes and is important in the removal of hydrogen peroxide generated in peroxisomes (microbodies) by oxidases involved in ß-oxidation of fatty acids, the glyoxylate cycle (photorespiration) and purine catabolism. Catalase was one of the first enzymes to be isolated in a highly purified state. All forms of the enzyme are tetramers in excess of 220,000 molecular weight. Multiple forms of catalase have been described in many plants. These forms have been cloned from maize and homologous genes has been cloned from several other plants. Maize has three isoforms termed cat-1, cat-2 and cat-3, that are on separate chromosomes and are differentially expressed and independently regulated. Cat-1 and cat-2 are localised in peroxisomes and the cytosol, whereas cat-3 is mitochondrial. Careful examination of the structure of beef liver catalase has shown four NADPH binding sites per catalase tetramer, but these sites were not in close association with the hydrogen peroxide reaction centre. Instead, NADPH functions in animal catalase to protect against inactivation by hydrogen peroxide. The only plant catalase examined, potato, does not contain NADPH. It is interesting in this regard to note that catalase is very sensitive to light and has a rapid turnover rate similar to that of the D1 protein of PSII. This may be a result of light absorption by the heme group or perhaps hydrogen peroxide inactivation. Regardless, stress conditions which reduce the rate of protein turnover, such as salinity, heat shock or cold, cause the depletion of catalase activity. This may have significance in the plant's ability to tolerate the oxidative components of these environmental stresses.

Ascorbic Acid

L-ascorbic acid (vitamin C) is an important vitamin in the human diet and is abundant in plant tissues. Green leaves have the same amount of ascorbate as chlorophyll. Because of its nutritional importance, the distribution of ascorbate has been extensively quantified in plants; however, relatively little consideration has been given to its function in the plant. Ascorbate has been shown to have an essential role in several physiological processes in plants, including growth, differentiation and metabolism. Ascorbate functions as a reductant for many free radicals, thereby minimising the damage caused by oxidative stress but ascorbate may have other functions which remain undefined.

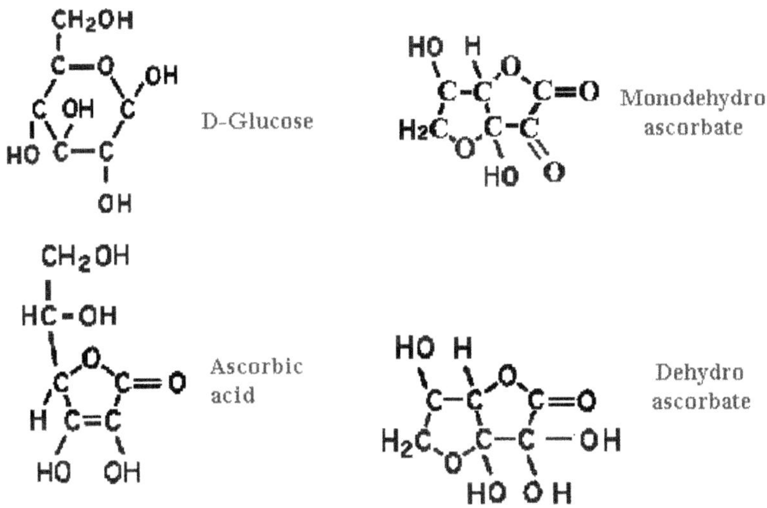

Figure 23.8 : Structure of ascorbic acid and its metabolites.

Apparently synthesis of ascorbate occurs in the cytosol because a specific ascorbate translocator has been identified on the chloroplast envelope. L-ascorbic acid is synthesised from hexose sugars in higher plants but controversy remains concerning some steps in its synthesis. Although two distinct pathways are possible, higher plants primarily convert D-glucose to ascorbate by a direct conversion that maintains the carbon chain in the same sequence. The pathway involves the oxidation of carbon-1 of D-glucose and enediol formation between carbons 2 and 3 :

(19) D-glucose D-glucosone L-sorbosone L-ascorbic acid

Ascorbate can directly scavenge oxygen free radicals with and without enzyme catalysts and can indirectly scavenge them by recycling tocopherol to the reduced form. By reacting with activated oxygen more readily than any other aqueous component, ascorbate protects critical macromolecules from oxidative damage. The reaction with the hydroxyl radical is limited only by diffusion.

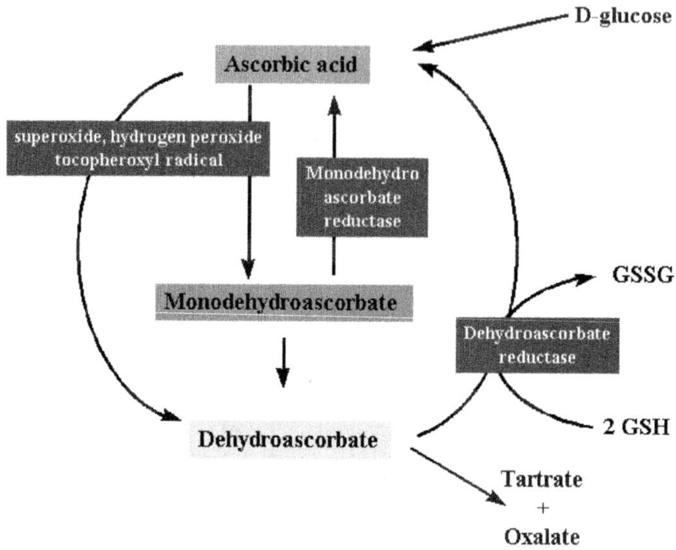

Figure 23.9 : Synthesis and degradation of L-ascorbic acid in plant tissues.

Ascorbic acid is synthesized from D-glucose. As an antioxidant, ascorbate will react with superoxide, hydrogen peroxide or the tocopheroxyl radical to form monodehydroascorbic acid and/or dehydroascorbic acid. The reduced forms are recycled back to ascorbic acid by monodehydroascorbate reductase and dehydroascorbate reductase using reducing equivalents from NAD(P)H or glutathione, respectively. Dehydroascorbate may decompose into tartrate and oxalate.

The reaction with superoxide may serve a physiologically similar role to SOD :

(20) $2\ O_2 + 2H^+$ + ascorbate $2H_2O_2$ + dehydroascorbate

The reaction with hydrogen peroxide is catalysed by ascorbate peroxidase :

(21) H_2O_2 + 2 ascorbate $2H_2O$ + 2 monodehydroascorbate

The indirect role of ascorbate as an antioxidant is to regenerate membrane-bound antioxidants, such as a-tocopherol, that scavenge peroxyl radicals and singlet oxygen, respectively :

(22) tocopheroxyl radical + ascorbate -tocopherol + monodehydroascorbate

The above reactions indicate that their are two different products of ascorbate oxidation, monodehydroascorbate and dehydroascorbate, that represent one and two electron transfers, respectively. The monodehydroascorbate can either spontaneously dismutate (reaction 23) or is reduced to ascorbate by NAD(P)H monodehydroascorbate reductase :

(23) 2 monodehydroascorbate ascorbate + dehydroascorbate

(24) monodehydroascorbate + NAD(P)H ascorbate + NAD(P)

The dehydroascorbate is unstable at pH greater than 6 decomposing into tartrate and oxalate. To prevent this, dehydroascorbate is rapidly reduced to ascorbate by dehydroascorbate reductase using reducing equivalents from glutathione (GSH) :

(25) 2 GSH + dehydroascorbate GSSG + ascorbate

Ascorbate has been found in the chloroplast, cytosol, vacuole and extracellular compartments of the cell. About 20-40% of the ascorbate in the mesophyll leaf cell is in the chloroplast. The chloroplast contains all the enzymes to regenerate reduced ascorbate from its oxidised products. Many of the details of this pathway and characterisation of the enzymes has been conducted by K. Asada in Japan. Consequently this sequence of reactions is referred to as the Halliwell-Asada pathway. Illuminated chloroplasts produce superoxide and hydrogen peroxide from sites on the thylakoids, most commonly PSI. Superoxide is converted into hydrogen peroxide by either spontaneous dismutation or by the SOD enzyme. Hydrogen peroxide is scavenged by ascorbate and the enzyme ascorbate peroxidase. The monodehydroascorbate has two routes of regeneration, one via monodehydroascorbate reductase, the other via dehydroascorbate reductase and glutathione. The terminal electron donor is NADPH. This pathway serves two functions. One is the detoxification of hydrogen peroxide that might otherwise participate is Fenton reactions, and the second in the oxidation of NADPH. The latter function is an apparently energy-consuming, wasteful process analogous to photorespiration. It might at first seem more logical that the chloroplast contain catalase because it would allow the dissipation of hydrogen peroxide without "wasting" NADPH. However, it should be realised that conditions favouring electron transfer from PSI to oxygen generally cause a high redox potential, *i.e.* high NADPH/NADP ratio. By reducing this redox potential through the Halliwell-Asada pathway, the tendency of PSI to reduce oxygen is minimised.

Although ascorbate metabolism has been studied in most detail in the chloroplast, it is likely that all enzymes for its regeneration also exist in the cytosol of both photo-synthetic and non-photo-synthetic cells. For example different isozymes for ascorbate peroxidase have been shown to exist in the cytosol and chloroplast compartments. The cell wall is also an important site of ascorbate metabolism because it contains mM concentrations of ascorbate. Here, ascorbate may have a role in cell wall bio-synthesis. The cell wall does not contain ascorbate peroxidase, but contains ascorbate oxidase. This enzyme contains 8-12 copper molecules per enzyme and catalyses the reaction :

(26) 2 ascorbate + O_2 + $2H^+$ 2 dehydroascorbate + $2H_2O$

Since the enzymes to recycle oxidised forms of ascorbate are not present in the cell wall, it has been proposed that the plasmalemma may have an ascorbate translocator to shuttle oxidised and reduced forms between the cytosol and cell wall.

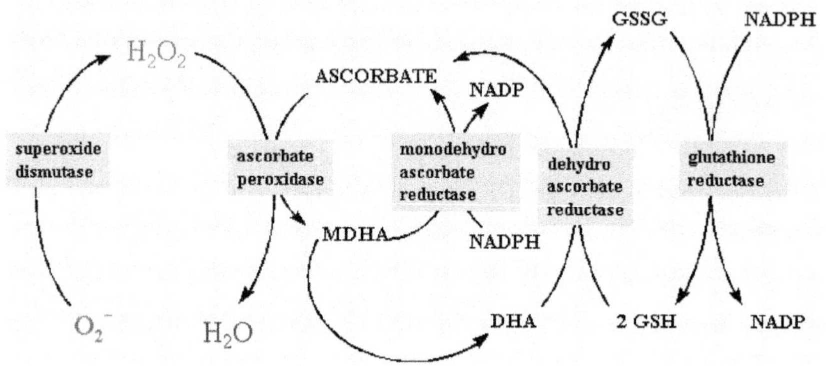

Figure 23.10 : The redox cycling of ascorbate in the chloroplast often referred to as the Halliwell-Asada pathway.

Glutathione

Glutathione (GSH) is a tripeptide (Glu-Cys-Gly) whose antioxidant function is facilitated by the sulphydryl group of cysteine. On oxidation, the sulphur forms a thiyl radical that reacts with a second oxidised glutathione forming a disulphide bond (GSSG). Some legumes contain homoglutathione (hGSH) that is a homologous tripeptide of Glu-Cys-Ala. Detailed reviews of GSH chemistry are available elsewhere. GSH has a redox potential of -340 mV that enables GSH to reduce dehydroascorbate to ascorbate or to reduce the disulphide bonds of proteins.

GSH is found in most tissues, cells and sub-cellular compartments of higher plants. Its levels declines with tissue age and vary among growth environments; for example, glutathione levels are higher in the light than in the dark. At the sub-cellular level, GSH concentration is highest in the chloroplast, averaging between 1 and 4 mM, but significant quantities also accumulate in the cytosol. GSH exists predominantly in the reduced form with estimates varying from 70% in barley chloroplasts to 90% in pea chloroplasts.

GSH can function as an antioxidant in many ways. It can react chemically with singlet oxygen, superoxide and hydroxyl radicals and therefore function directly as a free radical scavenger. GSH may stabilise membrane structure by removing acyl peroxides formed by lipid peroxidation reactions. As detailed in section 2.4.3, GSH is the reducing agent that recycles ascorbic acid from its oxidised to its reduced form by the enzyme dehydroascorbate reductase. GSH can also reduce dehydroascorbate by a non-enzymatic mechanism at pH > 7 and GSH concentrations greater than 1 mM. This may be a significant pathway in chloroplasts whose stromal pH in the light is about 8 and GSH concentrations may be as high as 5 mM.

There are alternative functions for GSH in cellular metabolism independent of its antioxidant properties. It may have a significant role in the transport of reduced sulphur from leaves to sink tissues such as the root. GSH also participates in the detoxification of xenobiotics as a substrate for the enzyme glutathione-S-transferase. The well documented tolerance of maize to the triazine herbicides is the result of conjugation of GSH to the herbicide. GSH is also the precursor of the phytochelatins that act as heavy metal binding peptides in plants.

The synthesis and degradation of GSH occurs continuously through the glutamyl cycle that has been well characterised in animals and at least portions have been confirmed in plants. The first step in GSH synthesis is the combination of glutamate and cysteine to form glutamylcysteine by the enzyme glutamylcysteine synthetase. The subsequent step involves the addition of glycine by the enzyme glutathione synthetase. In the legumes that accumulate hGSH, this second step involves the addition of alanine by the enzyme homoglutathione synthetase.

(27) Glu + Cys Glu-Cys

(28) Glu-Cys + Gly Glu-Cys-Gly

(29) Glu-Cys + Ala Glu-Cys-Ala

The degradation of GSH involves first the cleavage of the bond between glutamate and cysteine by glutamyl transpeptidase and the transfer of the glutamate residues to an acceptor amino acid :

Subsequently the Cys-Gly dipeptide is degraded by dipeptidases and Glu-aa by glutamylcyclotransferase :

(30) Glu-Cys-Gly + aa Glu-aa + Cys-Gly

(31) Cys-Gly Cys + Gly

(32) Glu-aa 5-oxoproline + aa

Alternatively, plants may contain an additional pathway to that described in animals for degradation of GSH involving a GSH-specific carboxypeptidase and glutamylcyclotransferase :

(33) Glu-Cys-Gly Glu-Cys + Gly

(34) Glu-Cys 5-oxoproline + Cys

The presence of the alternative degradative pathway in plants is significant, because Glu-Cys is a common intermediate in the synthesis and degradation of GSH and thus may be a site for regulating GSH content in plant cells. The final step in the degradation is the conversion of 5-oxoproline to glutamate by 5-oxoprolinase :

(35) 5-oxoproline Glu

The enzymes that catalyse the synthesis and degradation of GSH have been characterised in both the chloroplastic and cytosolic compartments. The numerous inhibitors of these enzymes have been used in animal systems to characterise the function of GSH in cellular metabolism but their use in plant systems has not been extensive. Similarly, although a number of environmental stresses have been shown to cause GSH accumulation, the mechanism involved has not been defined.

The reduction of GSSG to GSH is catalysed by the enzyme glutathione reductase (GR) which has been purified from a number of plant tissues, and cDNA for GR has been cloned from pea. The amino acid sequences and structures as determined by x-ray crystallography are very similar among the human and *E. coli* GR so far examined, suggesting that this is a very highly conserved enzyme. Nonetheless, the plant literature has conflicting reports of the sub-unit composition of GR; heterotetramer or homodimer forms have been reported. This is not totally unexpected. Although animals have only one form of GR, plants have multiple forms of the enzyme. For example, pea leaf GR has been resolved into eight isoenzymes by two dimensional gel electrophoresis. These isozymes are most likely associated with different subcellular compartments; GR is associated mainly with the chloroplast but significant activity is also found in the cytosol and a lesser amount in the mitochondria. The different isozymes quite probably represent the products of different genes with slightly or perhaps significantly different sequences, and with differing regulation and responses to environmental signals.

Animals contain a selenium containing enzyme, glutathione peroxidase, that reduces hydrogen peroxide forming GSSG and thereby serves as an alternative means of detoxifying activated oxygen. This enzyme was thought to be absent from higher plants but recently there have been reports of glutathione peroxidase in cultured cells. A plant cDNA showing homology to animal glutathione peroxidase has been isolated from *Nicotiana sylvestris*.

Tocopherol

The tocopherols, specifically a-tocopherol (vitamin E), have been studied extensively in mammalian research as membrane stabilisers and multifaceted antioxidants, that scavenge oxygen free radicals, lipid peroxy radicals, and singlet oxygen. This role is related to its fully substituted benzoquinone ring and fully reduced phytyl chain. Its hydrophobic nature dictates that a-tocopherol is located exclusively in cell membranes and is oriented with the benzoquinone ring in close association with the carbonyl of the glycerol component of the phospholipid, and with the phytyl chain associated with the fatty acids in the hydrophobic inner regions of the membrane bilayer. The ring oxygen is near the surface of the membrane bilayer but remains exposed to the lipid environment.

Because of its dietary importance tocopherol levels have been documented extensively in plant tissue. Tocopherol concentrations vary among plant tissues from 200 ng g^{-1} fresh weight (fw) in potato tubers to 5 mg g^{-1} fw in oil palm leaflets. Tocopherol has been found in all higher plants, in both photo-synthetic and non-photo-synthetic tissues. Although tocopherol is a well documented component of chloroplast membranes, there are no quantitative estimates of its distribution among other plant cell membranes.

Tocopherol is actually a family of antioxidants that includes four tocols that have a phytyl chain and analogous tocotrienols that have a geranylgeranyl chain. a-Tocopherol is generally considered to be the most active form of the tocols; the other forms may be bio-synthetic precursors. The relative synthesis of the tocols and tocotrienols may depend on the relative availability of phytyl and geranylgeranyl precursors. a-Tocopherol synthesis occurs in plastids with the aromatic ring formed by the shikimic acid pathway and the phytyl chain synthesized from geranylgeranyl pyrophosphate through the terpenoid pathway on the plastid envelope. The latter forms a common link among the synthesis of chlorophyll, tocopherol, and carotenoids.

a-Tocopherol is well established as a membrane stabilising agent. Although some of this activity is due to its influence on membrane lipid organisation, at least part of this activity is the result of its ability to complex free fatty acids. Free fatty acids act as detergents in membranes causing disruption of the bilayer, membrane aggregation and fusion. The association between the carboxyl group of the fatty acid and the ring of tocopherol reduces this destabilisation.

The antioxidant properties of tocopherol are the result of its ability to quench both singlet oxygen and peroxides. Tocopherol is a less efficient scavenger of singlet oxygen than ß-carotene and therefore in the thylakoid membrane it may function to break carbon radical chain reactions by trapping peroxyl radicals :

(36) ROO + tocopherol ROOH + tocopherol

The tocopheroxyl radical is stabilised by the fully substituted benzoquinone ring and therefore does not propagate the radical reaction. This is in effect a termination reaction making tocopherol an effective free radical trap.

Because the active oxygen of the a-tocopherol is located near the surface of the bilayer and because it readily diffuses laterally in the plane of the bilayer, tocopherol can react with peroxyl radicals formed in the bilayer as they diffuse to the aqueous phase. This position also allows the tocopheroxyl radical to be reduced by ascorbate in the aqueous phase to regenerate a-tocopherol. Glutathione reduces tocopheroxyl radicals in alcohol solutions, but there is only limited evidence for this in biological systems.

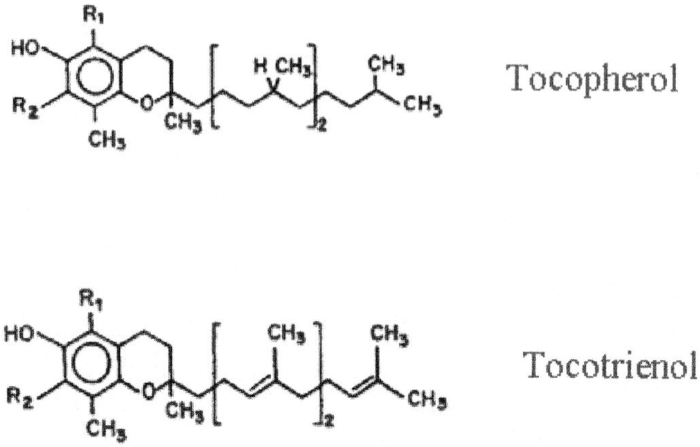

Figure 23.11 : Structure of tocopherols and tocotrienols commonly found in plants.

Carotenoids

Carotenoids are C_{40} isoprenoids and tetraterpenes that are located in the plastids of both photo-synthetic and non-photo-synthetic plant tissues. In chloroplasts, the carotenoids function act as accessory pigments in light harvesting, but perhaps a more important role is their ability to detoxify various forms of activated oxygen and triplet chlorophyll that are produced as a result of excitation of the photo-synthetic complexes by light.

There are two classes of carotenoids. The carotenes are hydrocarbons; the xanthophylls are carotene derivatives that contain one or more oxygen atoms. Carotenoids are synthesised from geranylpyrophosphate from the isoprenoid pathway in plastids, and thus have common precursors to chlorophyll and tocopherol. ß-carotene is formed by the cyclisation of lycopene and the xanthophylls are formed by mixed function oxidases that introduce hydroxyl groups to the carotene molecule. Certain classes of herbicides function by the inhibition of carotenoid bio-synthesis.

The carotenoids can exist in a ground state or in one of two excited states after the absorption of light energy. In terms of its antioxidant properties carotenoids can protect the photosystems in one of four ways : by reacting with lipid peroxidation products to terminate chain reactions; by scavenging singlet oxygen and dissipating the energy as heat; by reacting with triplet or excited chlorophyll molecules to prevent formation of singlet oxygen; or by the dissipation of excess excitation energy through the xanthophyll cycle.

Carotenoids may augment a-tocopherol in scavenging peroxy radicals :

(37) ß-car + ROO ß-car + ROOH

(38) ß-car + ROO inactive products

These reactions therefore act as chain terminations but unlike tocopherol a mechanism to recycle ß-carotene to the reduced form has not been described.

The main protective role of ß-carotene in photo-synthetic tissue may be through its direct quenching of triplet chlorophyll, which would prevent the generation of singlet oxygen, and therefore completely avoid oxidative stress.

(39) ^{3}Chl + ^{1}ß-car ^{1}Chl + ^{3}ß-car

(40) ^{3}ß-car ^{1}ß-car + heat

In reactions 39 and 40, energy is transferred from chlorophyll to a carotenoid which subsequently dissipates the energy in a non-radiative form (heat). Thus, the carotenoids act as a competitive inhibitor for the formation of singlet oxygen, and this is aided considerably by their proximity to chlorophyll in the light harvesting complex. This method of protection is especially critical as light intensity increases above saturating levels. Another carotenoid, zeaxanthin, has been implicated in the dissipation of thermal energy, but the precise mechanism has not been resolved. Zeaxanthin apparently facilitates the conversion of triplet to singlet chlorophyll in more efficient manner than ß-carotene. The xanthophyll cycle involves the reversible conversion of the xanthophylls between two forms, violaxanthin and zeaxanthin. A de-epoxidase enzyme catalyses the de-epoxidation of violaxanthin to zeaxanthin in the presence of excess light, and an epoxidase catalyses the reverse reaction in darkness or low light. Zeaxanthin therefore accumulates under light intensities that exceed photo-synthetic capacity. The de-epoxidase has a low pH optimum, whereas the epoxidase has a high pH optimum. Reduced ascorbate serves as the electron donor for the de-epoxidase, whereas NADPH supplies reducing equivalents for the epoxidase.

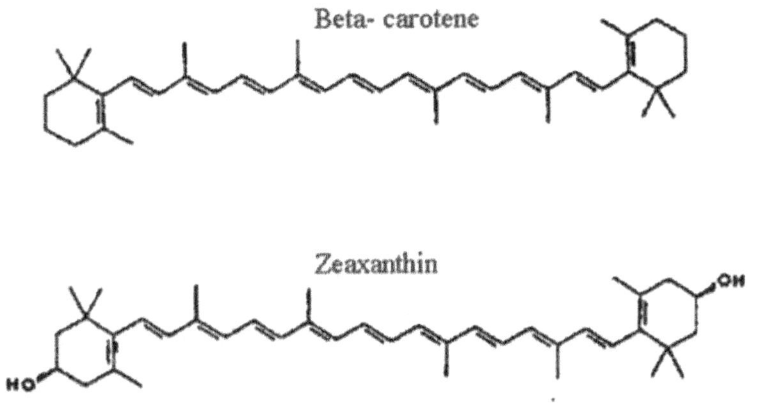

Figure 23.12 : Structure of two common carotenoids found in plants, ß-carotene and zeaxanthinin.

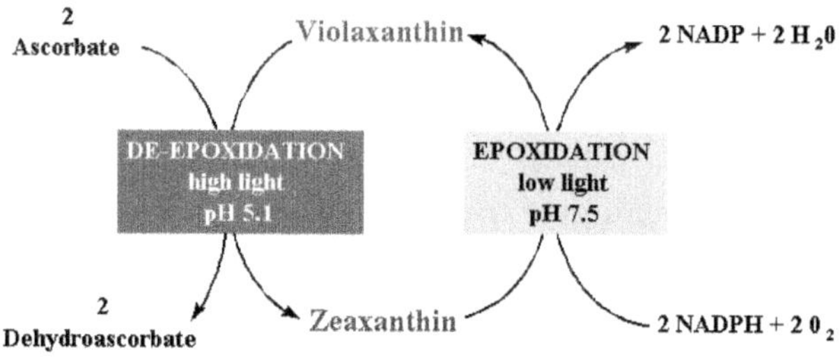

Figure 23.13 : The Xanthophyll cycle for the cycling of violaxanthin and zeaxanthin.

The de-epoxidation reaction converting violaxanthin to zeaxanthin is favoured under high light and pH 5.1, whereas the epoxidation reaction is favoured under low light and pH 7.5. The enzymes for both reactions are in the chloroplast thylakoid lumen and therefore during periods of photo-synthesis the lumen is acidified, and zeaxanthin accumulates. The reverse reactions occur in the dark and violaxanthin accumulates.

HERBICIDE TOLERANCE

A number of herbicides that are widely used on many agricultural crops act by the production of oxygen free radicals. A few plants have evolved or have been selected for tolerance to these herbicides. The mechanism of tolerance in some plants is related to the plants' ability to scavenge activated oxygen and thereby avoid oxidative stress. Two examples will be discussed in detail.

Paraquat

Paraquat and diquat are bipyridylium (viologen) herbicides that are effective as non-selective herbicides when applied to leaves. Symptoms include rapid wilting and desiccation of the leaf, followed by necrosis with 24 hours. Both light and chlorophyll are required for rapid response to the herbicide, which suggested to the early researchers that the herbicide was acting on photosynthesis. In the dark and in etiolated seedlings the symptoms appear more slowly. At the ultra-structural level, the first injury symptoms are chloroplast swelling, followed by breakdown of the tonoplast and the plasmalemma, and then breakdown of the thylakoids and chloroplast envelope. These observations indicated that cellular organelles differ in their sensitivity to the toxic agent and that the toxic agent is able to diffuse within the cell from its site of production in the chloroplast to the tonoplast and plasmalemma where it has its action promotes the visible symptoms of wilting and necrosis.

That the toxic agent is some form of activated oxygen was suggested by the first experiments of Mees who postulated that hydrogen peroxide was produced from the photo-synthetic electron transport system. This was based on the observations that monuron, an inhibitor of the Hill reaction, slowed the phytotoxic symptoms. Calderbank proposed that the hydroxyl radical was produced by these reactions, and subsequently Farrington *et. al.* suggested the involvement of superoxide.

Paraquat has a redox potential of -446 mV, which is critical to its function as a free radical generator. Any reducing agent with sufficient energy can denote an electron to the bipyridylium divalent cation, paraquat^{2+}, to form a free radical, paraquat^{1+}. The oxidation of the bipyridylium radical to form the original paraquat^{2+} results in the transfer of the electron to oxygen and the formation of superoxide. Subsequent Haber-Weiss and Fenton reactions yield toxic hydroxyl radicals. The paraquat^{2+} ion is again ready for another cycle, and thus the herbicide functions as a catalyst to transfer reducing equivalents to oxygen. In the leaf, the major reducing agent with sufficient energy to reduce paraquat is the primary electron acceptor in PSI, ferredoxin, but other reducing agents may be effective. Consequently, bipyridylium herbicides are also quite toxic to non-photo-synthetic organisms, including humans.

Resistance to paraquat has appeared in a number of weedy species after prolonged use of the herbicide, in *Lolium perenne* after breeding and selection, and in genetically engineered plants, animal cells, and microbes. In paraquat-sensitive plants that have been selected in the field by prolonged exposure paraquat, the mechanism(s) of resistance is ambiguous. In biotypes of *Conyza baraciensis* and in perennial ryegrass, *Lolium perenne,* resistance was correlated with increased activity of superoxide dismutase and other free radical scavenging enzymes. However, in other resistant plants there is reduced translocation of the herbicide to its site of action in the chloroplast, and in other instances, a correlation with superoxide dismutase activity was not observed.

Unfortunately, the situation is just as ambiguous in genetically engineered organisms that have modified expression of superoxide dismutase or glutathione reductase. Gruber *et. al.* expressed the *E. coli* Mn-superoxide dismutase in the cyanobacterium *Anacystis nidulans,* and observed that the transformants had less damage when treated with paraquat. Similarly, Bowler *et. al.* expressed a Mn-superoxide dismutase cDNA from *Nicotiana plumbiginfolia* in *N. tabacum* and showed reduced cellular damage when the transformed plants were treated with paraquat. However, tomato and tobacco plants that were transformed with a Cu/Zn-superoxide dismutase cDNA from Petunia did not have enhanced paraquat resistance.

Superoxide dismutase is unlikely to be the only enzyme involved in providing protection against activated oxygen produced by paraquat. Glutathione peroxidase has been associated with paraquat resistance in a human cell line, HL60. A transgenic N. *tabacum* plant that expresses the glutathione reductase gene from *E. coli* had reduced susceptibility to paraquat compared to the control plants based on visual leaf damage. Malan *et. al.* observed that in maize inbreds that tolerance of paraquat (and also drought) was not significantly correlated with the activity of either superoxide dismutase or glutathione reductase individually, but it was correlated with the two enzyme activities collectively. Glutathione reductase reduces production of reduced oxygen by isolated chloroplasts in the presence of paraquat presumably by restoration of the NADP pool.

There are also developmental changes in the plant's tolerance to paraquat. For example, in winter wheat, mesophyll protoplasts, excised leaves, and seedlings have increased tolerance to paraquat after acclimation of the seedlings at 2°C. This was associated with an increased quantity of lipid soluble antioxidants that would minimise the deleterious chain reactions associated with lipid peroxidation.

The above discussion illustrates a generalisation which will be a common theme in the chapters to follow. Most often our experimental research involves simple experiments that have focused on individual components of stress tolerance, in this case herbicide tolerance. The plant is not so simple and as a general rule will not develop a single method of surviving a potentially catastrophic stress, if two or more can be developed in parallel. In some instances discussed later these mechanisms of tolerance may be mutually exclusive, but very often they are complimentary and serve as backup systems, insurance or additive protection. Looking at the case in point, tolerance of the free radical generating herbicide paraquat, it is obvious that a number of mechanisms may be involved : (a) reduced translocation of the herbicide to its site of action; (b) altered redox potential of PSI primary electron acceptor so that the paraquat is not an efficient electron acceptor and will therefore not generate superoxide; (c) enhanced scavenging of superoxide produced by PSI; and (d) enhanced protection of membrane components from lipid peroxidation reactions. All the above mechanisms except (b) have been described in tolerant plants. Presumably mechanism (b) is not viable because changes in redox potential would also change the efficiency of electron transfer between ferredoxin and NADP, and this would reduce photo-synthetic efficiency and competitiveness or yield. The actual mechanism(s) selected will undoubtedly vary among plants.

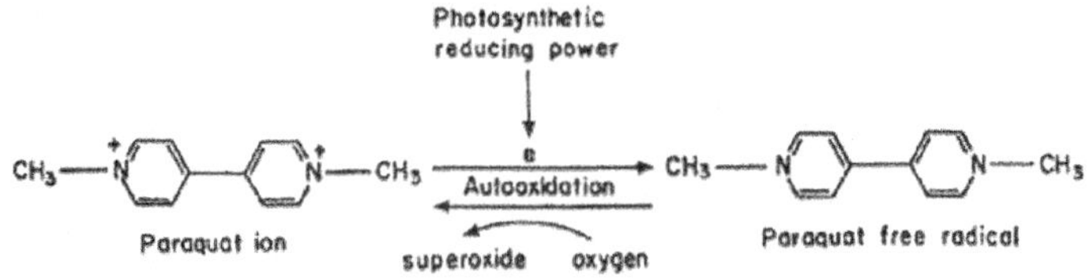

Figure 23.14 : The redox cycling reaction of paraquat that causes superoxide production.

Photo-sensitising Herbicides

Several herbicide formulations promote the accumulation of metabolic intermediates of chlorophyll, namely tetrapyrroles. In the light, photons are absorbed by the tetrapyroles and the energy used to create singlet oxygen that kills the plant. Two classes of photobleaching herbicides, the p-nitrodiphenyl ethers and aminolevulinic acid-based modulator, differ in the precise mechanisms of tetrapyrrole accumulation.

The p-nitrodiphenyl ethers, exemplified by acifluorfen, are used as selective herbicides in soybeans and require light for herbicidal activity. Early symptoms of injury include wilting, desiccation, and necrosis. Ultra-structural studies have shown that the plasmalemma and tonoplast are the first organelles to be disrupted, not the chloroplast, which explains why enhanced electrolyte leakage is one of the first indicators of injury from a photo-sensitising herbicide. Acifluorfen selectively inhibits protoporphyrinogen oxidase (Protox) which is the penultimate step in heme and Mg-protoporphyrin IX synthesis on the chloroplast envelope (Matrinage *et. al.*, 1989; Witkowski and Halling, 1989). This blockage results in the accumulation of its substrate proto-porphyrinogen which at the time was thought to oxidise to its product protoporphyrin IX by a "spontaneous" reaction. The loss of enzymatic control of protoporphyrin IX synthesis was associated with lower heme levels and increased carbon flow into the pathway. Because heme is a feedback inhibitor of aminolevulinic acid synthesis, this change in regulation of the pathway was thought to lead to the 100 fold accumulation of protoporphyrin IX but a detailed evaluation of the data from these early experiments on acifluorfen action were inconsistent with the herbicide acting solely via tetrapyrrole accumulation and other mechanisms have been proposed. Subsequently, *in vivo* localisation studies have shown that tetrapyrroles accumulate at sites outside of the chloroplast such as the plasmalemma, which may explain why photo-sensitising damage is not seen initially in the chloroplast.

The presumed "spontaneous" oxidation of protoporphyrinogen IX to protoporphyrin IX is apparently accomplished by a protoporphyrinogen oxidising enzyme outside of the chloroplast, probably on the plasmalemma. This enzyme is presumably less sensitive than the chloroplast enzyme to the nitrodiphenyl ether inhibitors, and the presence of this second enzyme would explain many of the inconsistencies of the earlier data, the observed accumulation of protoporphyrin IX outside of the chloroplast, and the observed depletion of heme in the chloroplast. The major mechanism of action of acifluorfen and other diphenyl either herbicides is now thought to involve a blockage of the chloroplastic pathway shunting the protoporphyrinogen to an extra-chloroplastic pathway that first uncouples the regulation of the tetrapyrrole pathway on the chloroplast envelope by low heme levels, and secondly causes the accumulation of a photo-sensitising agent, protoporphrin IX in the cellular membranes.

The mode of action of the aminolevulinic acid-based herbicides is more clearly established because unlike other herbicides they were designed based on a knowledge of chlorophyll synthesis and free radical chemistry. All formulations are based on the concept of supplying aminolevulinic acid to the plant as a precursor of tetrapyrrole synthesis, in conjunction with a modulator that qualitatively and quantitatively alters the pattern of tetrapyrrole accumulation by selectively stimulating or inhibiting various steps in the biosynthetic pathway. The accumulated tetrapyrroles serve as photosensitizers that produce singlet oxygen in the light. The herbicide formulations are most effective if the sprayed plants remain in darkness for a few hours to allow tetrapyrrole accumulation before exposure to the light. After about 20 min illumination, the leaves develop isolated bleached areas that expand followed by wilting, desiccation and necrosis.

Although the photo-sensitising herbicides were assumed to be non-selective, laboratory and field trials have indicated that significant selectivity can be achieved by modifications of the formulations. This is caused by the different capacities of tissues and plants of accumulate tetrapyrroles and presumably to sequester or translocate the herbicides.

Some plants have apparently achieved increased resistance to acifluorfen by an enhanced scavenging capacity of oxygen free radicals. Bean leaves treated with acifluorfen responded by higher levels of reduced glutathione and higher activity of glutathione reductase. A tobacco line that was resistant to paraquat was also co-tolerant of acifluorfen. This tolerance was associated with a greater inducibility of several components of the oxygen free radical scavenging system. Winter wheat (*Triticum aestivum*) seedlings that develop tolerance of freezing and of paraquat as they acclimate at 2°C, also develop increased resistance to acifluorfen. Alfalfa plants that express a transgene of the Mn-SOD cDNA from *Nicotiana* have an increased LD_{50} for acifluorfen.

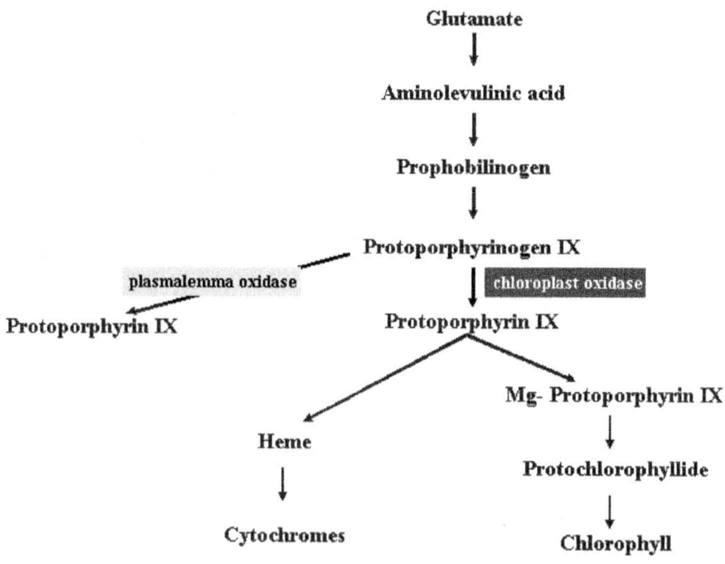

Figure 23.15 : The bio-synthetic pathway of chlorophyll and heme.

The proposed sites of action of diphenylether herbicides is the protoporphyrinogen oxidase on the chloroplast envelope. Inhibition of this enzyme has two effects :

a) reduces heme levels, which increases carbon flow into the pathway by reducing its feedback inhibition of the pathway.
b) diverts protoporphyrinogen IX to the plasmalemma where a second oxidase converts it to protoporphyrin IX.

Aminolevulinic acid-based herbicides stimulate the pathway by supplying excess substrate and modulators promote the accumulation of chlorophyll precursors that act as photosensitizers.

CONCLUSIONS

Oxygen free radicals or activated oxygen has been implicated in diverse environmental stresses in plants and animals and appears to be a common participation in most, if not all, degenerative conditions in eukaryotic cells. The peroxidation of lipids, the cross-linking and inactivation of proteins and mutations in DNA are typical consequences of free radicals, but because the reactions occur quickly and often are components of complex chain reactions, we usually can only detect their "footprints". Activated forms of oxygen are important in the bio-synthesis of "complex" organic molecules, in the polymerisation of cell wall constituents, in the detoxification of xenobiotic chemicals and in the defence against pathogens. Thus, the plant's dilemma is not how to eliminate the activation of oxygen, but how to control and manage the potential reactions of activated oxygen. Complex systems of scavenging activated oxygen therefore exist in plant cells with complimentary and interdependent strategies. Some components such as the carotenoids prevent the formation of activated oxygen by competing for the energy leaked from the photosystems. Other components are lipid soluble and reside in the membrane bilayer to terminate the lipid peroxidation chain reactions. Still others, ascorbate and glutathione, are aqueous scavenger that detoxify activated oxygen directly or serve to recycle other protective components back to their reduced state. The enzymes that catalyse the synthesis, degradation and recycling of these antioxidants are essential to viability. Consequently they are highly conserved among plants, and exist in multiple forms in different sub-cellular compartments and different tissues to allow precise regulation.

There are numerous sites of oxygen activation in the plant cell, which are highly controlled and tightly coupled to prevent release of intermediate products. Under stress situations, it is likely that this control or coupling breaks down and the process "dysfunctions" leaking activated oxygen. This is probably a common occurrence in plants especially when we consider that a plant has minimal mobility and control of its environment. These uncoupling events are not detrimental provided that they are short in duration and that the oxygen scavenging systems are able to detoxify the various forms of activated oxygen. If the production of activated oxygen exceeds the plant's capacity to detoxify it, deleterious degenerative reactions occur, the typical symptoms being loss of osmotic responsiveness, wilting, and necrosis. At the sub-cellular level, disintegration of membranes and aggregation of proteins are typical symptoms. Therefore it is the balance between the production and the scavenging of activated oxygen that is critical to the maintenance of active growth and metabolism of the plant and overall environmental stress tolerance.

Chapter 24

Impact of Cold Stress

THE ENVIRONMENTAL AND PHYSIOLOGICAL NATURE OF STRESS

Successful adaptation of a crop species is dependent upon the programming of critical growth stages so that the plant can capitalize on favourable weather periods during the growing season. Plants have evolved a variety of adaptive mechanisms that allow them to optimize growth and development while coping with environmental stresses. These mechanisms include seed and bud dormancy, photoperiod sensitivity, and low-temperature response. Seed dormancy delays germination until after the embryo has gone through an after-ripening period. The over-winter survival of buds of many temperate zone trees and shrubs is dependent on a dormancy stage that starts in the late summer or early fall and ends after exposure to an extended period of cold or increasing day length in the spring. In addition to trees, many other dicots and grasses have a photo-period response that can advance or delay flowering. Vernalization is a requirement for growth at low temperatures before a plant will flower. Most winter annual and biennial plants have a vernalization requirement. Low-temperature acclimation is an ability of plants to cold acclimate when exposed to gradually decreasing temperatures below a specific threshold. This is the most common mechanism that plants have evolved for adapting to low-temperature stress and examples of plants with the capacity to cold harden can be found in most species.

Types of Low Temperature Injury

There are two types of injuries a plant can sustain through exposure to low temperatures. The first is chilling injury that occurs from approximately 20 to 0°C. The resultant injuries may include a variety of physiological disruptions in germination, flower and fruit development, yield, and storage life. Minor chilling stress at non-lethal temperatures is normally reversible.

Exposure to gradually decreasing temperatures above the critical range can also result in hardening of plants that may reduce or eliminate injury during subsequent exposure to low temperatures.

The second type of injury is called freezing injury. This type of injury occurs when the external temperature drops below the freezing point of water. Some varieties of plants that are susceptible to chilling injury can be killed by the first touch of frost. At the other extreme, many plants that are native to cold climates can survive extremely low temperatures without injury. Plants may experience intracellular freezing and/or extra-cellular freezing. Intracellular freezing damages the protoplasmic structure and the ice crystals kill the cell once they grow large enough to be detected microscopically. In extra-cellular freezing, the protoplasm of the plant becomes dehydrated because a water-vapour deficit is created as cellular water is transferred to ice crystals forming in the inter-cellular spaces. In some cases, water can remain liquid as low as -47°C without nucleating and forming ice. When nucleation of this supercooled water does occur, intracellular ice forms suddenly resulting in death of the plant.

Types of Plants

Plants can be grouped into three different classes according to their low-temperature tolerance. The first group includes frost tender plants that are sensitive to chilling injury and can be killed by short periods of exposure to temperatures just below freezing. They cannot tolerate ice in their tissues and readily exhibit frost injury symptoms that include a water soaked flaccid appearance with loss of turger followed by rapid drying upon exposure to warm temperatures. Beans, corn, rice, and tomatoes are examples of plants in this category.

Low-temperature acclimation of plants in the second group allows them to tolerate the presence of extra-cellular ice in their tissues. Their frost resistance ranges from the broad-leafed summer annuals, which are killed at temperatures slightly below freezing, to perennial grasses that can survive exposure to -40°C. As temperatures decrease the outward migration of intracellular water to the growing extra-cellular ice crystal causes dehydration stress that will eventually result in irreversible damage to the plasma membrane, which is the primary site of low-temperature injury. If ice nucleation does not occur at -3 to -5°C, supercooling may result in intracellular freezing and death of individual cells.

The final group is made up of very cold hardy plants that are predominantly temperate woody species. Like the plants in the previous group, their lower limits of cold tolerance are dependent on the stage of acclimation, the rate and degree of temperature decline, and the genetic capability of tissues to accommodate extra-cellular freezing and the accompanying dehydration stress. Deep supercooling allows certain tissues in plants from this group to survive low temperatures without the formation of extra-cellular ice. However, the most cold hardy species do not rely on supercooling and can withstand temperatures of -196 °C.

PLANT CHILLING STRESS AND ITS REPERCUSSIONS

Introduction

Most crops of tropical origin as well as many of sub-tropical origin are sensitive to chilling temperatures. This limits production areas and causes potential damage during storage if they are exposed to low temperatures. The temperature below which chill injury can occur varies with species and regions of origin, ranging from 0 to 4°C for temperate fruits, 8°C for sub-tropical fruits, and about 12°C for tropical fruits such as banana. Amongst the highest volume world food crops, maize (*Zea mays*) and rice (*Oryza sativa*) are sensitive to chilling temperatures.

Their growth and development can be adversely effected by temperatures below 10°C resulting in yield loss or crop failure. Christiansen and St. John estimated annual losses of $60 million to the cotton industry due to chilling temperature immediately following field planting. Chilling during the seedling stage in cotton can reduce plant height, delay flowering and adversely affect yield and lint quality. Seedlings can also suffer water stress and leaf desiccation at chilling temperatures, floral initiation is inhibited at 7°C and seed set is inhibited at 15°C. Other crops suffering stand loss, delayed maturity, and reduced yield as a result of chilling after planting include soybean (*Glycine max* L.), lima bean (*Phaseolus lunatus* L.), cucurbits (*Cucurbita* sp.), tomato (*Lycopersicon esculentum* Mill.) pepper *Capsicum annuum* L.), eggplant (*Solanum melongena* L.), okra (*Abelmoschus esculentus* L.), and various cereal crops.

Physiological age, seedling development, and pre-harvest climate can also influence chilling sensitivity. Freshly imbibed seeds of chill-sensitive species tend to be very sensitive, as does the pollen development stage. Fruits maturing at high temperature are more susceptible than those maturing at lower temperatures. Post-harvest storage at lower temperatures is commonly used to extend the storage life of fruits and vegetables. Tropical and sub-tropical plants however are often subject to physiological damage and loss of quality due to chill injury under these storage conditions. The severity of injury to chill-sensitive tissues tends to increase with decreasing temperatures and with length of low-temperature exposure.

Chilling has been found to change the entire metabolic system of the cell with some processes recovering quickly and others only slowly. Chilling affects the entire internal environment of each cell and each molecule within the cells. Enzymatic reactions, substrate diffusion rates, and membrane transport properties are all affected. Chilling injury is therefore likely a direct consequence of these effects.

Amelioration of Chilling Injury

Avoidance : To avoid chilling injury, planting dates can be altered though this is often difficult because of its effect on later development of the plant. To overcome this problem, cultivars have been bred for early vigour and maturity. In the case of stored fruits and vegetables, maintenance of appropriate storage temperatures is essential to avoid chilling injury. Investigations have also been undertaken to examine synthetic plant growth regulators for the protection of chilling sensitive crops.

Temperature conditioning : Low-temperature 'hardening' allowing tolerance to chilling temperatures appears to have little effect although some sensitivity to 'slight chilling' can be reduced by exposure to temperatures slightly above the chilling range. It also appears that chilling injury to stored fruits and vegetables can be ameliorated by warm temperatures if they are imposed before tissue degeneration becomes advanced. Other treatments such as waxing, fungicides, hormones, and antioxidants have produced variable results that have been dependent upon the species and treatment conditions.

Duration : Ultra-structural-chilling injury increases with time and with prolonged exposure the injury becomes irreversible. It is therefore important to minimize the time of chilling temperature exposure.

Relative humidity : High relative humidity has been found to protect chloroplasts from chill injury, an effect that is enhanced by darkness.

Theories of Chilling Injury

Early research focused on chilling causing an imbalance in plant physiological processes. Chilling was found to affect O_2 evolution, organic acids, sugars, polyphenols, phospholipids, protein, and ATP. Research indicates that chilling stress in sensitive plants changes most chemical entities. There is evidence of accumulation of toxins such as ethanol and acetaldehyde. Although many altered processes involve key metabolites; it is difficult to separate the critical chilling-sensitive metabolic processes from those that are byproducts of metabolic disruptions or of ultra-structural breakdown. Ion leakage due to membrane permeability changes has often been reported in chill sensitive plants. Phase transition of the lipid portion of the cellular membranes has also received considerable attention as the *primary* response to chilling temperatures.

Ultra-structural changes : On an ultra-structural level, several changes have been associated with chilling injury. Although there are a number of variables affecting chill injury, the ultra-structural symptoms are very similar across species. Ultra-structural symptoms of chilling injury become evident before obvious physical symptoms are visible. These include changes to chloroplasts, mitochondria and membranes associated with these organelles and the vacuoles. The symptoms include swelling and disorganization of the chloroplasts and mitochondria, reduced size and number of starch granules, dilation of thylakoids and unstacking of grana, formation of small vesicles of chloroplast peripheral reticulum, lipid droplet accumulation in chloroplasts, and condensation of chromatin in the nucleus.

Chloroplasts are the first and most severely affected organelle. Irradiance during chilling greatly exacerbates the resulting injury. Chilled plants in darkness have been found to remain green and, except for starch depletion, chloroplasts appear normal. In the presence of light, however, chlorophyll becomes bleached, lipid droplets accumulate, and thylakoids degenerate. Mitochondria appear more resistant to chilling temperature but an immediate effect of low temperature on chilling-sensitive species is a suppression of mitochondrial activity. Electron micrographs of chilled sweet potato roots revealed that the mitochondria had a swollen appearance due to the release of phospholipids from the inner and outer membranes during storage at chilling temperatures. The capacity to bind phospholipids was also greatly decreased.

Membrane permeability and phase transition : Measures of solute leakage or ion permeability have provided evidence of increased membrane permeability in response to chilling. The plasma membrane is often considered the primary site of freezing injury and electrolyte leakage. Early work indicated that plants originating in warm climates tend to have more saturated fatty acids in their membrane lipids. More recent work on mitochondrial membranes has shown that membranes do undergo a physical phase transition from a flexible liquid-crystalline to a solid-gel structure at 10 to 12°C, which coincides with the temperature sensitivity range of species of tropical origin. Fruits of several apple cultivars have been observed to undergo phase transition in the 3 to 10°C range suggesting the same mechanism of chilling injury as found in tropical species. The correlation between fatty acid composition and temperature induced phase transition is, however, not precise. It may be that other membrane components such as sterols also play a role.

It is possible that the phase transition of cellular membranes could account for the entire range of physiological and metabolic changes associated with chilling injury. Increased membrane permeability could lead to an altered ion balance and also to the ion leakage observed from chilling of sensitive tissues. Phase transition could result in conformational changes in membrane bound enzymes and account for the observed discontinuities in the function of many enzyme systems. This may cause an imbalance between membrane bound and non-membrane bound systems. Over time the cells inability to cope with increased concentrations of metabolites could result in injury. Different tolerances to these metabolites could explain why some cultivars are more resistant to damage while still undergoing phase transition. Imbalances in metabolism, accumulation of toxic compounds, and increased permeability could all be the result of temperature-induced phase transition.

The contribution of unsaturated fatty acids in cell membrane lipids has been discussed for many years in relation to chilling sensitivity. Nishida and Murata have shown that chilling injury can be manipulated by modulating levels of unsaturation of fatty acids by the action of acyl-lipid desaturases and glycerol-3-phosphate acyltransferase. Lyons proposed that temperature induced phase transition of membrane lipids may play a primary role in chilling sensitivity of plants. Continued exposure to chilling temperatures would result in phase separated membranes becoming incapable of maintaining ionic gradients resulting in metabolic disruption and eventual cell death. A positive correlation has been found between chilling sensitivity of herbaceous plants and the level of saturated and *trans*-monounsaturated molecular species of phosphatidylglycerol (also termed high-melting-point molecular species) in thylakoid membranes. However, there is still a question of how directly these high-melting-point molecular species relate to chilling sensitivity in plants. Growth at low temperature generally increases the degree of unsaturation of membrane lipids, which compensates for the decrease in fluidity caused by the lower temperature. This increased unsaturation is also correlated with the sustained activity of membrane-bound enzymes at low temperature. The unsaturation of membrane lipids is therefore considered critical for the functioning of biological membranes and the survival of plant cells at low temperature. However, since low temperature is also known to induce or alter the expression level of a large number of genes it is not clear if the association between membrane lipid unsaturation and chilling tolerance is a cause or effect relationship.

Recently the role of unsaturation of membrane lipids in chilling tolerance and in response to low temperature has been reexamined using mutant and transgenic lines. In this way unsaturated fatty acids can be manipulated independent of temperature so that their individual effects can be evaluated. Tobacco was transformed with squash and *Arabidopsis* phosphatidylglycerol (PG) species found in thylakoid membranes. Squash has low levels of *cis*-unsaturated PG while *Arabidopsis* has relatively high levels of *cis*-unsaturated PG. It was found that tobacco transformed with squash PG was more chilling sensitive and tobacco transformed with *Arabidopsis* PG was the most chilling resistant, as measured by photo-synthesis at 1°C under strong illumination. These results indicate that chilling sensitivity can be manipulated by altering the level of unsaturated PG in the chloroplasts. These and other experiments have shown that unsaturation of membrane lipids protect the photo-system II complex from low-temperature photoinhibition by accelerating recovery from the photoinhibited state. However, it is likely that other factors such as accumulation of polyols and amino acids, or their derivatives, contribute to chilling sensitivity in plants. Some specific proteins may also be responsible for chilling tolerance.

Alteration of intra-cellular pH : Yoshida *et. al.* noted that intracellular pH was, in part, actively controlled by H^+-transport from the cytoplasm to the vacuole catalyzed by H^+-ATPase located on the vacuolar membrane in mung bean (Vigna radiata L.), which is a very chilling-sensitive species. The vacuolar H^+-ATPase is extremely sensitive to low temperature and is preferentially inactivated upon exposure to chilling temperatures. This inactivation occurs much earlier than the symptoms of cell injury and the decrease in enzyme activity associated with plasma membranes, endoplasmic reticulum, and mitochondria. Cold-induced inactivation of H^+-ATPase also occurs in chilling sensitive rice. Cold-induced suppression of proton transport disrupts cytoplasmic homeostasis and causes a change in the pH. The chilling sensitivity of cultured mung bean cells changed markedly during the growth cycle and a close relationship was found between sensitivity of the cells and of H^+-ATPase to the cold. Cold-induced inactivation of the vacuolar H^+-ATPase was closely linked to acidification of the cytoplasm and the corresponding alkalization of the vacuoles suggesting a passive release of H^+ ions across the vacuolar membrane. The susceptibility of vacuolar H^+-ATPase to low temperature *in vivo* was found to be markedly different between chilling-sensitive and chilling-resistant species. In contrast to the H^+-ATPases of chilling-sensitive species like mung bean and kidney bean (Phaseolus vulgaris), the H^+-ATPases of the chilling-tolerant species such as pea (Pisum sativum) and broad bean (Vicea faba) were very stable over long periods of low-temperature exposure. The molecular structures of the 16 kDa proteolipids from the two types of H^+-ATPase appeared to be very different. Low-temperature-induced pH reduction of the cytoplasm caused by inactivation of vacuolar H^+-ATPase may therefore be the cause of extreme chilling-sensitivity.

Conclusion

Plants are uniquely adapted to their native environment through developmental programming and the particular composition and conformation of their molecular components is optimized within each species for maximum competitive ability. These differences in adaptation result in the wide range of cellular disturbances that have been observed when these plants are moved to cooler environments. Changes in enzyme reactions, substrate diffusion rates, membrane properties, and cytoplasmic pH affect the entire metabolic system of cells subjected to chilling stress. The resulting injury depends on the duration of exposure and on the individual species, or variant, being observed.

THE PHYSIOLOGICAL AND AGRONOMIC REPERCUSSIONS OF FREEZING STRESS

Low-Temperature Response Mechanisms

Plants have adapted two mechanisms to protect themselves from damage due to below freezing temperatures. Supercooling is a low-temperature tolerance mechanism that is usually associated with acclimated xylem parenchyma cells of moderately hardy woody plants. When sources of ice nucleation are absent, pure water can supercool or remain unfrozen to its homogeneous nucleation point of approximately -40°C. The initiation of freezing at the limit of supercooling occurs suddenly and is accompanied by an exotherm that can be detected by thermal analyses of plant tissues. Plant tissues suffer irreversible damage once ice nucleation of supercooled water occurs and the distribution in nature of tree species with the ability to deep supercool is normally restricted to regions where winter temperatures are warmer than -40°C.

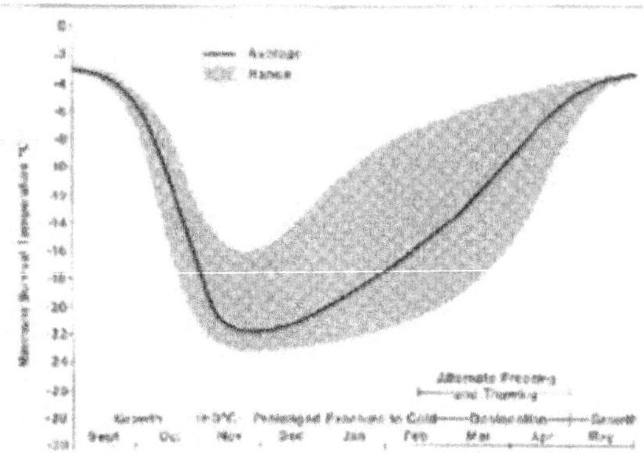

Figure 24.1 : Changes in cold hardiness of Norstar winter wheat for the period September to May. The primary factors responsible for these changes are shown at the bottom of the graph.

The second and most common low-temperature response mechanism is acclimation. Low-temperature acclimation is a gradual process during which there are changes in just about every measurable morphological, physiological, and bio-chemical characteristic of the plant. These changes are determined by genotype x environment interactions that are quite complex and not clearly understood. They have been studied most extensively in cereals where a wide range in genetic potential and the availability of unique cytogenetic stocks has allowed for novel approaches to investigations at the molecular and whole plant level. Potential gene donors have been evaluated for use in interspecific transfers and the control of alien (donor species) low-temperature gene expression has been studied in a variety of backgrounds. A survey of the published research in these areas has allowed us to construct a field validated winter survival model that successfully simulates the over winter changes in low-temperature tolerance of a wide range of genotypes. Consequently, this review will focus mainly on the genetic systems that winter cereals have evolved for low-temperature adaptation, the regulation of these systems, and their complex interaction with the environment.

Low-Temperature Acclimation in Winter Cereals

When growth starts in the early fall, winter cereal plants will not survive sub-freezing temperatures much better than spring cereal plants. However, winter cereals grown under cool fall temperatures will cold acclimate or 'harden off'. For example, in Saskatchewan (western Canada), the minimum survival temperature for 'Norstar' winter wheat is normally near -3°C at the beginning of September and -19°C or lower by the end of October.

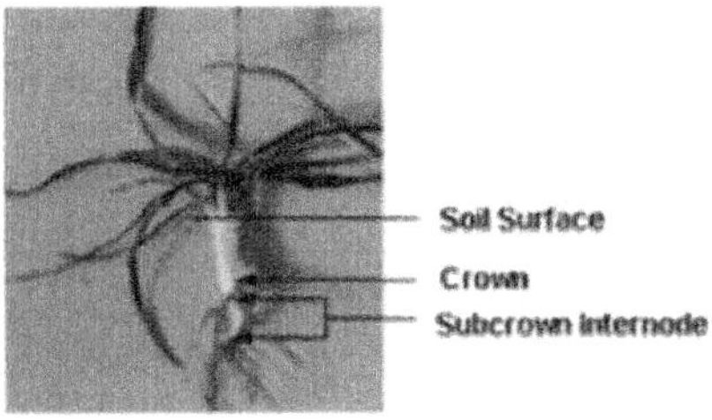

Figure 24.2.

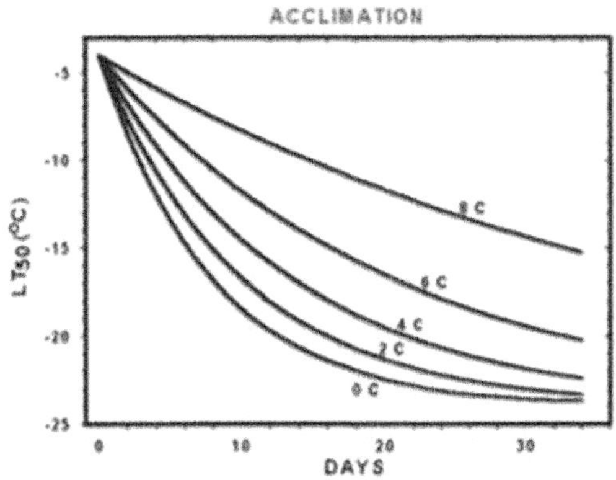

Figure 24.3 : Relationship between average daily crown temperature (0 to 8oC) and low-temperature tolerance of Norstar winter wheat during acclimation at a constant temperature before vernalization saturation (LT50 for Norstar = - 24oC when fully acclimated).

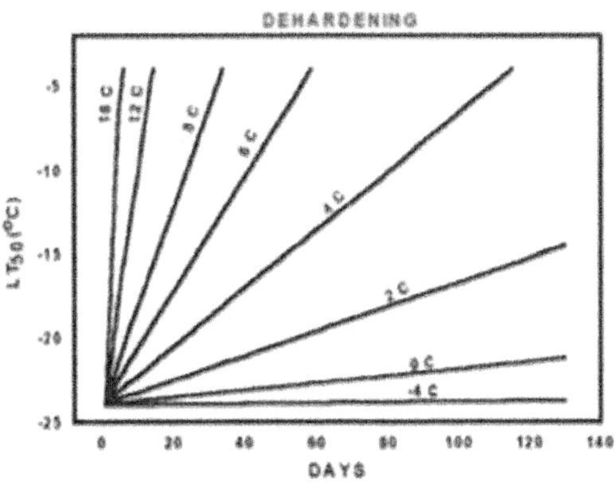

Figure 24.4 : Relationship between average daily crown temperature (- 4 to 18°C) and low-temperature tolerance of Norstar winter wheat during dehardening at a constant temperature (LT50 for Norstar = - 24°C when fully acclimated = day 0).

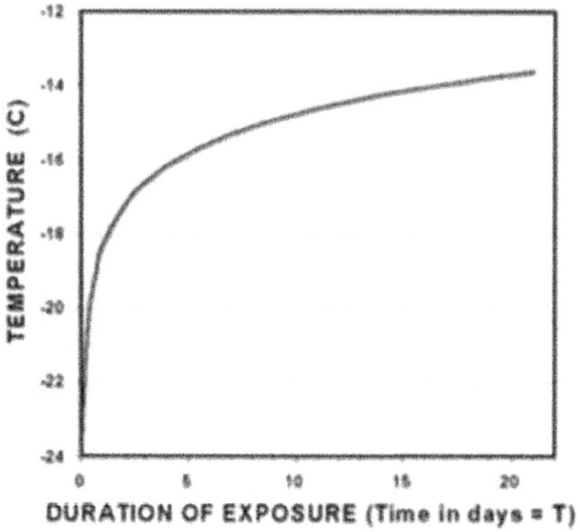

Figure 23.5 : The effect of exposure to a constant temperature on low-temperature tolerance of Norstar winter wheat (Norstar LT50 = - 24°C when fully acclimated at day 0).

Under field conditions in western Canada, eight to 12 weeks of fall growth is usually required for the full development of cold hardiness in winter cereals. The first four to five weeks is a period of active growth that takes place when average daily soil temperatures at crown depth are above 9°C. Both the cold acclimation process and winter survival require energy and this period of warm temperature allows for the establishment of healthy vigorous plants. Plants with well-developed crowns before freeze-up are in the best position to withstand the rigors of winter and regenerate roots and leaves in the spring. However, plants that enter the winter with two to three leaves are usually not seriously disadvantaged.

Cold acclimation of winter wheat plants begins once fall temperatures drop below approximately 9°C. A translocatable substance that promotes cold acclimation is not produced when winter wheat plants are exposed to acclimating temperatures. Consequently, the cold-hardiness level of different plant parts, such as leaves, crowns and roots, is dependent upon the temperature to which each part has been exposed. Because the crown contains tissues that are necessary for plant survival, it is the soil temperature at crown depth that determines critical cold-acclimation rates.

Plant growth slows considerably at temperatures that promote cold acclimation. In the field, soil temperatures gradually decrease as winter approaches and four to seven weeks at temperatures below 9°C is usually required to fully cold-harden plants. Cold acclimation during this period is dependent upon crown temperatures and the rate of acclimation increases dramatically as temperature drop from 9 to 0°C. Exposure of winter wheat crowns to soil temperatures above 9°C during this period results in a rapid loss of cold hardiness. The rate of dehardening is dependent upon the temperature to which the crown is exposed. At this stage, plants that have been exposed to crown temperatures above 9°C will resume cold acclimation once they return to temperatures below 9°C.

Winter wheat normally does not realize its maximum cold hardiness potential until after the soil is frozen in the late fall. In Saskatchewan, full acclimation is usually achieved by the middle to the end of November. Once cold acclimation has been completed, winter wheat can maintain a high level of cold hardiness provided crown temperatures remain below freezing. In the fall, winter wheat will cold acclimate when exposed to crown temperatures colder than 9°C. However, prolonged exposure of acclimated plants to winter temperatures above freezing results in the transition of the plant from the vegetative to the reproductive phase and a gradual loss of cold hardiness. The warmer the crown temperature during the winter, the shorter the period that maximum levels of cold hardiness can be maintained and, once started, the more rapid the rate of decline in cold hardiness.

Death of the crown tissue will result if the soil temperature falls below the plants minimum survival temperature. Exposure of winter cereal plants to crown temperatures that are 2 to 3°C warmer than their minimum survival temperature will cause immediate damage and a reduction in cold hardiness. Longer periods of exposure to temperatures approaching the minimum survival temperature can quickly reduce the plant's ability to tolerate cold stress. The expected LT_{50} for different exposure times (T) to constant temperature can be calculated from Equation 1.

$$LT50_{(T)} = LT50_{(0)} + 5.72 + 1.53 * \ln(T) \qquad \text{[Eq. 1]}$$

Where T is the number of days that plants are exposed to a constant low-temperature stress. $LT50_{(0)}$ is determined using a series of test temperatures where the low-temperature stress is removed as soon as the crown samples are exposed to a predetermined minimum temperature. For example, fully acclimated Norstar winter wheat will normally survive to -24.0°C in a controlled-freeze test where plant samples are gradually cooled at a rate of 2 to 6°C hr^{-1} and removed as soon as they reach a predetermined temperature. However, two days exposure to -17.2°C in a controlled environment will reduces the minimum survival temperature of fully hardened Norstar winter wheat from -24.0 to -17.2°C, a cold hardiness loss of 6.8°C.

Once vernalization saturation is complete and the plant enters the reproductive stage, it loses its ability to cold acclimate and it will start to deharden at temperatures warmer than approximately -4°C. This means that winter wheat will eventually completely deharden once plant growth resumes in the spring. Growth rate and rate of dehardening are both temperature dependent and because frozen soils warm slowly in the spring, several weeks of warm air temperatures are required to re-establish and completely deharden winter cereal plants that have survived without serious winter damage.

SYMPTOMS OF DEFICIENCY IN ESSENTIAL MINERALS

Introduction

Visual nutrient deficiency symptoms can be a very powerful diagnostic tool for evaluating the nutrient status of plants. One should keep in mind, however, that a given individual visual symptom is seldom sufficient to make a definitive diagnosis of a plant's nutrient status. Many of the classic deficiency symptoms such as tip burn, chlorosis and necrosis are characteristically associated with more than one mineral deficiency and also with other stresses that by themselves are not diagnostic for any specific nutrient stress. However, their detection is extremely useful in making an evaluation of nutrient status. In addition to the actual observations of morphological and spectral symptoms, knowing the location and timing of these symptoms is a critical aspect of any nutrient status evaluation. Plants do not grow in isolation, they are part of the overall environment and as such they respond to environmental changes as that affect nutrient availability. Also, plants do influence their environment and can contribute to environmental changes, which in turn can affect the nutrient status of the plant.

Sources of Visual Symptoms

Stresses such as salinity, pathogens, and air pollution induce their own characteristic set of visual symptoms. Often, these symptoms closely resemble those of nutrient deficiency. Pathogens often produce an interveinal chlorosis, and air pollution and salinity stress can cause tip burn. Although at first these symptoms might seem similar in their general appearance to nutrient deficiency symptoms, they do differ in detail and/or in their overall developmental pattern. Pathological symptoms can often be separated from nutritional symptoms by their distribution in a population of affected plants. If the plants are under a nutrient stress, all plants of a given type and age in the same environment tend to develop similar symptoms at the same time. However if the stress is the result of pathology, the development of symptoms will have a tendency to vary between plants until a relatively advanced stage of the pathology is reached.

Environmental Associations

Plants remove substantial amounts of nutrients from the soil during their normal growth cycle and many long-term environmental changes occur as a result of this process. Effects on the soil go considerably beyond the straight removal or depletion of nutrients. Charge balance must be maintained in the plant-soil system during nutrient uptake. Charge balance is usually achieved by the excretion of proton and/or hydroxyl ions by the plant to replace the absorbed nutrient cations or anions. For example when plants are fertilized with ammonia, they acquire most of their nitrogen in the form of the ammonium cation, rather than from the usual nitrate anion. Because nitrate is the only anion used by the plant in large amounts, the net result of this change is that during normal nutrient uptake the proton excretion will far exceed that of hydroxyl ions. In the case of vigorously growing plants, the amount of excreted protons can be sufficiently large as to decrease the pH of the soil by several pH units. Changes in soil pH of such magnitude can have large implications for a number of soil processes such as soil structure, nutrient availability and leaching of nutrients. The immediate effect on the soil may be favourable for some plants, especially acid-loving plants, in that it tends to make iron more available. However, in the long run, lowering the soil pH can be deleterious to plants in that the availability of nutrients will change. A lower soil pH will allow micro-nutrients to be more readily leached from the soil profile, eventually resulting in deficiencies of nutrients such as Cu and Zn. Additionally, when the pH of the soil drops much below pH 5, the solubility of Al and Mn can increase to such an extent as to become toxic to most plant growth.

Plants are often thought of as passive in relation to the environment. However this is not always a valid assumption; for there are many plants that clearly manipulate their environment in a fashion that tends to makes certain nutrients more readily available. For example, iron is a limiting nutrient in many agricultural areas, but it comprises about 3% of the average soil which, if available, would be far in excess of the needs of the average plant. Some plants actively excrete protons, and the resulting decrease in pH increases the solubility of iron in their environment. In addition, other plants excrete phytosiderophores that chelate the soil iron rendering it a more available form for the plants.

Pathways of Symptom Development

At first glance, it would appear that the distinction of deficiency symptoms for the 13 known essential mineral nutrients should be relatively simple. But such an assumption is incorrect. In fact, the deficiency symptoms are quite complex because each nutrient has a number of different biological functions and each function may have an independent set of interactions with a wide range of environmental parameters. In addition, the expression of these symptoms varies for acute or chronic deficiency conditions. Acute deficiency occurs when a nutrient is suddenly no longer available to a rapidly growing plant. Chronic deficiency occurs when there is a limited but continuous supply of a nutrient, at a rate that is insufficient to meet the growth demands of the plant.

Most of the classic deficiency symptoms described in textbooks are characteristic of acute deficiencies. The most common symptoms of low-grade, chronic deficiencies are a tendency towards darker green leaves and stunted or slow growth. Typically most published descriptions of deficiency symptoms arise from experiments conducted in greenhouses or growth chambers where the plants are grown in hydroponics or in media where the nutrients are fully available. In these conditions, nutrients are readily available while present, but when a nutrient is depleted, the plant suddenly faces an acute deficiency. Thus, hydroponic studies favor the development of acute deficiencies.

In experiments designed to study micro-nutrient deficiency symptoms, micro-nutrients are usually omitted from the nutrient solution. Micro-nutrients are often present in the seed or as contaminants in the environment, so a plant of adequate size will exhaust these trace amounts of micro-nutrient and develop characteristic acute deficiency systems. When deficiency symptoms of macronutrients are sought, the macronutrient is removed suddenly from a suitable sized rapidly growing plant. Alternatively the plant can initially be given a one-time supply of the nutrient that is sufficient for a limited amount of growth. Because macronutrients are continuously required in relatively large amounts by rapidly growing plants, the available nutrients will be rapidly depleted, resulting in an acute deficiency.

In natural systems, the plant encounters many degrees and types of stresses that result in different types of symptoms occurring over time. Perhaps the most common nutrient deficiency in natural environments is the case of a limited nutrient supply that is continuously renewed at a low rate from soil weathering processes. In such cases, the limited nutrient availability results in chronic nutrient deficiency symptoms.

Effect of Nutrient Mobility on Symptom Development

The interaction between nutrient mobility in the plant, and plant growth rate can be a major factor influencing the type and location of deficiency symptoms that develop. For very mobile nutrients such as nitrogen and potassium, deficiency symptoms develop predominantly in the older and mature leaves. This is a result of these nutrients being preferentially mobilized during times of nutrient stress from the older leaves to the newer leaves near the growing regions of the plant. Additionally, mobile nutrients newly acquired by the roots are also preferentially translocated to new leaves and the growing regions. Thus old and mature leaves are depleted of mobile nutrients during times of stress while the new leaves are maintained at a more favorable nutrient status.

The typical localization of deficiency symptoms of very weakly mobile nutrients such as calcium, boron, and iron is the opposite to that of the mobile nutrients; these deficiency symptoms are first displayed in the growing regions and new leaves while the old leaves remain in a favorable nutrient status. (This assumes that these plants started with sufficient nutrient, but ran out of nutrient as they developed). In plants growing very slowly for reasons other than nutrition (such as low light) a normally limiting supply of a nutrient could, under these conditions, be sufficient for the plant to slowly develop, maybe even without symptoms. This type of development is likely to occur in the case of weakly mobile nutrients because excess nutrients in the older leaves will eventually be mobilized to supply newly developing tissues. In contrast, a plant with a similar supply that is growing rapidly will develop severe deficiencies in the actively growing tissue such as leaf edges and the growing region of the plant. A classic example of this is calcium deficiency in vegetables such as lettuce where symptoms develop on the leaf margins (tip burn) and the growing region near the meristems. The maximal growth rate of lettuce is often limited by the internal translocation rate of calcium to the growing tissue rather than from a limited nutrient supply in the soil.

When moderately mobile nutrients such as sulfur and magnesium are the limiting nutrients of the system, deficiency symptoms are normally seen over the entire plant. However the growth rate and rate of nutrient availability can make a considerable difference on the locations at which the symptoms develop. If the nutrient supply is marginal compared to the growth rate, symptoms will appear on the older tissue, but if the nutrient supply is very low compared to the growth rate, or the nutrient is totally depleted, the younger tissue will become deficient first.

Plant Competition and Induced Deficiencies

When the observed symptoms are the direct result of a nutrient deficiency, the actions needed for correction are relatively straight-forward. However symptoms are often the result of interactions with other environmental factors limiting the availability of the nutrient whose symptoms are expressed. The classic instance is that of iron deficiency induced by an excess of heavy metals in the environment. Transition metals such as Cu, Zn Cr and Ni compete with Fe and each other for plant uptake. Competition for uptake is not specific to Fe and heavy metals but is true for all mineral nutrients that are chemically similar and have similar uptake mechanisms. For example, if the availability of Cu or Zn is relatively less than that of Fe, then excessive concentrations of some other metal such as Ni or Cr will induce a deficiency of one of these nutrients rather than Fe. In the case of the macronutrients, excessive amounts of Mg will compete with K for uptake and can possibly induce a K deficiency. The barrenness of serpentine soils is the result of such competition, with the high Mg of these soils inducing a Ca deficiency. The toxicity of a low pH soil is another example of a basic nutrient deficiency. Low pH has a two-fold effect on soil nutrients : It enhances the leaching of cations, reducing their availability in the soil, and the relatively abundant protons in the soil compete with Ca and other cations for uptake. Thus, nutrient deficiencies can be induced by a number of different mechanisms often working in concert to limit the availability of a nutrient.

Nutrient Demand and Use Efficiency

Although all plants of the same species respond similarly to nutrient stress, plants of similar species will often show significant differences in their nutrient use efficiency. This results from differences in growth rate, root distribution, phase of development, and efficiency of nutrient uptake and utilization. This implies that in any given location, plants from one species may become nutrient-deficient, while those from another species growing in the same environment right next to them, may not show any deficiency symptoms.

Growth rate also affects nutrient status. When the nutrient supply is barely inadequate for growth under existing environmental conditions, many plants adjust their growth rate to match that supported by the available nutrient supply without displaying typical visual deficiency symptoms.

Agricultural systems differ from natural systems in that crop plants have been selected primarily for rapid growth under low stress conditions. This rapid growth rate results in a high nutrient demand by these plants and a higher incidence of nutrient deficiency unless supplemental fertilizers are supplied. It is not uncommon to find agricultural crops showing severe signs of nutrient stress, with native plants growing in the same area showing little or no indication of nutrient stress. In agriculture systems chronic deficiency symptoms develop mostly in crops with little or limited fertilization. Acute nutrient deficiency symptoms most often occur when new crops with a higher nutrient demand are introduced, or less productive lands are brought under cultivation for the production of rapidly growing crop plants.

Uniformity of Nutrient Status

Not all tissues of a plant are at the same nutrient status during times of stress. Leaves on the same plant that are exposed to different environmental conditions, (such as light), or those of different ages may have considerable differences in nutrient status. Mineral nutrients are for the most part acquired by the roots and translocated throughout the plant. The distance of any part of the plant to the roots will influence nutrient availability, particularly in the case of the less mobile nutrients. In plants recovering from nutrient deficiency, the root and conductive tissues recover first. For example, in the case of recovery from Fe deficiency, it is common to see the veins regreen while the interveinal tissue remains chlorotic and Fe-deficient.

In order to maintain rapid, optimal growth, all plant tissues must have a favourable nutrient status. Although a plant may be marginally low in a number of nutrients, only one nutrient at a time will limit overall growth. However, if the supply of that limiting nutrient is increased even slightly, the resulting increase in growth will increase the demand for all other nutrients and another nutrient, the next lowest in availability, will become limiting.

Other Diagnostic Tools

Although visual diagnostic symptoms are an extremely valuable tool for the rapid evaluation of the nutrient status of a plant, they are only some of the tools available. Other major tools include microscopic studies, spectral analysis, and tissue and soil analysis. These methods all vary in their precision, rapidity and their ability to predict future nutrient status. Because of the close interaction between plant growth and the environment, all predictions of future nutrient status must make assumptions about how the environment will change in that time frame.

The principle advantage of visual diagnostic symptoms is that they are readily obtained and provide an immediate evaluation of nutrient status. Their main drawback is that the visual symptoms do not develop until after there has been a major effect on yield, growth and development.

Tissue analysis is nutrient-specific but relatively slow; tissues must be sampled, processed, and analyzed before the nutrient status can be determined. An analysis of the mineral nutrient content of selected plants tissues, when compared against Critical Level values, can be used to evaluate the plant nutrient status at the time of sampling with a relatively high degree of confidence and can be extrapolated to project nutrient status at harvest. Soil analysis is similar to tissue analysis but evaluates the potential supplying power of the soil instead of plant nutrient status. Plant analysis provides information as to what the plant needs, while soil analysis provides information about the status of the nutrient supply.

Spectral analysis of nutrient status is still in its infancy and is presently used primarily in the inventory of global resources and in specialized studies. Microscopic studies are most valuable in looking at the physiological aspects of nutrient stress rather than the evaluation of plant nutrient status on a whole plant or crop basis.

Symptom Descriptions

It is unusual to find any one leaf or even one plant that displays the full array of symptoms that are characteristic of a given deficiency. It is thus highly desirable to know how individual symptoms look, for it is possible for them to occur in many possible combinations on a single plant. Most of the terms used below in the description of deficiency symptoms are reasonably self evident; a few however have a distinct meaning in the nutrient deficiency field. For example, the term *chlorotic,* which is a general term for yellowing of leaves through the loss of chlorophyll, cannot be used without further qualification because there may be an overall chlorosis as in nitrogen deficiency, interveinal, as in iron deficiency, or marginal, as in calcium deficiency. Another term used frequently in the description of deficiency symptoms is *necrotic,* a general term for brown, dead tissue. This symptom can also appear in many varied forms, as is the case with chlorotic symptoms.

Nutrient deficiency symptoms for many plants are similar, but because of the large diversity found in plants and their environments there is a range of expression of symptoms. Because of their parallel veins, grasses and other monocots generally display the affects of chlorosis as a series of stripes rather than the netted interveinal chlorosis commonly found in dicots. The other major difference is that the marginal necrosis or chlorosis found in dicots is often expressed as tip burn in monocots.

Magnesium. The Mg-deficient leaves show advanced interveinal chlorosis, with necrosis developing in the highly chlorotic tissue. In its advanced form, magnesium deficiency may superficially resemble potassium deficiency. In the case of magnesium deficiency the symptoms generally start with mottled chlorotic areas developing in the interveinal tissue. The interveinal laminae tissue tends to expand proportionately more than the other leaf tissues, producing a raised puckered surface, with the top of the puckers progressively going from chlorotic to necrotic tissue. In some plants such as the Brassica (*i.e.*, the mustard family, which includes vegetables such as broccoli, brussel sprouts, cabbage, cauliflower, collards, kale, kohlrabi, mustard, rape, rutabaga and turnip), tints of orange, yellow, and purple may also develop.

Figure 24.6 : A Magnesium deficiency symptoms in tomato.

Manganese. These leaves show a light interveinal chlorosis developed under a limited supply of Mn. The early stages of the chlorosis induced by manganese deficiency are somewhat similar to iron deficiency. They begin with a light chlorosis of the young leaves and netted veins of the mature leaves especially when they are viewed through transmitted light. As the stress increases, the leaves take on a gray metallic sheen and develop dark freckled and necrotic areas along the veins. A purplish luster may also develop on the upper surface of the leaves. Grains such as oats, wheat, and barley are extremely susceptible to manganese deficiency. They develop a light chlorosis along with gray specks which elongate and coalesce, and eventually the entire leaf withers and dies.

Figure 24.7 : Manganese deficiency symptoms in tomato.

Molybdenum. These leaves show some mottled spotting along with some interveinal chlorosis. An early symptom for molybdenum deficiency is a general overall chlorosis, similar to the symptom for nitrogen deficiency but generally without the reddish colouration on the undersides of the leaves. This results from the requirement for molybdenum in the reduction of nitrate, which needs to be reduced prior to its assimilation by the plant. Thus, the initial symptoms of molybdenum deficiency are in fact those of nitrogen deficiency. However, molybdenum has other metabolic functions within the plant, and hence there are deficiency symptoms even when reduced nitrogen is available. In the case of cauliflower, the lamina of the new leaves fail to develop, resulting in a characteristic whiptail appearance. In many plants there is an upward cupping of the leaves and mottled spots developing into large interveinal chlorotic areas under severe deficiency. At high concentrations, molybdenum has a very distinctive toxicity symptom in that the leaves turn a very brilliant orange.

Figure 24.8 : Molybdenum deficiency symptoms in tomato.

Nitrogen. The chlorotic symptoms shown by this leaf resulted from nitrogen deficiency. A light red cast can also be seen on the veins and petioles. Under nitrogen deficiency, the older mature leaves gradually change from their normal characteristic green appearance to a much paler green. As the deficiency progresses these older leaves become uniformly yellow (chlorotic). Leaves approach a yellowish white color under extreme deficiency. The young leaves at the top of the plant maintain a green but paler color and tend to become smaller in size. Branching is reduced in nitrogen deficient plants resulting in short, spindly plants. The yellowing in nitrogen deficiency is uniform over the entire leaf including the veins. However in some instances, an interveinal necrosis replaces the chlorosis commonly found in many plants. In some plants the underside of the leaves and/or the petioles and midribs develop traces of a reddish or purple color. In some plants this coloration can be quite bright. As the deficiency progresses, the older leaves also show more of a tendency to wilt under mild water stress and become senescent much earlier than usual. Recovery of deficient plants to applied nitrogen is immediate (days) and spectacular.

Figure 24. 9 : Nitrogen deficiency symptoms in tomato.

Figure 24. 10 : Phosphorus deficiency symptoms in tomato.

Phosphorus. These phosphorus-deficient leaves show some necrotic spots. As a rule, phosphorus deficiency symptoms are not very distinct and thus difficult to identify. A major visual symptom is that the plants are dwarfed or stunted. Phosphorus deficient plants develop very slowly in relation to other plants growing under similar environmental conditions but without phosphorus deficiency. Phosphorus deficient plants are often mistaken for unstressed but much younger plants. Some species such as tomato, lettuce, corn and the brassicas develop a distinct purpling of the stem, petiole and the under sides of the leaves. Under severe deficiency conditions there is also a tendency for leaves to develop a blue-gray luster. In older leaves under very severe deficiency conditions a brown netted veining of the leaves may develop.

Sulfur. This leaves show a general overall chlorosis while still retaining some green color. The veins and petioles show a very distinct reddish colour. The visual symptoms of sulfur deficiency are very similar to the chlorosis found in nitrogen deficiency. However, in sulfur deficiency the yellowing is much more uniform over the entire plant including young leaves. The reddish colour often found on the underside of the leaves and the petioles has a more pinkish tone and is much less vivid than that found in nitrogen deficiency. With advanced sulfur deficiency brown lesions and/or necrotic spots often develop along the petiole, and the leaves tend to become more erect and often twisted and brittle.

Figure 24.11 : Sulfur deficiency symptoms in tomato.

Zinc. This leaves show an advanced case of interveinal necrosis. In the early stages of zinc deficiency the younger leaves become yellow and pitting develops in the interveinal upper surfaces of the mature leaves. Guttation is also prevalent. As the deficiency progress these symptoms develop into an intense interveinal necrosis but the main veins remain green, as in the symptoms of recovering iron deficiency. In many plants, especially trees, the leaves become very small and the internodes shorten, producing a rosette like appearance.

Figure 24.12 : Zinc deficiency symptoms in tomato.

Boron. These boron-deficient leaves show a light general chlorosis. The tolerance of plants to boron varies greatly, to the extent that the boron concentrations necessary for the growth of plants having a high boron requirement may be toxic to plants sensitive to boron. Boron is poorly transported in the phloem of most plants, with the exception of those plants that utilize complex sugars, such as sorbitol, as transport metabolites. In a recent study, tobacco plants engineered to synthesize sorbitol were shown to have increased boron mobility, and to better tolerate boron deficiency in the soil.

Figure 24.13 : Boron deficiency symptoms in tomato.

In plants with poor boron mobility, boron deficiency results in necrosis of meristematic tissues in the growing region, leading to loss of apical dominance and the development of a rosette condition. These deficiency symptoms are similar to those caused by calcium deficiency. In plants in which boron is readily transported in the phloem, the deficiency symptoms localize in the mature tissues, similar to those of nitrogen and potassium. Both the pith and the epidermis of stems may be affected, often resulting in hollow or roughened stems along with necrotic spots on the fruit. The leaf blades develop a pronounced crinkling and there is a darkening and crackling of the petioles often with exudation of syrupy material from the leaf blade. The leaves are unusually brittle and tend to break easily. Also, there is often a wilting of the younger leaves even under an adequate water supply, pointing to a disruption of water transport caused by boron deficiency.

Calcium. These calcium-deficient leaves show necrosis around the base of the leaves. The very low mobility of calcium is a major factor determining the expression of calcium deficiency symptoms in plants. Classic symptoms of calcium deficiency include blossom-end rot of tomato (burning of the end part of tomato fruits), tip burn of lettuce, blackheart of celery and death of the growing regions in many plants. All these symptoms show soft dead necrotic tissue at rapidly growing areas, which is generally related to poor translocation of calcium to the tissue rather than a low external supply of calcium. Very slow growing plants with a deficient supply of calcium may re-translocate sufficient calcium from older leaves to maintain growth with only a marginal chlorosis of the leaves. This ultimately results in the margins of the leaves growing more slowly than the rest of the leaf, causing the leaf to cup downward. This symptom often progresses to the point where the petioles develop but the leaves do not, leaving only a dark bit of necrotic tissue at the top of each petiole. Plants under chronic calcium deficiency have a much greater tendency to wilt than non-stressed plants.

Figure 24.14 : Calcium deficiency symptoms in tomato.

Chloride. These leaves have abnormal shapes, with distinct interveinal chlorosis. Plants require relatively high chlorine concentration in their tissues. Chlorine is very abundant in soils, and reaches high concentrations in

Figure 24.15 : Chloride deficiency symptoms in tomato.

saline areas, but it can be deficient in highly leached inland areas. The most common symptoms of chlorine deficiency are chlorosis and wilting of the young leaves. The chlorosis occurs on smooth flat depressions in the interveinal area of the leaf blade. In more advanced cases there often appears a characteristic bronzing on the upper side of the mature leaves. Plants are generally tolerant of chloride, but some species such as avocados, stone fruits, and grapevines are sensitive to chlorine and can show toxicity even at low chloride concentrations in the soil.

Copper. These copper-deficient leaves are curled, and their petioles bend downward. Copper deficiency may be expressed as a light overall chlorosis along with the permanent loss of turgor in the young leaves. Recently matured leaves show netted, green veining with areas bleaching to a whitish gray. Some leaves develop sunken necrotic spots and have a tendency to bend downward. Trees under chronic copper deficiency develop a rosette form of growth. Leaves are small and chlorotic with spotty necrosis.

Figure 24.16 : Copper deficiency symptoms in tomato.

Iron. These iron-deficient leaves show strong chlorosis at the base of the leaves with some green netting. The most common symptom for iron deficiency starts out as an interveinal chlorosis of the youngest leaves, evolves into an overall chlorosis, and ends as a totally bleached leaf. The bleached areas often develop necrotic spots. Up until the time the leaves become almost completely white they will recover upon application of iron. In the recovery phase the veins are the first to recover as indicated by their bright green colour. This distinct venial re-greening observed during iron recovery is probably the most recognizable symptom in all of classical plant nutrition. Because iron has a low mobility, iron deficiency symptoms appear first on the youngest leaves. Iron deficiency is strongly associated with calcareous soils and anaerobic conditions, and it is often induced by an excess of heavy metals.

Figure 24.17 : Iron deficiency symptoms in tomato.

Potassium. Some of these leaves show marginal necrosis (tip burn), others at a more advanced deficiency status show necrosis in the interveinal spaces between the main veins along with interveinal chlorosis. This group of symptoms is very characteristic of K deficiency symptoms.

Figure 24.18 : Potassium deficiency symptoms in tomato.

The onset of potassium deficiency is generally characterized by a marginal chlorosis progressing into a dry leathery tan scorch on recently matured leaves. This is followed by increasing interveinal scorching and/or necrosis progressing from the leaf edge to the midrib as the stress increases. As the deficiency progresses, most of the interveinal area becomes necrotic, the veins remain green and the leaves tend to curl and crinkle. In some plant such as legumes and potato, the initial symptom of deficiency is white speckling or freckling of the leaf blades. In contrast to nitrogen deficiency, chlorosis is irreversible in potassium deficiency, even if potassium is given to the plants. Because potassium is very mobile within the plant, symptoms only develop on young leaves in the case of extreme deficiency. Potassium deficiency can be greatly alleviated in the presence of sodium but the resulting sodium-rich plants are much more succulent than a high potassium plant. In some plants over 90% of the required potassium can be replaced with sodium without any reduction in growth.

Chapter 25

Hydrogen Peroxide Functions as a Stress Signal in Plants

INTRODUCTION

In nature, plants are exposed to various stresses, which affect their physiology, morphology, and development. Every year, environmental stress causes considerable losses in the productivity of many crops. Among these stresses, the fluctuation of temperature, the water status of soil, and the intensity of light are the most crucial signals affecting plant growth. In addition to external stimuli, a variety of internal signals, such as hormones and nutrient conditions, also modify a plant's metabolism, growth, and development. Hence, rapid and precise perception of and response to various stimuli are important as plants adapt to changing natural environments. However, how a plant perceives environmental changes and how it subsequently triggers signals to activate the physiological response are yet to be explored.

Extensive study on oxidative stress has demonstrated that exposure of plants to adverse environmental conditions induces the overproduction of ROS, such as superoxide radical (O_2^-), H_2O_2, and hydroxyl radical ($HO^\cdot$) in plant cells. ROS are highly reactive to membrane lipids, protein, and DNA; they are believed to be the major contributing factors to stress injuries and to cause rapid cellular damage, particularly when plants are exposed to low temperature. At low temperature, electron transport chains tend to form O_2^-, which dismutates to form H_2O_2. In chloroplasts, low temperature limits the light reactions, restricting the supply of $NADP^+$ and promoting the reduction of O_2 by photo-system I. In mitochondria, inhibition of ATP formation or electron flow through cytochrome *b* stimulates O_2^- formation by complex I and by ubiquinone. Therefore, more serious damage is observed when plants are exposed to low temperature in combination with high light intensities, drought, air pollutants (*e.g.* ozone or sulphur dioxide), ultraviolet light, and herbicides (*e.g.* paraquat).

Plants have evolved both enzymatic and non-enzymatic mechanisms to scavenge the rapidly evolving ROS under low temperature or other stresses. Enzymes, including superoxide dismutase (SOD), catalase (CAT), peroxidase (POX), ascorbate peroxidase (APX) and glutathione reductase (GR), and non-enzymatic antioxidants such as tocopherols, ascorbic acid (AsA), and glutathione (GSH) work in concert to detoxify ROS. Among the antioxidant mechanisms AsA is a key antioxidant for elimination of ROS, especially H_2O_2. The reaction of AsA with H_2O_2 can occur directly or it can be catalysed by APX. APX, a key H_2O_2-scavenging enzyme, is found in the cytosol, chloroplasts,, and mitochondria of higher plants. The regeneration of AsA is catalysed directly by reduced ferredoxin in photo-system I and by monodehydroascorbate reductase (MDHAR), dehydroascorbate reductase (DHAR), and GSH. APX, MDHAR, DHAR, GSH/glutathione disulfide (GSSG), NADPH, and GR together form the AsA-GSH cycle, an H_2O_2-scavenging pathway. In accordance with its role in scavenging H_2O_2, APX expression is induced in response to chilling. Glutathione is a crucial antioxidant associated with regenerating AsA in the AsA-GSH cycle, and thus GSH is also involved in the regulation of H_2O_2 concentrations and control of the redox state in plant cells.

Our present understanding about the stress signaling in plants is that the changes of GSH/GSSG, following the synthesis of H_2O_2, constitute an early stress signal leading to the physiological response, usually activation of antioxidant mechanisms or modification of gene expression. It is suggested that GSH and GR are important factors participating in the redox state regulation of plant cells. The ratio of GSH to GSSG decreases under stress conditions. GSH is oxidized for the reduction of excess H_2O_2 in the AsA-GSH cycle upon stress.

The increase in H_2O_2 level and subsequent changes in the GSH/GSSG ratio during chilling seem to be involved in redox signaling, which activates specific transcription factors and antioxidative enzymes. Indeed, it has been shown that pretreatment of seedlings with H_2O_2 induces chilling tolerance (Yu *et. al.*, 2002; Yu *et. al.*, 2003). More studies have provided evidence that H_2O_2 itself is a key signal molecule mediating a series of responses. Other studies suggest that H_2O_2 is a signal mediator for programmed cell death (PCD) of plants in response to pathogens, elicitors, and hormones (Levine *et. al.*, 1994; Tenhaken *et. al.*, 1995; Levine *et. al.*, 1996; Desikan *et. al.*, 1998; Mittler *et. al.*, 1999; Solomon *et. al.*, 1999; Bethke and Jones, 2001). Furthermore, a number of studies indicate that H_2O_2 is synthesized in response to exogenous abscisic acid (ABA) and that H_2O_2 mediates, at least in part, ABA responses, including stomatal closure and gene expression. Using cDNA micro-array technology to carry out a transcriptomic analysis, Desikan *et. al.* provided further evidence of H_2O_2 as a central signaling mediator. Their study showed that the expression of some genes is up-regulated by H_2O_2. In addition, some are repressed. Classified by their potential biological functions, these genes include heat shock proteins, heat shock transcription factors, mediators for calcium signal transduction, such as calmodulin, important signaling enzyme protein tyrosine phosphatases (PTPs), a blue copper-binding protein which is an essential catalyst for redox reactions, a mitochondrial uncoupling protein, pyruvate decarboxylase, and a *myb*-related transcription factor. In addition, some of the H_2O_2-sensitive genes could also be involved in plant hormone signaling. It is interesting to note that various genes encoding potential transcription factors were induced by H_2O_2, suggesting that these transcription factors mediate further downstream H_2O_2 responses and finally induce physiological changes to promote adaptation to stresses.

Our knowledge about the scope of stress-induced signaling in plants leading to modification of gene expression has increased in recent years. A number of signaling molecules have been demonstrated to be involved in the stress signaling pathway.

THE GENERATION AND REMOVAL OF H_2O_2

Stress often leads to the production of ROS such as O_2^- and H_2O_2 in plant tissues. ROS are highly active molecules that can easily damage membrane and other cellular components. Hence, it is important to remove ROS to avoid chilling or other stress-induced injuries. Anti-oxidation mechanisms of the cell include the enzymatic ROS-scavenging system and non-enzymatic antioxidants. Scavenging mechanisms for ROS involve these enzymes : SOD, CAT, APX, MDHAR, DHAR, glutathione peroxidase (GPX), and GR.

H_2O_2 can be directly decomposed through CAT. On the other hand, by combining APX with GR, H_2O_2 can also be removed via recurrent oxidation-reduction reactions promoted by GSH, hence preventing cell damage.

Non-enzymatic antioxidants include pigments, vitamin E, tocopherol, AsA, and GSH. Among these, GSH is most important. Glutathione is a common cellular component. Its biological functions include : (1) adjusting gene performance; (2) acting as a precursor in the synthesis of phytochelatin; (3) serving as a substrate of glutathione S-transferase (GST), and thus assisting in clearing the harmful components in the cell; (4) relating to cell cycle control; and (5) playing the key role in the AsA-GSH cycle. This last one is generally considered the most important biological function of GSH. Therefore, elevating the total quantity of GSH within a plant becomes the critical step in reducing stress injury.

The fluctuation of the GSH level also could be a good indicator of the redox status within plant cells. Therefore, it is interesting to monitor this fluctuation in stressed plant cells. For example it was observed that mung bean seedlings receiving various stresses—chilling, ozone, and heat shock treatments—showed 2.8, 3.9, and 1.9-fold increases, respectively, in transient GSH level over the level in the control plants. Salt stress and flooding treatments produced moderate GSH increases of 36% and 60%, respectively. Interestingly, the GSH content of drought stressed mung bean plants dropped to 52% of the control level. Using buthionine sulfoximine (BSO) to inhibit reduced GSH synthesis, our previous studies also demonstrated that GSH is the "key" molecule in the acclimation of mung beans to chilling stress. Thus, the gene coding for gamma-glutamylcysteine synthetase (g-ECS) (the rate limiting enzyme in the synthesis of GSH) must be a critical downstream target gene of H_2O_2 triggered signal transduction.

Together, these defensive responses imply that one consequence of many stresses is an increase in the concentration of ROS. ROS are generated via electron transport reactions both in mitochondria and chloroplasts and are speedily converted to H_2O_2. In cells, H_2O_2 is also generated via several enzyme-mediated reactions, such as glycollate oxidase during photorespiration, acyl CoA oxidase during *b*-oxidation, oxalate oxidase that converts oxalate and oxygen to H_2O_2 and CO_2, and NADPH oxidase (a plasma membrane located enzyme) that transfers electrons from NADPH to molecular oxygen to generate O_2^- (subsequently converted to H_2O_2). The effects of various stresses on H_2O_2 generation have not yet been completely revealed, nor is it known which if any of these enzymes is involved in stress-induced H_2O_2 synthesis.

Glutathione was shown to regulate the redox status of cells. Disturbance of cellular redox homeostasis may activate transcription factors and gene expression of those involved in the protection of plants against damage during chilling and cold acclimation. This may be explained by the fact that the stress-induced oxidized state of the cell may lead to the disulfide bridge formation in proteins, including transcription factors. The modification of proteins may lead to conformational change and consequently activate transcription factors, which will in turn control the expression and accumulation of certain antioxidants, as described for *E. coli*.

H_2O_2 AS A SIGNALING MOLECULE

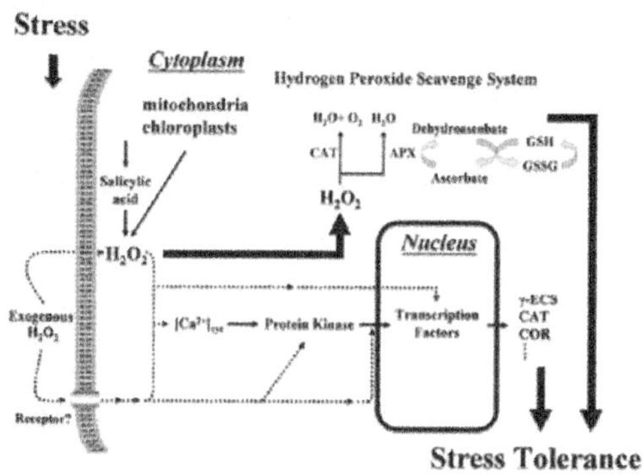

Figure 25.1.

ROS are extremely reactive molecules that have high affinity to membranes, DNA, or proteins in plant cells. It is not surprising that until recently ROS were still viewed mainly as toxic cellular metabolites in plants. However, increasing lines of evidence support the idea that ROS also function as signaling molecules that mediate responses to various stimuli. Among ROS, H_2O_2 seems best suited to play the role of signaling molecule due to its higher stability and longer half-life. If H_2O_2 serves as a stress signal, the fluctuation of H_2O_2 level in plants should spatially and temporally reflect changes in the environment. Indeed, an oxidative burst is a common response to both biotic and abiotic stresses. These stresses include pathogens, elicitors, wounding, heat, low temperature, ultra-violet (UV) light, and ozone. For example, in winter wheat (*Triticum aestivum* L.) leaf, chilling at 4°C caused the H_2O_2 level to increase to threefold that of control within 1 min. Similarly, ROS levels shot up during chilling of non-acclimated maize seedlings.

It is further reported that maize seedlings pretreated with H_2O_2 or menadione, an $O_2^{\cdot-}$-generating compound, acquire additional chilling tolerance as compared with control plants. These findings suggest that chilling-induced ROS appears as a message to induce antioxidant systems in cells. In addition to plants, H_2O_2 has been also demonstrated to act as a secondary messenger in mammalian cells, mediated through the activation of transcription activators, such as nuclear factor kB (NF-kB) and AP-1 (activator protein-1; a complex composed of *jun* and *fos* gene products). Thus H_2O_2 seems to serve as a common stress signal in organisms.

For H_2O_2 to really be a specific signaling molecule, a mechanism must exist to perceive the elevation of H_2O_2 in cells. H_2O_2 can interact with cysteine residues within proteins. This redox modulation of protein could potentially alter protein conformation, affecting protein activity, and therefore initiating subsequent cellular responses.

The yeast *Saccharomyces cerevisiae* has been a useful model for exploring the eukaryotic response to oxidative stress. The modifications of thiol residues leading to conformational change of protein by H_2O_2 have been demonstrated in vitro for the yeast transcription factor YAP1. Coleman *et. al.* have shown that controlling the activity of the positive transcriptional regulator YAP1 is a key step in inducing normal tolerance of oxidative stress elicited by the redox-active agents diamide and H_2O_2. A cysteine-rich domain of YAP1 was required for H_2O_2 resistance, which influenced the nuclear localization of YAP1 when it was activated. Two cysteines in this domain are essential for YAP1 oxidation.

Information on how higher organisms perceive the change of the cellular redox state is still limited. Our previous studies showed that H_2O_2 applied to the abaxial surface of a *Phalaenopsis* leaf increased chilling tolerance effectively; however, adaxial surface application was not effective. Although H_2O_2 can freely diffuse across membranes, no changes in the leaves' endogenous H_2O_2 levels were observed after exogenous application of H_2O_2. These observations suggest that receptors on the plant surface, interacting with H_2O_2

molecules, might initiate the signal for development of chilling tolerance. Alternatively, the difference between the abaxial and adaxial treatment effects might have been due to the easier access of H_2O_2 molecules to receptors located in the mesophyll through the abaxial stomata. In either case, the receptor seems to be isolated from the major pool of endogenous H_2O_2.

H_2O_2 was also clearly demonstrated to play a prominent role in the transduction of ABA signals in Arabidopsis by regulating the activity of phosphatase. In this pathway, Arabidopsis protein phosphatase 2C (PP2C) enzymes ABI1 and ABI2 serve as direct targets for H_2O_2 modification of cysteine residues in vitro. ABI1 and ABI2 are type 2C serine/threonine protein-phosphatases that are negative regulators of ABA signaling. Upon H_2O_2 challenge in vitro, ABI1 and ABI2 are inactivated rapidly, with an IC_{50} value of 140 mM for ABI1 and 50 mM for ABI2 in the presence of physiological concentrations of GSH. Based on these findings, ABI1 and ABI2 represent a likely receptor for the H_2O_2 signal in higher plants.

It is reported that, in animals, H_2O_2 activates the transcription factor NF-kB. NF-kB usually resides in the cytoplasm of the cell in association with an inhibitor, Ik-B. Addition of H_2O_2 to cells results in the dissociation of NF-kB from Ik-B and translocation of activated NF-kB into the nucleus, thus triggering related gene expression. Other transcriptions affected by exogenous H_2O_2 include AP-1, Myb, and Ets.

ROLE OF CA^{2+} IN H_2O_2 SIGNALING PATHWAY

Cold signal transduction can be defined as a process that starts with the perception of low temperature and ends with the expression of genes encoding proteins that lead to increased chilling tolerance of plants. Ca^{2+} provides favourable features as a secondary messenger and serves as one of the major components in the cellular signaling pathways of many organisms. Cytosolic Ca^{2+} signals can be viewed as the result of two opposing factors : influx into and efflux out of the cytosol. Under ambient conditions, delicate control is required to maintain homeostasis; and upon instances of stress, monitoring the fluctuation of cytosolic Ca^{2+} concentration, $[Ca^{2+}_{cyt}]$, is necessary for interpreting the signal. Many studies have revealed that transient $[Ca^{2+}_{cyt}]$ fluctuation plays an essential signaling role in many stress responses. Nevertheless, how various stress signals are mediated through Ca^{2+} to subsequently affect the expression of different downstream genes is still unclear. At least we know that many factors—including duration, frequency, and location of the Ca^{2+} signal, as well as interactions with other cellular components—affect the downstream response.

Cold shock evoked transient increases in $[Ca^{2+}_{cyt}]$ level in tobacco and Arabidopsis (Knight *et. al.*, 1991; Knight *et. al.*, 1996; Polisensky and Braam, 1996; Lewis *et. al.*, 1997). These findings suggested that calcium influx plays a major role in the cold shock response and also that an intra-cellular calcium source might be involved. Further evidence indicated that H_2O_2 and Ca^{2+} were both involved in a signaling cascade leading to the closure of stomata in Arabidopsis. In this study, H_2O_2-activated Ca^{2+} channels mediated both the influx of Ca^{2+} in protoplasts and increases in $[Ca^{2+}_{cyt}]$ in intact guard cells. In addition to serving as a link in a signaling cascade, the fluctuation of $[Ca^{2+}_{cyt}]$ could be one of the mechanisms that lead plants to remember what they have suffered. This inference comes from the observation that Arabidopsis treated with either sublethal cold or H_2O_2 modifies its calcium signature in response to subsequent cold stress as compared to an untreated control. Based on these findings, it is proposed that a Ca^{2+} channel may be employed as a possible putative cold sensor in higher plants.

More lines of evidence concerning the relationship between H_2O_2 and Ca^{2+} signals were provided by the study of H_2O_2 homeostasis in Arabidopsis. Yang and Poovaiah indicated that a close interaction exists between intracellular H_2O_2 and cytosolic Ca^{2+} in response to biotic and abiotic stresses. This study indicated that an increase in cytosolic Ca^{2+} boosted the generation of H_2O_2.

In summary, treatment with H_2O_2 could activate Ca^{2+} channels to elevate $[Ca^{2+}_{cyt}]$ level. For a certain period of time, this could be autocatalytic, because H_2O_2 production in the oxidative burst requires a continuous Ca^{2+} influx, which activates the plasma membrane-localized NADPH oxidase. Later, the $[Ca^{2+}_{cyt}]$ elevation activates the calcium sensor calmodulin and subsequently passes the signal to a downstream target "CAT." This finally down-regulates H_2O_2 levels by stimulating the plant CAT activity. Together, these results provide evidence indicating that Ca^{2+} has dual functions in regulating H_2O_2 homeostasis, which in turn influences redox signaling in response to environmental signals in plants.

THE ROLE OF KINASES AND PHOSPHATASES IN THE H_2O_2 SIGNALING PATHWAY

Reversible protein phosphorylation and dephosphorylation, catalysed by protein kinases and protein phosphatases (PTPs), respectively, is a common molecular mechanism that passes on information within cells. Substrate specificity for protein phosphatases can be classified into two major groups, serine/threonine phosphatase and tyrosine phosphatase. Ca^{2+}-regulated protein phosphatase plays a role in cold signal transduction. The expression of Ca^{2+} binding calcineurin-B-like protein (AtCBL1), an effector of calcium signaling, is highly up-regulated by cold, drought and salt stresses in Arabidopsis. Notably, the expression of receptor-like PTPase, AtPTP1, is transiently down-regulated by cold treatment.

Studies of protein kinases have shown that mitogen-activated protein kinases (MAPKs) are activated by H_2O_2 in both animals and plants, which could lead to the modulation of gene expression. Whether H_2O_2 has a direct effect on MAPK or activates upstream effectors is unclear.

In addition to activating MAPK, H_2O_2 inhibits phosphatase activities, probably by the direct oxidation of cysteine in the active site of these enzymes. Hence, instead of positively transmitting a Ca^{2+} signal, some types of protein phosphatase such as PP2C act as a negative regulator of ABA signaling. The Arabidopsis with null mutations in *abi1* and *abi2* (genes coding for a PP2C) loses its susceptibility to ABA and shows a diminishing of ABA-induced $[Ca^{2+}_{cyt}]$ elevation. The *abi1* mutation has been shown to diminish cold-inducible gene expression, affecting the development of cold-induced freezing tolerance in *abi1*-knockout Arabidopsis. These observations could explain at least part of the role of H_2O_2 in the ABA-triggered signal transduction pathway. However, the function of PTPase and its interaction with H_2O_2 in cold signal transduction remain to be elucidated.

Besides PTPase, the signal transduction in many eukaryotic cells is also controlled by protein phosphorylation by MAPKs. In the MAPK cascade, mitogen-activated protein kinase kinase kinases (MAPKKKs) are the primary signal receiver, activated via phosphorylation by mitogen-activated protein kinase kinases (MAPKKs). In recent years various groups have shown that H_2O_2 is able to activate the MAPK cascade in plants and thus provide the linkage between an upstream H_2O_2 signal and downstream gene expression. Activation of MAPKs by H_2O_2 is a critical step in mediating cellular responses to multiple stresses. The ANP class of MAPKKKs from Arabidopsis can be induced specifically by H_2O_2, and this then activates a specific class of stress-induced MAPKs. These MAPKs activated by ANP1 include AtMPK3 and AtMPK6 of Arabidopsis. By sequence comparison, Arabidopsis ANP1 is classified as a *Nicotiana* protein kinase-1 (NPK-1) like protein kinase. Transgenic plants over-expressing ANP1 show extra-tolerance to heat shock, freezing, and salt stress. Recently, a related study found that Arabidopsis nucleoside diphosphate kinase 2 (NDPK2) is an important upstream signaling component of the AtMPK3 and AtMPK6-mediated signaling cascade, which fortifies plants against cold, salt, and oxidative stress. This phenomenon seems to occur because the expression of AtNDPK2 antagonizes the elevation of cytosolic H_2O_2 induced by stresses. Together, all of these data suggest that various stresses induce a generation of cytosolic H_2O_2, which in turn induces a MAPK cascade that subsequently stimulates expression of antioxidant genes, reducing H_2O_2 levels and restoring cellular homeostasis.

TRANSCRIPTIONAL REGULATION OF GENE EXPRESSION IN RESPONSE TO H_2O_2

H_2O_2 modulates the expression of various genes, including those encoding antioxidant enzymes and modulators of H_2O_2 production. However, accessing the enzymes involved in the oxidative scavenging system can only provide pieces of a jigsaw puzzle, and it is difficult to visualize the whole picture of how plants respond to various environmental stimuli. Using cDNA micro-array technology, Desikan *et. al.* have undertaken a large-scale analysis of the Arabidopsis transcriptome during oxidative stress. Of the transcripts, 113 were induced and 62 were repressed by H_2O_2. A comparison of the micro-array results of oxidative stress with other stresses such as wilting, UV irradiation, and elicitor challenge showed overlapping expression of some of these genes. The H_2O_2-induced transcripts encoded proteins with functions such as metabolism, energy, protein destination and transport, cellular organization and biogenesis, cell rescue or defense, and transcription. Among these genes, the genes encoding potential transcription factors should be emphasized due to their capacity for activating the expression of downstream target genes. In Arabidopsis, a transcription factor CBF1 (C-repeat binding factor), binding to C-repeat and dehydration-responsive element (CRT/DRE), was first identified by Stockinger *et. al.*. CBF1

Belongs to the APETALA2/EREBP-family of transcription factors. Five additional CBF1 homologous genes, namely DRE binding proteins (DREBs) and two additional CBFs, called CBF2 and CBF3, were subsequently cloned from Arabidopsis. Tomato seedlings (*Lycopersicon esculentum*) over-expressing an Arabidopsis *CBF1* cDNA showed a higher tolerance for water deficit stress. Subtractive hybridization showed that CAT activity increased and H_2O_2 concentration decreased in the transgenic tomato plants compared with the wild-type plants with or without water deficit stress. According to this result, CBF seems to be able to indirectly regulate the cytosolic redox status by activating downstream genes encoding antioxidant enzymes, for example CAT. Nevertheless, there still is no evidence connecting the H_2O_2 signal with CBF accumulation or activation.

The strongest evidence showing H_2O_2 itself activates transcription factor is provided by the study on yeast transcription factor YAP1. Recently, using a functional selection strategy, Arabidopsis cDNAs were overexpressed in the fission yeast *Schizoaccharomyces prombe,* and the transformants were screened for an enhanced ability to tolerate diamide-induced oxidative stress. The gene selected encodes a transcription factor named OXS2, which is related to oxidative stress tolerance in plants. This gene belongs to a five-member family of C_3H zinc finger protein genes. OXS2 protein can auto-activate its own promoter, and this correlates with the elevated accumulation of *OXS2* mRNA during oxidative stress. The regulation mechanism activating OXS2 is similar to that of YAP1. The inactivation of Exportin-mediated export of OXS2 and hence nuclear accumulation of OXS2, and the auto-activation of its own gene, are all caused by oxidative stress and followed by the activation of a stress-induced pathway. Apart from this study, the direct influence of H_2O_2 on transcription factor activity is, to date, still far from clear.

CONCLUSIONS

Currently, research data show that H_2O_2 can play a dual role in cells. During oxidative stress, H_2O_2 is a strong toxic oxidant causing cell damage or even cell death. At the same time it serves conversely as a signaling molecule to activate a rescue/defense system for restoring the redox homeostasis in plant cells. In addition, H_2O_2 is involved in mediating biological processes, including PCD, ABA-mediated stomatal closure, auxin-regulated gravitropic responses, mechanical wounding response, systematically acquired resistance (SAR), plant pathogen interaction, and some other processes relating to both biotic and abiotic stresses.

So far it is still unclear whether H_2O_2 is situated at a common center for the signaling pathways providing responses to various environmental stimuli. Extensive work has indicated that H_2O_2 indeed plays a role in diverse cellular signals, especially signals triggered by stresses. These observations suggest the existence of "cross-talk" between different stresses. Hence, elevation of H_2O_2 concentration in cells is a physiological response to many stimuli and results in activation of downstream target genes. However, the same target gene induced by different stimuli always showed a dissimilar expression profile, an observation that needs to be clarified. Based on current data, a model for an H_2O_2 signaling pathway is proposed here. An H_2O_2 signal may be perceived by a receptor and then result in elevated $[Ca^{2+}{}_{cyt}]$. Next, Ca^{2+} may activate a signaling protein such as a protein kinase or phosphatase to trigger a cascade that then alters the activity of a transcription factor by phosphorylation or dephosphorylation. In addition, H_2O_2 may activate transcription by directly oxidising H_2O_2-responsive transcription factors via oxidation of thiols of cysteine residues in protein. In either case, having moved into the nucleus, the modified (activated) transcription factor interacts with its corresponding *cis*-acting element on target promoters to regulate gene expression.

In the future, developments in post-genomic technologies will clarify the biological significance of H_2O_2 and provide more clues to how plants sense and respond to environmentally adverse conditions.

ABSCISIC ACID : A SEED MATURATION AND STRESS-RESPONSE HORMONE

THE STRUCTURE OF LUNULARIC ACID FROM LIVERWORTS

It was at first believed that ABA was present in all plants except the liverworts, and that the inhibitory compound lunularic acid served a function similar to ABA in the bryophytes. However, with improved detection techniques it was possible to show that ABA is also present in liverworts at the same concentrations as in other plants.

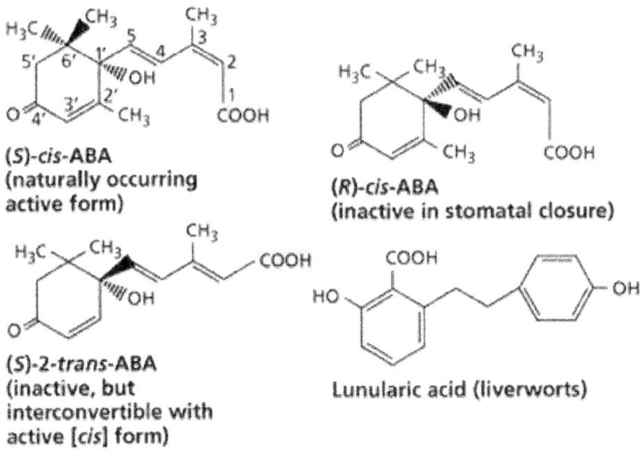

Figure 25.1. A The chemical structures of the *S* (counterclockwise array) and *R* forms of *cis*-ABA, the (*S*)-2-*trans* form of ABA, and lunularic acid. The numbers in the diagram of (*S*)-*cis*-ABA identify the carbon atoms.

ABA MAY BE AN ANCIENT STRESS SIGNAL

Although long considered a plant-specific signal, several recent reports have documented a role for ABA in regulating stress responses in animals ranging from sea sponges to mammals. ABA was first detected in animals over twenty years ago, where it was found to accumulate to the highest levels (~15-20 nM) in mammalian brain tissue. Despite evidence indicating that it was unlikely that the ABA could be coming from the animals' diets, no function for the ABA was proposed or tested.

Subsequent studies in diverse organisms (cyanobacteria, unicellular parasites, sponges, hydroids, and mammals) have suggested that ABA may serve as a "universal Ca^{2+} agonist" in a variety of stress responses. In cyanobacteria, ABA is produced in response to salt stress and promotes formation of heterocysts, thick-walled cells specialized for nitrogen fixation. The protozoan parasite *Toxoplasma gondii* uses ABA to control Ca^{2+} release, which stimulates egress and therefore transmission of the parasite from its host.

Over the past decade, Zocchi and coworkers have worked their way up the evolutionary tree, investigating ABA responses from sponges to mammals. Sponges respond to a high temperature stress by an ABA-induced transient increase in water filtration, mediated by cADPR-stimulated release of intra-cellular Ca^{2+}. Hydroids are capable of regenerating entire sections of their bodies following damage; this process is stimulated by light, in an ABA-dependent process that is again mediated by cADPR-stimulated increases in Ca^{2+}. In humans, ABA signaling has been demonstrated in pathogen response, where it acts as a pro-inflammatory cytokine to stimulate phagocytosis, cell migration, ROS, and NO production. ABA also mediates response to high glucose levels in pancreatic β-cells, resulting in insulin secretion, and promotes stem cell proliferation and mobility in response to specific growth factors and inflammatory cytokines. In all of these examples, ABA is produced and active at nanomolar to sub-micromolar levels. Also similar to ABA signaling in plants is the implication of G-protein involvement and evidence of binding at the cell surface, but the specific receptors involved in animal recognition of ABA have not yet been identified.

However, the observations that ABA production by diverse organisms is sensitive to common inhibitors of carotenoid synthesis, and that ABA signaling is mediated by mechanisms involving G-protein complexes, cADPR activation of Ca^{2+} release, and in some cases ROS and NO production suggest that this is an evolutionarily ancient signal.

STRUCTURAL REQUIREMENTS FOR BIOLOGICAL ACTIVITY OF ABSCISIC ACID

The features shown to be essential for biological activity include the carboxyl group, the tertiary hydroxyl group, the 2-*cis*, 4-*trans*-pentadienoic side chain, a 42 -ketone, and a double bond in the cyclohexane ring. Catabolic products of ABA that are present in the tissue and are missing any of these groups are biologically inactive.

THE BIOASSAY OF ABA

Coleoptile growth, the classic bioassay devised for auxins, is also used for ABA detection in plant extracts by measurement of coleoptile growth inhibition. This bioassay has adequate sensitivity (minimum detectable level is 10^{-7} *M*) and shows a linear response in the range of 10^{-7} to 10^{-5} *M*, but it has some disadvantages. Plant extracts have other promoters and inhibitors, which decrease the specificity of the bioassay and make preliminary purification steps necessary.

Other available bioassays for ABA include inhibition of germination, inhibition of gibberellic acid- induced α-amylase synthesis in cereal aleurone layers, induction of leaf abscission, and stomatal closure. Stomatal closure is highly specific for ABA because it is affected little by other plant growth regulators. Additional advantages of this bioassay include a fast response of guard cells to ABA, high sensitivity (minimum detectable level 10^{-9} *M*), and a linear response over a wide range of concentrations.

EVIDENCE FOR BOTH EXTRA-CELLULAR AND INTRA-CELLULAR ABA RECEPTORS

Evidence for an extra-cellular ABA receptor is provided by studies showing that ABA-protein conjugates that are unable to cross the membrane can activate both ion channel activity and gene expression.

The presence of an extra-cellular receptor is also suggested by the failure of micro-injected ABA to alter stomatal opening in the spiderwort *Commelina*, or to inhibit GA-induced α-amylase synthesis in barley aleurone protoplasts.

In addition, multiple small patches of possible ABA receptors were visualized on guard cell protoplast surfaces by treating the protoplasts with biotinylated ABA, then with fluorescent avidin, a protein that binds to biotin, providing histochemical evidence for extra-cellular receptors.

Other experiments, however, support an intra-cellular location for the ABA receptor :

- Direct delivery of ABA to the cytosol via a "patch pipette" inhibited inward K^+ channels, which are required for stomatal opening.
- Similarly, micro-injection of an inactive, molecularly caged form of ABA into guard cells of *Commelina* resulted in stomatal closure after the stomata were treated briefly with UV irradiation to activate the hormone; that is, release it from its molecular cage.

Furthermore, extra-cellular application of ABA was nearly twice as effective at inhibiting stomatal opening at pH 6.15, when it is fully protonated and readily taken up by guard cells, as it was at pH 8, when it is largely dissociated to the anionic form that does not readily cross membranes.

(A)

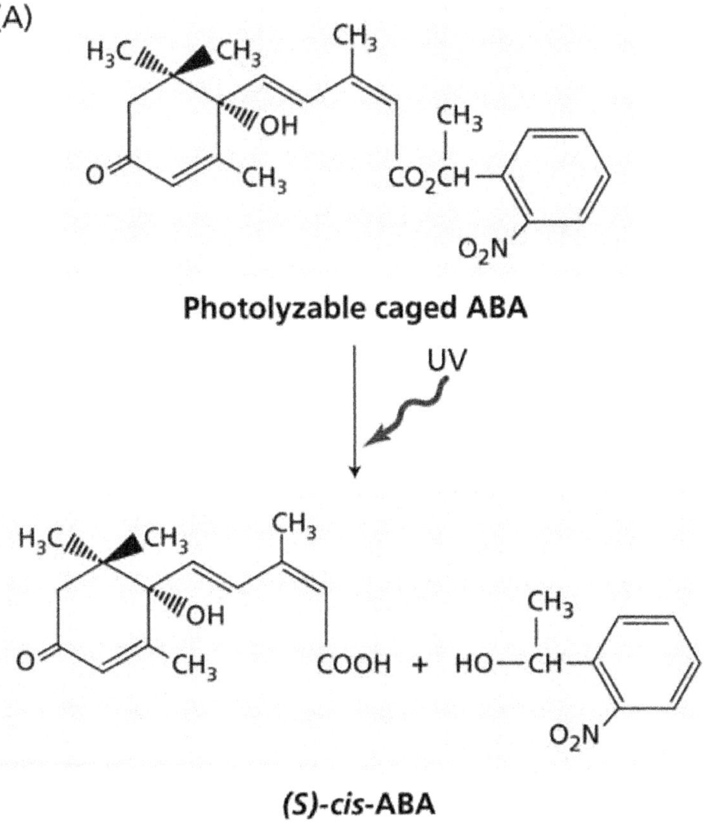

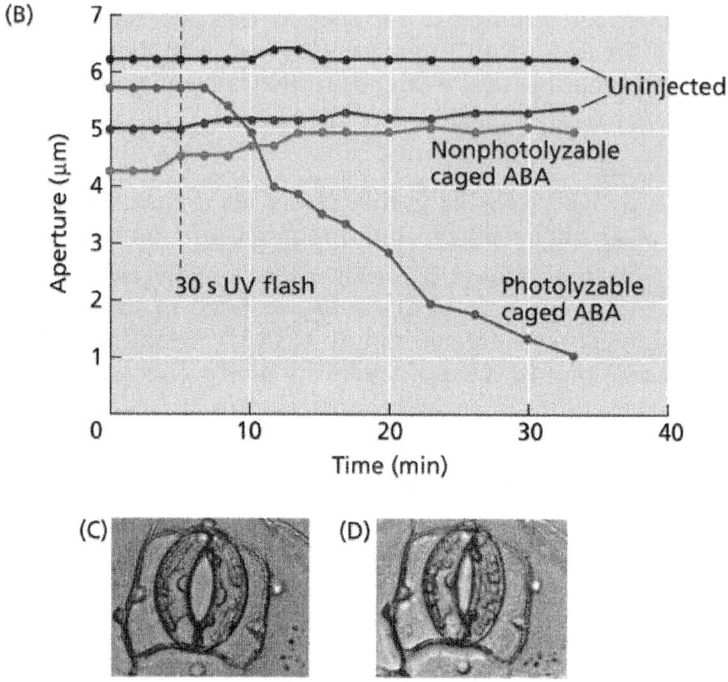

Figure 25.3 : Stomatal closure induced by UV photolysis of caged ABA in the guard cell cytoplasm implies intra-cellular recognition of ABA. Single guard cells in stomatal complexes of *Commelina* were micro-injected with caged ABA. (A) Photolysis reaction induced by UV irradiation. (B) The stomatal apertures recorded before and after a 30-second exposure of the cells to UV. (C, D) Light micrographs of the same stomatal complex in which the right-hand guard cell was loaded with the photolyzable cages ABA 10 minutes before UV photolysis (C) and 30 minutes after photolysis (D). (A and B from Allan *et. al.* 1994; C and D courtesy of A. Allan, from Allan *et. al.* 1994; © American Society of Plant Biologists, reprinted with permission.)

THE EXISTENCE OF G PROTEIN-COUPLED ABA RECEPTORS IS STILL UNRESOLVED

A classic signal transduction pathway of animal cells begins with ligand binding to a receptor on the plasma membrane, leading to the activation of a heterotrimeric GTP-binding protein (G protein), composed of α, β, and γ sub-units. The Arabidopsis genome encodes single copies of the genes for the Gα and Gβ sub-units, and two copies of the gene for the Gγ sub-unit. Mutations in the Gα and Gβ sub-units result in defective ABA response, but this is seen as *decreased* response in stomatal regulation yet *hypersensitivity* in seed and seedling growth. This indicates that G-protein signaling may either promote or inhibit a given ABA response, depending on the cell and tissue type.

Because both pharmacological and genetic evidence have implicated G-proteins in mediating ABA response, researchers have sought to identify an appropriate G-protein coupled receptor (GPCR) to link ABA to these responses. The first two candidates, the Arabidopsis genes *G-protein coupled receptor (GCR)1* and *GCR2*, encode proteins with significant sequence similarity to non-plant GPCRs, but seem unlikely to be ABA receptors. Although GCR1 knockout mutants exhibit *hypersensitivity* to ABA, indicating that the protein functions as an inhibitor or negative regulator of Gα-mediated ABA responses, the ligand for GCR1 is unknown. On the other hand, GCR2 was initially reported to bind ABA, interact with the Gα sub-unit and mediate multiple ABA responses, but subsequent reports have contradicted both its role in ABA signaling and its identity as a GPCR. Thus, although these GCRs may function in ABA response, they are probably not receptors.

The GPCR-type G (GTG) proteins, were identifed by an alternate bio-informatic approach searching for distantly related GPCRs. A novel feature of these GPCR-related proteins is their intrinsic GTPase activity. Functional characterization of mutant and overexpression lines demonstrated redundancy of the two GTG proteins, and a fairly mild effect on ABA responses such as inhibition of both germination and root growth. Preparations of recombinant GTG proteins specifically bind the naturally occuring (+)-ABA enantiomer, and this binding is more efficient in the presence of GDP than GTP.

Both GTG proteins are localized at the plasma membrane and interact with GPA1, the sole Gα sub-unit in Arabidopsis, but it is not clear whether they bind intra-cellular or extra-cellular ABA. The model proposed from these results is that GPA1 modulates the GTPase activity of the GTGs, thereby altering their ability to bind ABA and initiate a signaling cascade via unknown intermediates. This mechanism is unusual in that the GDP-bound form is more active than the GTP-bound GTGs, and the classic relationship between GPCR and G-protein is reversed : the G-protein regulates the receptors, not *vice versa*.

THE YEAST TWO-HYBRID SYSTEM

Many cellular proteins interact with a variety of other proteins to perform their physiological functions. Identification of interacting proteins can be accomplished by bio-chemical assays, such as co-purification of a complex, but these require relatively stable interactions and the ability to identify components of the complex immunologically or based on amino acid sequence and composition. Alternatively, the two-hybrid assay is a genetic approach that takes advantage of the modular nature of proteins to identify interacting domains. Although there are several variations, in all cases two fusion proteins are co-expressed in a cell line that will activate expression of a reporter gene if the fusion proteins interact. For example, the yeast transcription factor GAL4 can be separated into distinct DNA binding (BD) and transcription activation (AD) domains. If hybrid genes are constructed such that they encode fusion proteins BD-protein X and AD-protein Y, interactions between proteins X and Y will bring the GAL4 domains close enough to reconstitute a functional transcription factor that can trans-activate a GAL4-responsive promoter. The cell lines often contain several GAL4-responsive promoter-reporter fusions, whose activity can be scored either qualitatively or quantitatively. Qualitative scoring (+/- growth) makes it possible to screen a cDNA library for fusions encoding interacting proteins by requiring expression of a gene whose activity can complement an auxotrophy. Quantitative scoring (*e.g.* β-galactosidase activity from a *lacZ* reporter) permits comparisons of the relative strength of interactions among a variety of protein pairs.

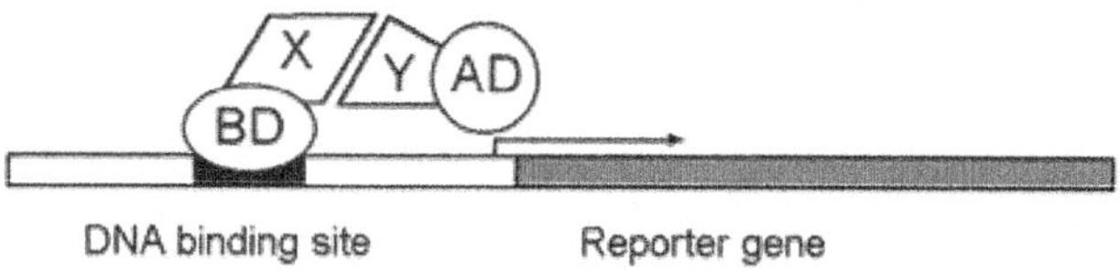

Figure 25.5 : Two fusion proteins containing the GAL4 DNA binding (BD) and transcription activation (AD) domains can reconstitute a functional transcription factor that can trans-activate a GAL4-responsive promoter if the interactions between the two proteins, X and Y, will bring the GAL4 domains close enough together.

YELLOW CAMELEON : A NONINVASIVE TOOL FOR MEASURING INTRA-CELLULAR CALCIUM

Cytoplasmic free calcium ($[Ca^{2+}]_{cyt}$) acts as a secondary messenger in a variety of signaling pathways and it appears that the frequency, magnitude and localization of $[Ca^{2+}]_{cyt}$ oscillations are important determinants of specificity. "Traditional" techniques for measuring Ca^{2+} have made use of micro-injected Ca^{2+} sensitive indicator dyes, but these methods are technically difficult and invasive. An alternative strategy makes use of Ca^{2+} binding proteins that can be transgenically expressed throughout an organism, eliminating the need for micro-injection. The Ca^{2+} indicator yellow cameleon is a fusion protein composed of four domains : two modified green fluorescent proteins (GFPs) linked via calmodulin (CaM) and a CaM-binding peptide ("M13"). The GFPs are modified to shift their emission wavelengths toward blue (enhanced cyan fluorescent protein, ECFP) or yellow (enhanced yellow fluorescent protein, EYFP). When the CaM domain binds to $[Ca^{2+}]_{cyt}$ it associates with the CaM-binding peptide, producing a conformational change that brings the modified GFP domains closer together. Following exposure to a laser, excitation energy from ECFP (the higher-energy fluorescent protein) is passed to EYFP resulting in increased emission by this lower-energy fluorescent protein. Consequently, increased $[Ca^{2+}]_{cyt}$ results in increased emission at 535 nm relative to that at 480 nm, and the 535/480 emission ratio is proportional to the $[Ca^{2+}]_{cyt}$.

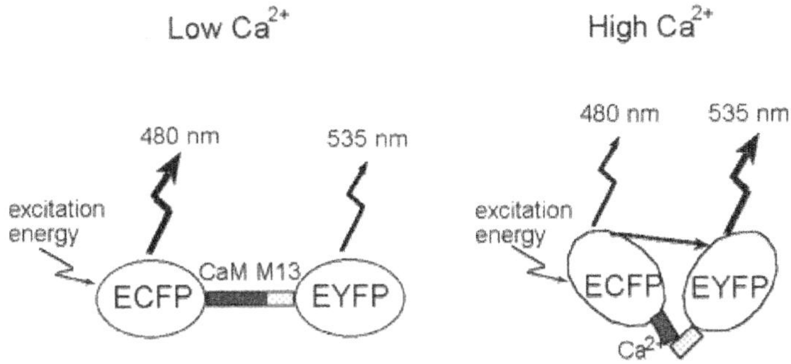

Figure 25.5 : The binding of calcium brings the two ends of the yellow cameleon protein together, altering the ratio of 535 nm to 480 nm emission.

PHOSPHATIDIC ACID MAY STIMULATE SPHINGOSINE-1-PHOSPHATE PRODUCTION

Phosphatidic acid may stimulate sphingosine-1-phosphate production, thereby promoting G protein-mediated ABA signaling leading to inhibition of stomatal opening. PLD also binds directly to the Gα sub-unit of heterotrimeric G protein, thereby regulating stomatal opening. PA catabolism by lipid phosphate phosphatase (LPP) results in DAG production, but mutants with reduced LPP activity over-accumulate PA and display ABA hypersensitivity at germination. Finally, PLD and PA also appear to mediate both the production of and the response to reactive oxygen species and phosphoinositides.

THE ABA SIGNAL TRANSDUCTION PATHWAY INCLUDES SEVERAL PROTEIN KINASES

In addition to the SnRK2 family of stress-activated protein kinases, specific Ca^{2+}-dependent protein kinases (abbreviated as CPKs, CDPKs or CDKs) have been implicated in the ABA regulation of both stomatal aperture and seedling gene expression. The calcium-dependent kinases are another branch of the protein superfamily containing the SnRKs. Genetic studies have shown that some CPKs function redundantly in guard cells, affecting ABA regulation of Ca^{2+} channels and Ca^{2+} activation of one class of anion channels. Similar to the SnRK2s, other sets of CPKs act redundantly in multiple tissues and developmental stages and can phosphorylate members of the AREB/ABF family transcription factors *in vitro*. Although some CPK expression is stress-induced, most does not require ABA-induction, but depends on post-translational activation of these kinases.

The calcineurin B-like interacting protein kinases (CIPKs) are another branch of the SnRK superfamily of protein kinases that are activated by interaction with specific Ca^{2+}-binding calcineurin B–like proteins, CBLs. Some of the CBLs are membrane associated, which could result in localization of the

CIPKs near their phosphorylation targets or permit sensing of localized changes in cellular Ca^{2+} levels. Subsets of the CBL and CIPK families are induced by ABA and/or stress, and genetic studies have implicated some CBL/CIPK pairs in ABA or stress signaling, but have also revealed opposing effects of specific CBLs. Furthermore, a given CBL/CIPK pair can either repress or promote distinct ABA responses, depending on the response.

The mechanistic details leading to this complexity are not yet clear, but the presence in the cell of approximately ten Ca^{2+} sensors (the CBLs) that can regulate another set of about 25 protein kinases (the CIPKs) provides plant cells with a robust network of signaling pathways for transducing the information from internal Ca^{2+} signals into physiological responses. Although most of their targets are unknown, CIPKs regulate some transcription factors involved in ABA or stress signaling, possibly indirectly, and interact directly with protein phosphatases implicated in ABA response.

Yet another group of serine/threonine-selective protein kinases implicated in ABA signaling are those named for their homology to the mitogen-activated protein kinases, initially characterized in the context of cancerous cell growth. Each of the three tiers of MAPK cascades (MAPKs, MAPKKs, and MAPKKKs) are represented by multiple family members with specific functions. Guard cell specific MAPKs are ABA-induced, and an inhibitor of MAPKK activity inhibits ABA-induced gene expression and stomatal closing. Additional MAPK signaling regulates activity of a transcription factor mediating seedling ABA response.

THE *ERA*1 AND *ABH* GENES CODE FOR NEGATIVE REGULATORS OF THE THE ABA RESPONSE

Negative regulators of the ABA response (protein phosphatases) have been identified by isolation of dominant negative mutants such as *abi1* and *abi2* that result in ABA-insensitive phenotypes.

Other negative regulators have been identified through isolation of mutants exhibiting enhanced responses to ABA, based on either physiological criteria or reporter gene expression. Mutants showing increased sensitivity to ABA during germination include *enhanced response to ABA (era), ABA hypersensitive (abh),* and *ABA hypersensitive germination (ahg)*. The *era* and *abh* mutants both confer ABA hypersensitivity in both stomatal closing and germination, making these mutants resistant to wilting and mildly drought-tolerant. These characteristics have been engineered into the crop plant canola by drought-inducible expression of an *ERA1* antisense gene, resulting in significantly higher seed yields under drought conditions.

The *ERA1* gene encodes a sub-unit of the enzyme farnesyl transferase. Farnesyl transferases catalyze attachment of the 15-C isoprenoid intermediate farnesyl diphosphate to proteins that contain a specific signal sequence of amino acids. Many proteins that have been shown to participate in signal transduction are farnesylated, and *era1* mutants have pleiotropic defects in meristem function and pathogen response as well as ABA response. Farnesylation is a lipid modification and farnesylated proteins are anchored to the membrane via hydrophobic interactions between the farnesyl group and the membrane lipids. The identification of ERA1 as part of farnesyl transferase suggests that protein(s) that normally suppress the ABA response require farnesylation for anchoring to the membrane. One such protein is the monomeric GTPase ROP10, a negative regulator of ABA signaling affecting cytoskeletal organization whose plasma membrane localization is partly dependent on ERA1-mediated farnesylation.

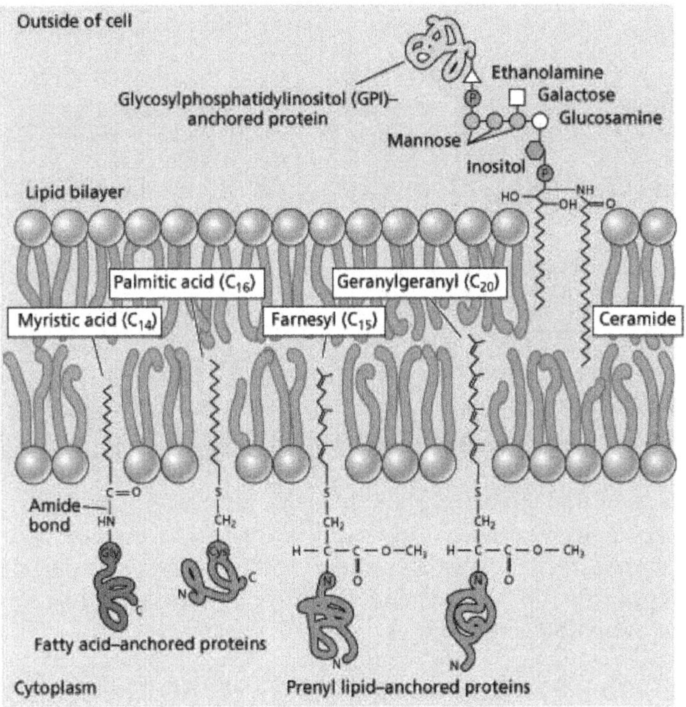

Figure 25.6 : Various anchored membrane proteins are attached to the membrane via GPI, fatty acids, and prenyl groups.

PROMOTER ELEMENTS THAT REGULATE ABA INDUCTION OF GENE EXPRESSION

The underlined DNA sequences are most highly conserved. Many of the DNA-binding proteins are members of large families, comprised of dozens (bZIP and B3 domain) to hundreds (Myb) of related proteins regulating diverse processes in plant growth. It is likely that the specificity of the response is controlled by the "context" of the conserved binding site, the specific protein(s) bound, and the other proteins present in a complex interacting with any given promoter.

Table 25.1.

Cis-acting Promoter Element Class	Consensus Sequence(s)	DNA-Binding Protein Class
Abscisic acid response elements (ABREs):		
ACGT- (G/ABRE)	GaCACGTG(G/t)C	Basic leucine zipper (bZIP) e.g. EmBP1, GBF3 and ABI5/DPBF/ABF/AREB family
Non ACGT- (C/ABRE)	CGCGTGgc	
DPBF	ACACNNG	
ABF	GNTGACGTG(G/t)C	
CE3-	RCGY(C/G)T(C/G)	
Sph/RY elements	CATGCATG	B3 domain e.g. VP1/ABI3, FUS3, LEC2
Myb recognition elements	YAAC(G/T)G	MYB e.g. AtMyb2
Myc recognition elements	CANNTG	MYC e.g. AtMyc1

REGULATORY PROTEINS IMPLICATED IN ABA-STIMULATED GENE TRANSCRIPTION

Studies aimed at identifying new ABA transcriptional regulators have shown that the protein ABI3/VP1 interacts physically with a variety of proteins, including ABI5 and its rice homolog (TRAB1). ABI5 also forms homodimers and heterodimers with other bZIP family members. At least one of these heterodimers actually represses ABA-inducible gene expression, illustrating another mechanism for post-transcriptional regulation of activity. Additional interactions have been identified between specific transcription factors and kinases, phosphatases, and 14-3-3 proteins. These latter are a class of acidic proteins that facilitate protein–protein interactions involved in signaling, transport, and enzyme activity. Another class of phosphatases regulates RNA polymerase II directly, specifically repressing transcription of ABA-inducible genes.

ABA GENE EXPRESSION CAN ALSO BE REGULATED BY mRNA PROCESSING AND STABILITY

Some mutations affecting basic aspects of RNA metabolism, such as splicing and addition of a "cap" of methylated guanosine at the 52 end, have surprisingly specific effects in stress responses, such as drought tolerance, by altering the sensitivity to ABA. For example, two mRNA 52 cap-binding proteins (CBPs) have been implicated in mRNA processing of negative regulators of ABA signaling : ABH1, which encodes a protein named CBP80 in animal systems, and CBP20. Comparison of transcript accumulation in wild-type and *abh1* plants showed a small number of mis-expressed genes in the mutant, including some encoding possible ABA signaling molecules such as an ABI-related PP2C. CBP80 and CBP20 also bind to the caps of primary micro-RNA (miRNA) transcripts and are required to complete their processing to miRNAs; mutants are hypersensitive to ABA and consequently drought tolerant.

The *SUPERSENSITIVE TO ABA AND DROUGHT 1 (SAD1)* locus encodes a member of a small nuclear ribonucleoprotein (snRNP) complex implicated in splicing, export, and degradation of RNAs. A point mutation in this locus is sufficient to confer hypersensitivity to ABA, but the specific targets of *SAD1* regulation have not been reported. At the 32 end of transcripts, processing of the poly(A) tail involves *AHG2*, which encodes a poly(A)-specific ribonuclease that appears to destabilize transcripts induced in response to ABA, abiotic stress, or salicylic acid.

MicroRNA (miRNA) is short, non-coding RNA that regulates expression of some genes. MiRNA production also depends on HYPONASTIC LEAVES 1 (HYL1), a double-stranded RNA-binding protein. The *hyl1* mutation has pleiotropic effects, including ABA hypersensitivity, that can be partially explained by altered MAP kinase signaling leading to over-accumulation of the ABI5 transcription factor. At least two small RNAs affect ABA response via effects on transcription factor levels. ABA signaling via ABI3 in stressed seedlings induces miR159 production, which promotes cleavage of transcripts for two MYB factors that promote ABA response, thereby attenuating the response. The targets of miR160 are a set of auxin response factors (ARFs); over-production of these ARFs results in ABA hypersensitivity at germination, reflecting cross-talk between auxin and ABA signaling.

ABA MAY PLAY A ROLE IN PLANT PATHOGEN RESPONSES

ABA has long been recognized as a mediator of response to abiotic stress, but it is now receiving more attention for its role in pathogen response. Several lines of evidence support this connection. Physiological studies have shown that increased ABA levels, due to stress or external application, are correlated with *increased* susceptibility to infection by some pathogens. However, the role of ABA is complex in that it also *increases* resistance to other pathogens, and mutations affecting ABA response may have opposite effects on susceptibility to different pathogens. It has been suggested that these apparent contradictions reflect differences in the mode of entry for different pathogens, which affects the timing of ABA participation in the defense response.

Attempts to define the molecular basis of ABA effects on pathogen response have used global comparisons of gene expression (transcriptomes); these studies show that nearly half the genes regulated by ABA are also regulated by certain bacterial virulence factors (Type III). This may partly reflect the importance of common secondary messengers acting in both abiotic and biotic stress. For example, Ca^{2+} and ROS regulate stomatal aperture, whether in the context of gas exchange or as a potential entry point for fungal or bacterial pathogens.

In general, ABA promotes early defense responses such as stomatal closure, may either increase or decrease callose deposition in phloem sieve elements depending on whether the pathogen is fungal or bacterial, and is more likely to inhibit late defense responses by suppressing salicylic acid-dependent responses and modulating jasmonic acid- and ethylene-dependent defenses.

PROTEINS REQUIRED FOR DESICCATION TOLERANCE

LEA protein sequences are highly conserved among a wide variety of species. They are related to two other groups of proteins, the RAB (*r*esponsive to *AB*A) and DHN (*deh*ydri*n*) proteins. These protein classes accumulate in a variety of tissues experiencing cellular dehydration, *e.g.* maturing embryos and pollen, and vegetative tissues exposed to drought stress, including extreme desiccation of "resurrection plants" (*Craterostigma plantagineum*). The LEA, RAB, and DHN proteins are all water soluble, basic, rich in glycine and lysine, and low in hydrophobic residues. As a result they are extremely hydrophilic and stable to boiling. These features led Dure and colleagues to propose that they function specifically in the protection of membranes and proteins against desiccation damage, possibly by binding water tightly or providing hydrophilic interactions in the absence of free water, and by preventing crystallization of cellular components through their ability to act as stabilizing "solvents," similar to the effects of sugars. One of these proteins confers increased osmotolerance in transgenic yeast.

THE TYPES OF COAT-IMPOSED SEED DORMANCY

1. *Prevention of water uptake.* This type of coat-imposed dormancy is common in plants found in arid and semi-arid regions, especially among legumes, such as clover (*Trifolium* spp.) and alfalfa (*Medicago* spp.). Waxy cuticles, suberized layers, and lignified sclereids all combine to restrict the penetration of water into the seed.
2. *Mechanical constraint.* The first visible sign of germination is typically the radicle (embryonic root) breaking through the seed coat. In some cases, however, the seed coat may be too rigid for the radicle to penetrate. For the seeds to germinate, the endosperm cell walls must be weakened by the production of cell wall–degrading enzymes.
3. *Interference with gas exchange.* Lowered permeability of seed coats to oxygen suggests that the seed coat inhibits germination partly by limiting the oxygen supply to the embryo.
4. *Retention of inhibitors.* The seed coat may prevent the escape of inhibitors from the seed.
5. *Inhibitor production.* Seed coats and pericarps may contain relatively high concentrations of growth inhibitors, including ABA, that can suppress germination of the embryo.

Some components of the seed coat can contribute to several aspects of dormancy. For example, pigments not only filter light, but become cross-linked to other wall constituents thereby increasing mechanical constraints and decreasing permeability to water, gases, and hormones.

TYPES OF SEED DORMANCY AND THE ROLES OF ENVIRONMENTAL FACTORS

During seed maturation, the embryo enters a quiescent phase in response to desiccation. Seed germination can be defined as the resumption of growth of the embryo of the mature seed; it depends on the same environmental conditions as vegetative growth does. Water and oxygen must be available, the temperature must be suitable, and there must be no inhibitory substances present.

However, in many cases a viable (living) seed will not germinate even though all the necessary environmental conditions for growth are satisfied. This phenomenon is termed seed dormancy. Seed dormancy introduces a temporal delay in the germination process that provides additional time for seed dispersal over greater geographical distances. It also maximizes seedling survival by preventing germination under unfavorable conditions.

Two types of seed dormancy have been recognized, coat-imposed dormancy and embryo dormancy. Coat-imposed dormancy is dormancy imposed on the embryo by the seed coat and other enclosing tissues, such as endosperm, pericarp, or extra-floral organs. The embryos of such seeds will germinate readily in the presence of water and oxygen once the seed coat and other surrounding tissues are either removed or damaged. There are five basic mechanisms of coat-imposed dormancy :

1. *Prevention of water uptake.* Prevention of water uptake by the seed coat is a common cause of seed dormancy in families found in arid and semiarid regions, especially among legumes, such as clover (*Trifolium* spp.) and alfalfa (*Medicago* spp.). Waxy cuticles, suberized layers, and lignified sclereids all combine to restrict the penetration of water into the seed.
2. *Mechanical constraint.* The first visible sign of germination is typically the radicle breaking through the seed coat. In some cases, however, the seed coat may be too rigid for the radicle to penetrate. Nuts with hard, lignified shells are examples of dormancy caused by mechanical constraint. Such shells must be broken by biotic or environmental forces for the seed to germinate. Even nonlignified tissues, such as the endosperm of lettuce seeds, can suppress expansion of the embryo. For the seeds to germinate, the endosperm cell walls must be weakened by the production of cell wall-degrading enzymes.
3. *Interference with gas exchange.* Some seed coats are considerably less permeable to oxygen than an equivalent thickness of water is— *e.g.*,

less by a factor of about 10^4 in seeds of cocklebur. This lowered permeability to oxygen suggests that the seed coat inhibits germination by limiting the oxygen supply to the embryo. In support of this idea, investigators can break the dormancy of such seeds either by making a small hole in the coat with a pin (without weakening the coat mechanically), or by treating the coat with concentrated oxygen. However, other studies suggest that the oxygen consumption of the embryos from such seeds is considerably less than the amount of oxygen able to penetrate the seed coats under normal aerobic conditions. Thus the role of oxygen impermeability in seed coat dormancy remains unresolved.

4. *Retention of inhibitors.* The seed coat may prevent the escape of inhibitors from the seed. For example, growth inhibitors readily diffuse out of isolated *Xanthium* embryos, but not from intact seeds.
5. *Inhibitor production.* Seed coats and pericarps may contain relatively high concentrations of growth inhibitors that can suppress germination of the embryo. ABA is a common germination inhibitor present in these maternal tissues. In certain cases where repeated washing (leaching) removes dormancy, the effect is thought to be due to the loss of such inhibitory compounds.

The second type of seed dormancy is embryo dormancy, which refers to a dormancy that is inherent in the embryo and is not due to any influence of the seed coat or other surrounding tissues. In some cases, embryo dormancy can be relieved by amputation of the cotyledons. Species in which the cotyledons exert an inhibitory effect include European hazel (*Corylus avellana*) and European ash (*Fraxinus excelsior*). A fascinating demonstration of the cotyledon's ability to inhibit growth is found in species (*e.g.*, peach) in which the isolated dormant embryos germinate but grow extremely slowly to form a dwarf plant. If the cotyledons are removed at an early stage of development, however, the plant abruptly shifts to normal growth.

Embryo dormancy is thought to be due to the presence of inhibitors, especially ABA, as well as the absence of growth promoters, such as GA (gibberellic acid). The loss of embryo dormancy is often associated with a sharp drop in the ratio of ABA to GA.

SEED DORMANCY MAY BE PRIMARY OR SECONDARY

Different types of seed dormancy also can be distinguished on the basis of the timing of dormancy onset rather than the cause of the dormancy.

- Seeds that are released from the plant in a dormant state are said to exhibit primary dormancy.
- Seeds that are released from the plant in a non-dormant state but which become dormant if the conditions for germination are unfavourable exhibit secondary dormancy. For example, seeds of *Avena*

sativa (oat) can become dormant in the presence of temperatures higher than the maximum for germination, whereas seeds of *Phacelia dubia* (small-flower scorpionweed) become dormant at temperatures below the minimum for germination. The mechanisms of secondary dormancy are poorly understood.

Environmental Factors Control the Release from Seed Dormancy

Various external factors release the seed from dormancy, and dormant seeds typically respond to more than one condition. Many seeds lose their dormancy when their moisture content is reduced to a certain level by drying. This method of breaking seed dormancy is called afterripening, and is usually performed in a special drying oven. On the other hand, if the seed becomes too dry (5% water content or less), the effectiveness of afterripening is diminished.

Another factor that can release seeds from dormancy is low temperature, or chilling. Many seeds require a period of cold (0 to 10°C) while in a fully hydrated (imbibed) state in order to germinate. In temperate-zone species, this requirement is of obvious survival value, since such seeds will germinate not in the fall, but only in the following spring. Chilling seeds to break their dormancy is a time-honored practice in horticulture and forestry and traditionally has been referred to as *stratification*. This term is derived from the old agricultural practice of allowing seeds with a chilling requirement to overwinter outdoors in layered mounds of moist soil. Today the seeds are simply stored in a refrigerator.

The third external factor that plays an important role in breaking seed dormancy is light. Many seeds have a light requirement for germination, which may involve only a brief exposure, as in the case of lettuce, an intermittent treatment (*e.g.*, succulents of the genus *Kalanchoe*), or even a specific photoperiod involving short or long days. Phytochrome is the main sensor for light-regulated seed germination. Interestingly, all light-requiring seeds exhibit seed coat dormancy, and removal of the outer tissues of the seed allows the embryo to germinate in the absence of light. The effect that light has on the embryo is thus to enable the radicle to penetrate the seed coat. This penetration often involves some enzymatic weakening of the enclosing tissues.

The Longevity of Seeds

How long can a seed remain dormant and still remain viable? Seed longevity is of practical importance because of ongoing efforts to preserve plant genetic resources for future agricultural crops by setting up seed gene banks. Spectacular claims of dormant seeds that remained viable for thousands of years have been made, but are considered highly controversial. Extreme examples of ancient seeds that have been reported to retain their viability include submerged lotus seeds found in a 3,000-years-old boat near Tokyo, barley seeds from the 3,000-year-old tomb of King Tutankhamen, and arctic lupine seeds associated with rodent burrows determined to be 14,000 years old. In most cases, however, a scientifically rigorous examination of the data has either disproved or raised serious doubts about the claimed antiquity of the seeds.

The most serious criticism is that the dates of the seeds have been inferred from the age of their immediate surroundings or associated artifacts rather than from direct measurements of the seeds themselves. This question raises the possibility that the seeds are much younger than the site at which they were found. In fact, microscopic examination of authenticated ancient cereal grains from various Middle Eastern sites has shown that cells and tissues of these seeds are no longer intact. Such seeds are clearly inviable.

Although the most sensational claims for seed longevity are almost certainly bogus, seeds of *Canna compacta* apparently can live for at least 600 years. Viable *Canna* seeds were obtained from inside a walnut in a tomb in Argentina. The *Canna* seeds had apparently been inserted into the immature seeds of a growing walnut fruit before the hard outer shell formed. Once the shell hardened and the nut dried out, the result was a rattle. Native people strung the rattles together to form a necklace. In this case, the seeds had to be at least as old as the walnut shell, and carbon dating of the shell indicated that it was about 600 years old.

Plant collections begun during the late eighteenth century in Europe—for example, at the British Museum in London and at the Museum of Natural History in Paris—have been reliable sources of seeds for viability determinations. *Cassia multijuga* (false sicklepod) seeds, collected in 1776, were still viable after being tested in 1934. In 1879, W. J. Beal initiated the longest-running experiment on seed longevity by burying the seeds of 21 different species in unstoppered bottles in a sandy hilltop near the Michigan Agricultural College in East Lansing. After 100 years, only one species, *Verbascum blattaria* (moth mullein), remained viable.

Why are some seeds able to survive in a dormant state for hundreds of years, whereas others lose viability in less than 5 years? Almost nothing is known about the mechanisms that determine the longevity of seeds. If these mechanisms could be understood, we might someday be able to greatly increase the seed longevity of agriculturally important species and varieties, thereby enhancing our ability to preserve plant genetic resources for generations to come.

Genetic Mapping of Dormancy : Quantitative Trait Locus (QTL) Scoring of Vegetative Dormancy Combined with a Candidate Gene Approach

Complex traits, *e.g.* dormancy, are affected by many different, often unlinked genes. Consequently, attempts to map them result in continuous, or quantitative, variation within a mapping population. Quantitative trait locus (QTL) mapping is used in these instances to map individual genes affecting the trait. In QTL mapping, inbred lines that differ in the trait of interest are crossed and an F_2 population is produced. The F_2 population has continuous variation for the trait of interest, and the goal of the analysis is to determine which chromosomal segments contain genes affecting the trait. Using marker loci along each chromosome (loci that have different alleles between the parental inbred lines), the genotypes of entire chromosomal segments in each F_2 individual can be inferred as being homozygous for one or the other parental lines or heterozygous. If the inheritance of a chromosomal segment is correlated with the trait of interest, then a gene affecting the trait is likely to be between the two marker loci. By analyzing chromosomal segments along each of the chromosomes, the number and locations of genes affecting the quantitative trait can be determined.

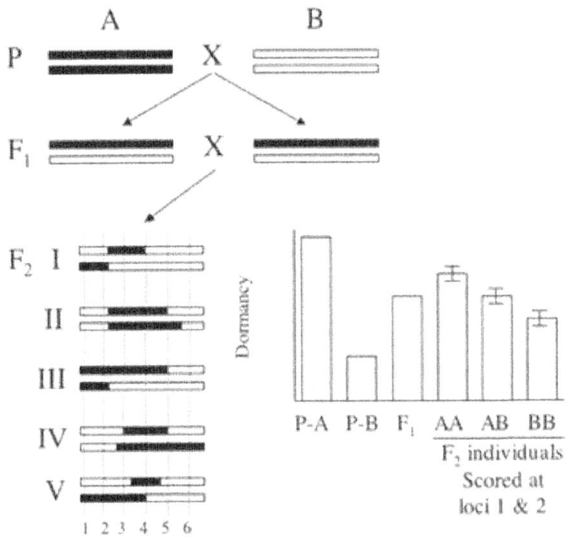

Figure 25.7 : A Two parental (P) inbred lines, A & B, which differ for a trait of interest (*e.g.* dormancy) are crossed to produce heterozygote F_1 progeny which have an intermediate degree of dormancy. Only one chromosome pair is depicted for simplicity. The F_1 progeny are crossed to produce a large segregating F_2 population (usually 100-500 individuals are analyzed, here only five individuals, I-V, are depicted). Because of recombination during F_1 meiosis, the F_2 individuals have varying genetic combinations of the original inbred lines (depicted by shading of chromosomes) and have continuous variation, from high to low, in degree of dormancy. In order to estimate whether segments of the chromosome are homozygous for one of the parental-lines or heterozygote, the F_2 individuals are scored for marker loci (1-6) along the chromosome. For example, loci 1 & 2 are identified as homozygous for A-line alleles in individual III, as heterozygous in individuals I & V and as homozygous for B-line alleles in individuals II & IV. The F_2 individuals are also scored for their dormancy. After scoring many F_2 individuals a correlation between the genotypes at locus 1 & 2 and phenotype is determined and therefore this segment likely contains at least one gene affecting dormancy (bar graph). The analysis is repeated for each chromosomal segment to identify which ones are likely to contain genes affecting dormancy.

Candidate genes are specific loci, identified by predicted bio-chemical function or homology to regulatory genes from other species, that are suspected of affecting the trait of interest. Molecular differences between the parental alleles at these loci can be used to determine whether they map to the same region as any of the QTLs, thereby supporting the hypothesis that a given gene contributes to regulation of the trait.

ABA-INDUCED SENESCENCE AND ETHYLENE

Ethylene has also been found to play a role in the senescence of oat leaf segments, but ABA appears to be the initiating agent, whereas ethylene appears to exert its effects at a later stage. To determine whether the ABA effects are mediated by ethylene, the effects of ABA and ethylene on senescence of leaf discs from wild-type and ethylene-insensitive mutants were compared. Chlorophyll loss from untreated leaf discs of the ethylene-insensitive mutant was much slower than that from the wild type, in the light and in the dark. Ethylene accelerated the chlorophyll loss from the wild-type leaf discs, but it had no effect on the yellowing of the ethylene-insensitive mutant discs. In contrast, treatment with ABA stimulated chlorophyll loss in *both* the wild-type and the mutant discs, whether they were held in the light or in the dark. These results indicate that the promotion of senescence by ABA is not mediated through its stimulation of ethylene production.

Chapter 26

Responses and Adaptations to Abiotic Stress

STOMATAL CONDUCTANCE AND YIELDS OF IRRIGATED CROPS

In the last few decades, the breeding of agricultural crops for higher yields has been very successful. Breeders usually select for high-yielding genotypes empirically, without particular attention to specific plant traits that might be conducive to higher yields. However, comparing old, low-yielding lines of any crop with advanced, high-yielding lines shows clearly that many morphological, physiological, and bio-chemical traits have been altered by the intense selection pressures for higher yields. These changes indicate that selection for high-yielding genotypes has generated indirect selection pressures on the altered traits. If we exclude effects of genes that regulate the expression of two or more unrelated traits (pleiotropic genes), the study of high-yielding lines might reveal specific traits and genes associated with the higher yield. With this information, breeders could explicitly select for yield-enhancing traits to improve yields further.

Changes in plant traits associated with yield increases can be studied conveniently in a historical series of a crop. Historical series include successive commercial releases within a breeding program, with each member of the series representing an incremental increase in yield. In studies aimed at characterizing relationships between specific traits and higher yields, all the members of the series are grown together under the same conditions, and the traits of interest, such as plant architecture or photo-synthetic rates, are measured in parallel with yields. Recent studies of this type focusing on historical series of Pima cotton (*Gossypium barbadense*) and bread wheat (*Triticum aestivum*) have shown a remarkable positive correlation between yield increases and increases in stomatal conductance.

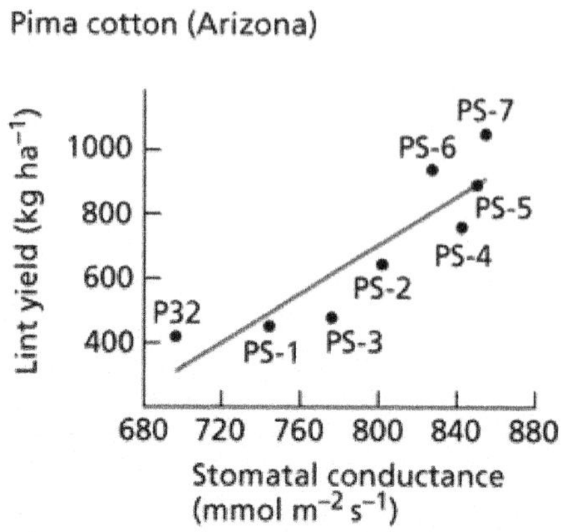

Figure 26.1 : Stomatal conductance has increased in parallel with agronomic yields in irrigated Pima cotton (Gossypium barbadense) selected for higher yields at a high temperature. The figure shows the relationship between lint yield and stomatal conductance in a historical series of Pima cotton grown in Arizona.

Pima cotton is a high-quality fiber cotton grown in hot environments under intensive irrigation. The historical series used in the Pima studies included eight members, released between 1949 and 1996, and encompassing a nearly threefold increase in lint yield. The yield increase attained in each commercial release was accompanied by a corresponding increase in stomatal conductance.

Genetic crosses between low- and high-yielding Pima strains have shown that stomatal conductance in Pima cotton has a clear-cut genetic component. The studies with stomata of Pima cotton are the first to show a clear-cut genetic regulation of stomatal conductance proper, of the stomatal response to temperature, and of proton pumping rates of isolated guard cells.

In the absence of explicit selection for higher stomatal conductance in the Pima breeding program, what could be the nature of the indirect selection pressures that caused the increases in conductance paralleling the increases in yield? Higher stomatal conductance increases CO_2 diffusion into the leaf and favours higher photo-synthetic rates. Higher photo-synthetic rates could in turn favour a higher biomass and higher crop yields. Advanced Pima lines show a higher photo-synthetic capacity than older, low-yielding lines, but photo-synthetic rates measured in the same leaves used to measure stomatal conductance were not positively correlated with yields. Thus, higher stomatal conductance appears to favour higher yields by a mechanism not directly related to photo-synthesis.

Evapotranspiration at the leaf surface lowers leaf temperature, and higher stomatal conductance enhances this leaf cooling. Optimal daytime temperatures for growth, photo-synthesis, and reproduction in Pima cotton are below 30°C, while afternoon air temperature in the hot Pima-growing areas often exceeds 40°C. Evaporative cooling of the leaves thus reduces the gap between optimal growth temperatures and air temperature, and provides heat resistance. Studies with the Pima historical series showed that, because of leaf cooling, leaf temperatures were several degrees lower than air temperature, and leaf and canopy temperature were lower in the advanced, high-yielding lines than in low-yielding lines.

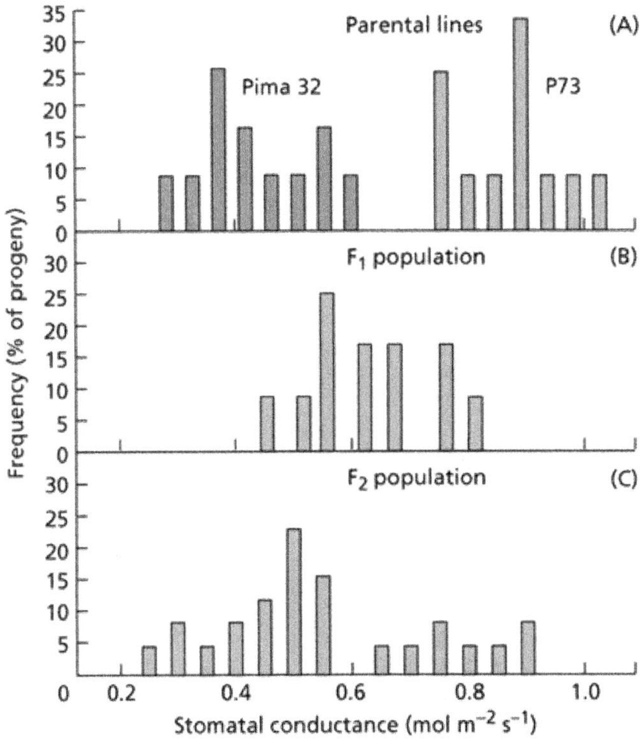

Figure 26.2 : Frequency distribution of stomatal conductance in parental populations of the high-yielding Pima cotton line, the old Pima line, and their F_1 and F_2 progeny. (A) Non-overlapping distribution of stomatal conductance in the two parental populations. (B) Distribution of stomatal conductance in an F_1 population from a cross between the two parents. (C) Distribution of stomatal conductance in an F_2 population derived from F_1 parents.

The same relationship among yields, stomatal conductance, and leaf temperature was found in a historical series of bread wheat grown in the warm Yaqui Valley of north-western Mexico. These studies indicate that selection for higher yields in irrigated crops grown at high temperatures imposes indirect selection pressures for high stomatal conductance that lowers leaf temperature and appears to reduce deleterious effects of heat stress on critical flowering and fruiting stages, thus resulting in higher crop yields.

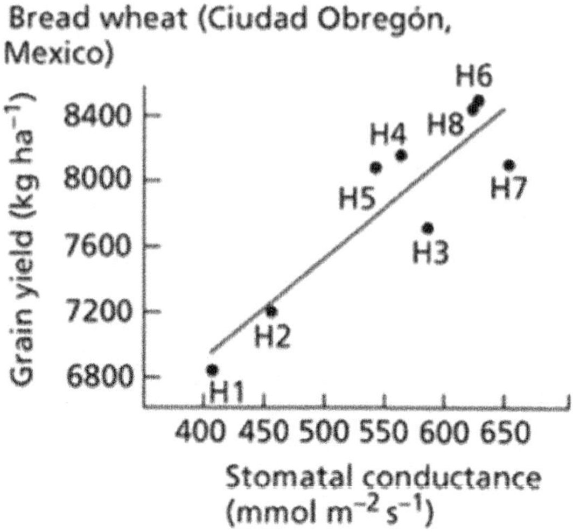

Figure 26.3 : The relationship between grain yield and stomatal conductance in a historical series of semi-dwarf bread wheat grown in Ciudad Obregón, Mexico. The abbreviations H1 through H8 designate successive commercial lines released by the International Maize and Wheat Improvement Center between 1962 and 1988.

If selection for higher yields generates indirect selection pressures for higher stomatal conductance, can breeders explicitly select for higher conductance and thus obtain lines with higher yields? Studies with Pima cotton have shown that selection of high-conductance F_2 progeny from a cross between high- and low-conductance parents produces high-conductance F_4 lines with higher lint yields than low-conductance lines have. These experiments indicate that high stomatal conductance could be used as a selection trait for high yields in irrigated crops grown at high temperature.

The observed increases in stomatal conductance in crops grown at high temperature also have valuable implications for models developed to predict global climate changes, which are expected to occur because of increases in atmospheric CO_2. Under atmospheric CO_2 concentrations that are twice as high as the present ones, evapotranspiration would decrease over the continents and air temperatures would increase significantly over the tropical land masses, amplifying the changes resulting from atmospheric radiative effects. However, these models have yet to take into consideration the effect of higher temperatures on the stomatal control of evapotranspiration. Incorporation of the stomatal response to temperature in climate change models is likely to improve the accuracy of the predictions generated by the models. These studies illustrate both the complex interactions between physical and biological factors in the biosphere, and the advantage of a thorough understanding of the physiological properties of plants for efforts aimed at increasing agronomic yields and at improving models that predict global climate changes.

MEMBRANE LIPIDS AND LOW TEMPERATURES

The importance of membrane lipids for tolerance to low temperatures has been demonstrated by work with mutant and transgenic plants in which the activity of particular enzymes led to a specific change in membrane lipid composition independent of acclimation to low temperature.

In one study, the genes used to modify the lipid composition of thylakoid membranes encoded glycerol-3-phosphate acyltransferases (GPATs).

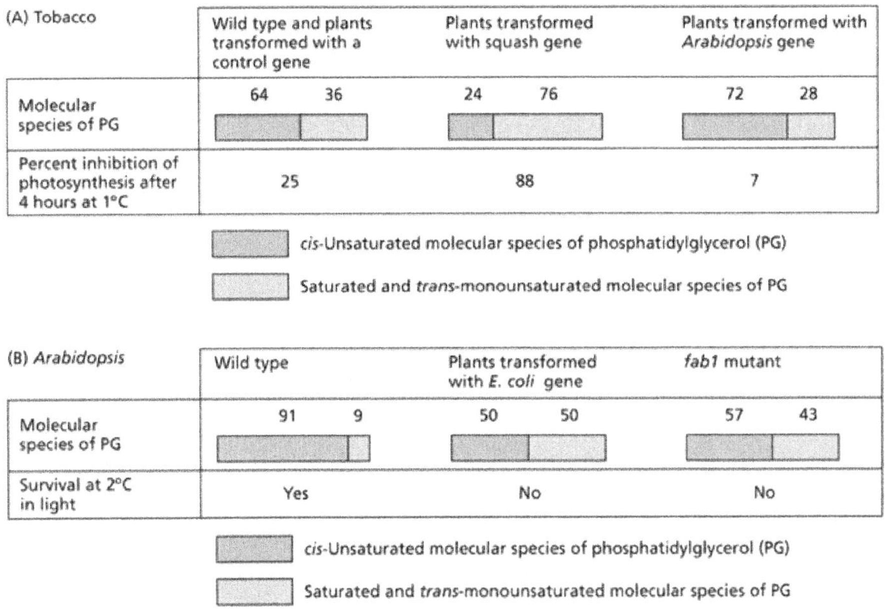

Figure 26.4 : How changes in relative levels of *cis*-unsaturated molecular species of phosphatidylglycerol (PG) in thylakoid membranes of (A) tobacco and (B) *Arabidopsis* affect sensitivity to chilling.

The degree of fatty acid saturation of phosphatidylglycerol (PG) is particularly important to the chilling resistance of chloroplasts. In one experiment, tobacco was transformed with an acyltransferase gene from either squash (chill sensitive) or *Arabidopsis* (chill tolerant) to modify the fatty acid saturation of PG in the thylakoid membranes of the chloroplasts.

In transgenic tobacco transformed with the squash gene, which increased the proportion of high-melting-point (trans-mono-unsaturated and saturated) molecular species of PG, chilling tolerance decreased (88% inhibition of photosynthesis following 4 hours at 1°C). Conversely, when tobacco was transformed with the *Arabidopsis* chill-tolerant gene, the proportion of high-melting-point molecular species of PG decreased, and tolerance of chilling was greater than in wild-type plants.

ICE FORMATION IN HIGHER-PLANT CELLS

During the initial phase of heat loss from tissue, when the temperature drops below the freezing point of the cytosol and the vacuole without ice forming, the liquid phase in the cytosol and vacuole is said to be supercooled. As the temperature drops further, ice forms in the inter-cellular spaces, and heat energy (0.33 kJ or 80 calories per gram) is released as a result of the latent heat of fusion of water. At this stage, the temperature of the tissue reflects the balance between heat gain from ice formation and heat loss to the environment. As a result, when ice first forms, the temperature rises rapidly, and it remains at that level until all the extra-cellular water is frozen. When that point is reached, heat release stops, and the temperature begins to fall again. The release of heat energy during ice formation is the basis for the common practice of spraying crops with water during frost : As long as the water continues to freeze extracellularly, it releases heat that prevents intra-cellular freezing.

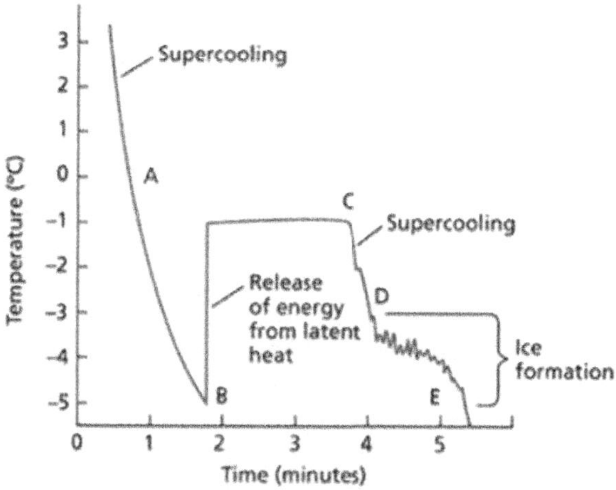

Figure 26.5 : Temperature of parenchyma cells in cucumber (*Cucumis sativus*) fruit during freezing. The temperature was recorded with an electronic device, a thermistor, inserted into a 5 × 20 mm cylinder of tissue and immersed in a coolant at –5.8°C. (A–B) Supercooling. (B–C) Release of heat during freezing in cell walls and inter-cellular spaces. (C–D) Supercooling. (D–E) Small heat spikes released during intra-cellular freezing of individual protoplasts.

The formation of ice within the inter-cellular spaces of cells that are sensitive to freezing is not lethal, but extended exposure to freezing temperatures causes water vapour to move through the plasma membrane from the unfrozen protoplast to the cell wall, causing ice crystals to grow within the inter-cellular spaces. This slow dehydration concentrates solutes within the protoplast, depressing the freezing point by 2 to 3°C. As the temperature continues to drop, a second phase of release of the heat of fusion of water is detectable. This phase reflects a series of small freezing events : Each "spike" of heat release is thought to represent the freezing of cell protoplasts and coincides with loss of viability. The formation of ice crystals on cell walls or in the protoplasm requires the presence of ice nucleation points on which crystals can be initiated and grow.

In some species, acclimation confers an ability to suppress ice nucleation in the protoplast, allowing deep supercooling to many degrees below the freezing point without ice formation. However, deep supercooling of intra-cellular water has a lower limit of about -40°C, the temperature at which ice forms spontaneously. (The homogeneous ice formation temperature of pure water droplets is -38.1°C, but the presence of solutes lowers this temperature.) At or below -40°C, ice crystals form without nucleation points, and intra-cellular freezing and cell death are unavoidable. Species that tolerate temperatures below -40°C under natural conditions do so not by supercooling, but by tolerating gradual dehydration.

Plants produce compounds that protect cells against intra-cellular ice formation. Some of these compounds, such as sugars, amino acids, and other solutes, induce supercooling in the plant's tissues. Overwintering plants also produce antifreeze proteins and other low molecular weight compounds. These compounds provide freeze tolerance by inhibiting ice crystal growth and the nucleation of ice crystals.

WATER-DEFICIT-REGULATED ABA SIGNALING AND STOMATAL CLOSURE

ABA accumulation mediates water deficit, and dark and day/night regulated stomatal closure. A plant response to water deficit (drought) includes the accumulation of ABA in leaves, which induces stomatal closure and inhibits opening. This is a fundamental process in the plant's capacity to maintain plant water status under conditions of low soil moisture content and high evaporative demand.

Water deficit induces activation of genes that encode enzymes for both ABA bio-synthesis and hydrolysis of ABA conjugates releasing the active hormone from an inactive pool. Water deficit induces expression of the gene encoding NCED (9-*cis*-epoxycarotenoid dioxygenase), which catalyzes a critical committed step in drought-induced ABA bio-synthesis. Water deficit induces, by transcriptional and post-transcriptional regulation, activity of the BG1 â-glucosidase, which specifically cleaves biologically inactive ABA-glucose ester (ABA-GE) to form ABA. Further, water deficit causes acidification of the leaf symplast and alkalinzation of the apoplast—both mechanisms enhancing accumulation of ABA in the leaf apoplast—leading to increased ABA arriving at the guard cells to initiate closure in response to water deficit.

ABA induces stomatal closure through a physiological series of steps. These begin with cytosolic Ca^{2+} transients, activation of anion efflux channels, and inhibition of plasma membrane H^{+} pump activity, which depolarize membranes and cause K^+ loss through outward channels. This leads to water loss, loss of turgor, and stomatal closing. Our current understanding is that ABA functions both inter- and intra-cellularly to cause stomatal closure, implicating the presence of receptors that sense ABA in both the apoplast and the symplast. GTG1 and GTG2 (G protein coupled receptor [GPCR]-type G protein) bind ABA and hydrolyze GTP.

GTG1 and GTG2 are hypothesized to function as ABA receptors through GPCR activity and as initiators of ABA signaling through their G protein activity. RCAR (regulatory component of ABA receptor)/PYR1/PYL (pyrabactin resistance 1/PRY-like) family proteins also bind ABA, which promotes interaction with phosphatase 2C (PP2C) proteins that represses their activity to inhibit ABA signaling. Repression of PP2C activity is associated with ABA-dependent regulation of stomatal closure.

There are numerous intermediates and secondary messengers of ABA signaling that results in stomatal closure. Most of these determinants function to regulate Ca^{2+} while some directly modulate ion channel activity. ABA induces ROS production either independent of or through activation of the OST (open stomata) kinase. ROS leads to production of NO that activates plasma membrane Ca^{2+} channels, resulting in cytosolic transients that activate anion efflux channels.

Endomembrane Ca^{2+} channels are activated by signaling that results in cytosolic Ca^{2+} accumulation from internal pools. S1P (sphingosine-1-phosphate) is a determinant of the G protein signaling pathway that activates anion channels. PP2Cs such as ABI1 and ABI2 are negative regulators of anion channels that cause membrane depolarization, while protein kinases are implicated in anion channel activity. ABA inhibits the plasma membrane H^+ pump which facilitates membrane depolarization and cytosolic alkalization. Alkalization activates K^+ efflux and anion channels.

GENETIC AND PHYSIOLOGICAL ADAPTATIONS REQUIRED FOR ZINC HYPERACCUMULATION

The recognition that plants may contain very high concentrations of certain metals such as zinc (Zn) dates back to 1885 when a German botanist was working near the border of Germany and Belgium. He noticed that leaves of *Thlaspi calaminare* (now classified as *Thlaspi caerulescens*) contained 1.7 % of their dry biomass as Zn, which is 100 times higher than Zn concentrations in leaves of non-accumulating plants growing nearby. The ecological role of this Zn hyperaccumulation is still not entirely clear, though it has been suggested that metal accumulation provides protection for the plants against fungal and insect attack.

Plant scientists have also become very interested in understanding the processes involved in Zn hyperaccumulation, as it may lead to the development of food crops with an enhanced capacity to accumulate the essential micro-nutrient zinc from soils. Such micro-nutrient efficient crops should help to alleviate some of the mineral deficiencies that plague a large proportion of the world's human populations. We might also expect to see transgenic plants with the capacity to remove pollutant metals from soils and accumulate them in their shoots for harvest and removal. This type of technology should help remediate some of the heavy metal pollution generated by various industrial processes.

In the previous case study on plant responses to elevated salinity, the excellent genetic tools available for *A. thaliana* allowed discoveries to be primarily driven by the forward genetics approach. However, because Zn hyperaccumulator plants such as *T. caerulescens* and *Arabidopsis halleri* lack well developed genetics tools, uncovering the mechanisms involved in Zn hyperaccumulation has required an elegant combination of comparative physiological, genetic, genomic, and reverse genetic approaches performed in several laboratories in the US and Europe over the last 15 years. These efforts have shown how these plants have evolved the unique Zn hyperaccumulation adaptation.

Zn^{2+} stored in vacuole. To identify the targets for research into the mechanism of Zn hyperaccumulation, it was important to first describe the physiological differences between the Zn hyperaccumulator and a related non-accumulator plant. Mitch Lasat and Leon Kochian at the United States Plant, Soil, and Nutrition Laboratory at Cornell University performed one of the first such studies using the radioactive isotope Zn to identify differences in Zn transport between the Zn hyperaccumulator *T. caerulescens* and the non-accumulator *T. arvense*. These studies identified that the hyperaccumulator has an enhanced ability for Zn uptake in roots and xylem loading for increased root-to-shoot transport, and also an increased ability to take up Zn into shoot cells. When Zn reaches the leaves it is detoxified by storage in the central vacuole of the leaf cells. Using x-ray absorption spectroscopy (XAS) at the Stanford Synchrotron Radiation Laboratory, David E. Salt, Ingrid Pickering, and colleagues determined that the Zn^{2+} stored in the vacuole is primarily bound to organic acids and also stored as the unbound Zn^{2+} ion.

Genes for Zn accumulation. In an elegant application of **classical genetics**, crosses were made between the Zn hyperaccumulator *A. halleri* and the non-accumulator *Arabidopsis petraea* to characterize the genetic basis underlying Zn hyperaccumulation. By measuring both Zn tolerance and shoot Zn accumulation in the progeny of the cross it was determined that Zn tolerance and hyperaccumulation are inherited independently, and are governed by a small number of different genes.

Using DNA micro-arrays and also targeted analysis of particular genes identified that the gene encoding the plasma membrane Zn pump protein HMA4 and the vacuolar membrane Zn^{2+}/H^{+} transporter MTP1 are highly expressed in the Zn hyperaccumulators compared to the non-accumulators. Furthermore, the research groups of Macnair, Saumiton-Laprade, Verbruggen, and Kramer showed that these genes are also linked to the genomic regions in *A. halleri* associated with Zn tolerance. By both knocking out expression of the gene encoding HMA4 using RNA-interference (RNAi) in *A. halleri,* and over expressing the same gene in the non-accumulator *A. thaliana,* Ute Kramer's laboratory was able to firmly establish that HMA4 plays a critical role in Zn hyperaccumulation.

Arabidopsis halleri contains three copies of the *HMA4* gene compared to a single copy in *A. thaliana.* This gene duplication, coupled with an enhancement of the strength of all three HMA4 promoters, is what drives the elevated expression of *HMA4* in *A. halleri.* This elevated expression leads to an increased accumulation of the HMA4 protein in the xylem parenchyma leading to elevated loading of Zn into the xylem for transport to the shoot. This increased Zn flux to the shoot also leads to localized Zn-deficiency in the root, enhancing expression of the plasma membrane Zn influx transporters *ZIP4* and *IRT3.* In this way altered copy number **(the number of genes encoding HMA4 in the genome)** and *cis*-regulation (regulation of gene expression driven by DNA sequences on the same DNA strand as the gene being regulated. These sequences are usually in the promoter region 5′ to the gene being regulated) of *HMA4* leads to the increased uptake and flux of Zn from roots-to-shoots first observed in Zn hyperaccumulators by Lasat and Kochian 10-years previously.

HMA4 only explains the increased Zn flux from roots-to-shoots observed in the hyperaccumulators. The vacuolar Zn^{2+}/H^{+} transporter MTP1 is known to transport Zn^{2+} into the vacuole, and has been shown to have enhanced expression in the hyperaccumulator due at least in part to a gene duplication, and it has also been shown to be genetically associated with Zn tolerance.

By expressing *MTP1* specifically in shoots of *A. thaliana,* to mimic the elevated *MTP1* expression observed in the hyperaccumulator, David E Salt's laboratory at Purdue University showed that increased accumulation of MTP1 in shoots increases Zn tolerance. This elegant grafting experiment also established that elevated expression of MTP1 specifically in shoots also enhances expression of genes encoding Zn-deficiency responsive plasma membrane Zn influx transporters *ZIP4* and *ZIP5* in shoots, and the expression of *ZIP3* and *ZIP9* in roots. Increased expression of these Zn transporters leads to an increased accumulation of Zn in shoots, explaining the elevated shoot uptake of Zn first observed in the hyperaccumulator by Lasat and Kochian.

Taken together this data provides a compelling story of how elevated accumulation of two proteins, HMA4 in roots and MTP1 in shoots, acting in concert, can explain the alterations in Zn uptake, root-to-shoot transport, shoot uptake and vacuolar storage required for the Zn hyperaccumulation adaptation. Genetic adaptations that elevate the expression of *HMA4* and *MTP1* lead to expression changes in various other genes, including various plasma membrane Zn transporters involved in Zn^{2+} uptake that together provide the necessary mechanisms to support enhanced Zn uptake and hyperaccumulation.

In some ways this mechanism can be viewed as HMA4 "pushing" Zn to the shoot, and MTP1 providing a sink that not only promotes storage of Zn in the shoot, but also provides an additional "pull" for Zn from the roots. Thought by no means the end of this story, what we already know provides an excellent example of how a small number of stable genetic changes can give rise to a physiologically complex phenotype such as Zn hyperaccumulation. If such genetic changes better adapt an individual to a particular environment these genetic changes will be selected for, giving rise to an adapted population, which may eventually give rise to a new, reproductively isolated species.

CELLULAR AND WHOLE PLANT RESPONSES TO SALINITY STRESS

Arabidopsis thaliana has proved to be a very powerful genetic model system for dissecting the molecular genetics and bio-chemical basis of a multitude of important biological processes in plants, including numerous studies on plant responses various environmental conditions including salinity (NaCl). In comparison to many other plants *A. thaliana* is quite sensitive to NaCl, with 75 mM NaCl causing a 50% reduction in plant biomass, compared to 200 mM in Alfalfa (*Medicago sativa*) and 400 mM in saltbush (*Atriplex amnicola*).

Even though *Arabidopsis thaliana* is quite sensitive to NaCl it is still able to deploy a sophisticated suite of molecular and bio-chemical responses to help alleviate the toxicity of NaCl. These responses are directed at controlling Na accumulation at both the cellular and whole plant levels. The details of these mechanisms and their experimental elucidation, provide an excellent study to help illustrate certain fundamental biological principles involved in plant responses to the environment, and also how the creative application of experimental techniques in the laboratory can lead to a deeper understanding of bio-chemical, cellular and whole plant processes.

Na^+ efflux and SOS proteins. Between 1996 and 2000, Jian-Kang Zhu and coworkers at the University of Arizona published a series of papers describing experiments that uncovered the first bio-chemical pathway in plants that regulates the cellular concentration of Na. This pathway contains three proteins which act in concert to enhance the efflux of Na from cells in response to exposure of the plant to NaCl. By screening over 250,000 *A. thaliana* plants for mutants that showed reduced root growth when exposed to elevated NaCl (**S**alt **O**verly **S**ensitive mutants), Jian-Kang Zhu and his colleges and students were able to identify three proteins called SOS1, SOS2 and SOS3 that play a critical role in allowing *A. thaliana* to respond to NaCl toxicity (Wu, Ding and Zhu 1996; Shi *et. al.* 2000; Zhu, Liu and Xiong 1998; Liu *et. al.* 2000; Liu and Zhu 1997; Liu and Zhu 1998).

At the cellular level SOS1 functions to pump Na^+ out of the cell across the plasma membrane. This movement of Na out of the cell against an **electrochemical gradient** requires energy which is derived from coupling Na transport to the inward movement of H^+ down an electro-chemical gradient In response to NaCl exposure SOS3, a Ca^{2+} binding protein, binds Ca^{2+} initiating SOS3 binding directly with SOS2. SOS2 is a **protein kinase** with the ability to phosphorylate other proteins.

Binding of SOS3 to SOS2 activates this phosphorylase activity, and the SOS2/SOS3 complex is then directed to the plasma membrane were it phosphorylates SOS1. The Na^+ pumping activity of SOS1 is enhanced after it is phosphorylated, allowing SOS1 to increase the efflux of Na^+ from cells.

The level of expression of the *SOS1* and *SOS2* genes also appears to be elevated after exposure to NaCl which would be expected to further enhance Na efflux. In this way SOS1, SOS2 and SOS3 act together to provide a Na^+ responsive cellular Na^+-efflux pathway that functions to maintain reduced cytoplasmic Na^+ levels. Loss of SOS1 function not only leads to enhanced Na sensitivity but also to increased accumulation of Na in both shoots and roots. Such accumulation suggests that SOS1's role in removing Na from the symplast is to both reduce Na^+ cytotoxicity and also to prevent Na moving symplastically to the xylem parenchyma to limit its loading into the xylem for transport to the shoot.

Na^+ efflux to vacuole. An alternative mechanism to reduce cytoplasmic Na^+ levels is to efflux cytoplasmic Na^+ into the large central vacuole that is present in most plant cells. The bio-chemical capacity to pump Na^+ into vacuoles in exchange for H^+ was first observed in isolated vacuolar membranes from red beet by Eduardo Blumwald when he was a post-doctoral researcher in Ron Poole's laboratory at McGill University. However, it took another 14 years and the release of the *A. thaliana* full genome sequence before the protein underlying this activity was characterized in *A. thaliana* as NHX1, a pump moving Na^+ across the tonoplast into the vacuole.

By both expressing *NHX1* in yeast and in *A. thaliana*, the NHX1 protein was established to pump Na into the vacuole, and plants expressing elevated levels of *NHX1* were able to tolerant higher levels of NaCl exposure. Furthermore, *A. thaliana* lacking a functional copy of the *NHX1* gene (*nhx1* mutant) was shown to have lost the Na^+/H^+ exchange activity at the vacuolar membrane in a process known as reverse genetics.

Na^+ movement in plant. It is not enough for plants to simply control cellular Na through efflux via SOS1 and NHX1. Systems are also required for controlling the movement of Na from the roots to the shoots. In an attempt to uncover the mechanisms involved in such whole-plant Na regulation, Berthomieu and colleagues measured shoot Na levels in thousands of *A. thaliana* plants and were able to identify a mutant *sas2* (sodium over-*a*ccumulation in *s*hoot) which showed elevated shoot Na and reduced root Na. Using forward genetics the mutation in *sas2* causing altered Na accumulation was determined to result in the loss-of-function of the Na transporter HKT1.

It was initially concluded that HKT1 functioned to load Na into the phloem for recycling of Na from the shoot to the root. However, further studies using an antibody against HKT1 revealed that HKT1 was in fact localized in the xylem parenchyma, and most likely involved in unloading Na from the xylem to limit Na movement from the roots to the shoots. Reciprocal grafting experiments by Rus *et. al.* directly established that HKT1 functions primarily in the root to regulate shoot Na, and using radioactive Na (^{23}Na) Davenport and colleagues directly measured the role of HKT1 in Na efflux from the xylem in roots.

Over the four years following Berthomieu's original discovery of the role of HKT1 in regulating shoot Na, various laboratories refined our understanding of the function of HKT1 using a variety of techniques including : immune localization, reciprocal grafting and ^{23}Na flux studies. We shouldn't think of science as being a simple series of discoveries but rather a series of interconnected episodes of creative thinking, experimentation, and refinement of our ideas. In this case such activities have led us to develop an elegant model described above of how *A. thaliana* is able to regulate both whole plant distribution of Na through control of Na entering the xylem transpiration stream, and control of the cellular levels of Na through the co-ordination of Na efflux at the plasma membrane and vacuolar membrane.

Dealing with excess NA^+ loss. The electro-chemical potential normally maintained by cells at the plasma membrane can facilitate Na^+ accumulation into the cytosol 100- to 1000-fold above the external concentrations. As a charged ion, Na^+ has very low permeability to the plasma membrane but it is transported by both low- and high-affinity transport systems, many of which are normally used for K^+ uptake.

External Ca^{2+} at millimolar concentrations (a typical physiological concentration of Ca^{2+} in the apoplast) increases the selectivity of the K^+ transporters and minimizes Na^+ uptake through mechanisms that reduce uptake and facilitate efflux of Na^+ to the apoplast. Influx of Na^+ occurs through non-selective, voltage-insensitive gated channels that passively transport cytotoxic cations into cells. When such channels are open, Na^+ diffuses into the cytosol, where it can be substantially concentrated. Physiological levels of Ca^{2+} cause closure of these channels, restricting Na^+ uptake.

SIGNALING DURING COLD ACCLIMATION REGULATES GENES THAT ARE EXPRESSED IN RESPONSE TO LOW TEMPERATURE AND ENHANCES FREEZING TOLERANCE

Cold acclimation is the process by which temperate plants are able to survive over-winter at freezing temperatures. Capacity for cold acclimation is an adaptation that determines the geographic limits for native species, and areas and seasonal periods in which crops are grown. Almost 25 years ago, Guy and colleagues determined that cold acclimation causes alterations in gene expression. Research over the years now makes it evident that cold acclimation includes the induction of signal transduction pathways that condition the process through regulation of expression of a suite of genes responsive to low temperature. A current hypothesis is that low temperatures alter physical properties of lipids, which increases membrane rigidity leading to mechanical activation of a, yet unknown, cold sensor initiating signal transduction pathways. These pathways activate expression of a suite of genes regulated by low temperature, which are necessary for cold acclimation and tolerance, and include genes that encode signaling and cyroprotective proteins, and enzymes that mediate lipid and sugar metabolism, compatible solute bio-synthesis, ROS scavenging and metabolic reprogramming (Guy 1990; Thomashow 1999; Yamaguchi-Shinozaki and Shinozaki 2006; Chinnusamy *et. al.* 2006; Survila *et. al.* 2009).

Cold acclimation signaling pathways are integrated into a network that interacts with signaling to other environmental responses. Hence, low temperatures induce expression of genes that are also induced by water deficit and salinity, often through common *cis*-regulatory elements and *trans*-acting factors.

ABA and Ca^{2+} in low temperature signaling. ABA, Ca^{2+}, specific Ca^{2+} decoding proteins, and other kinases and phosphatases are signaling intermediaries in low temperature signaling. Mutations of *A. thaliana* causing ABA deficiency (*e.g.*, *aba*) or insensitivity (*e.g.*, *abi1*) result in plants that are incapable of cold acclimation. ABA dependent signaling and transcriptional regulation is necessary but not sufficient for full cold acclimation. On the other hand, not all genes induced by low temperature are ABA dependent. Cold acclimation involves Ca^{2+} influx to the cytosol from the apoplast, and from endoplasmic reticulum and vacuolar pools resulting in cytosolic transients necessary for induction of some low temperature responsive genes and cold acclimation.

Ca^{2+}-decoding proteins implicated in cold acclimation include calmodulin and calmoudlin-like proteins, Ca^{2+}-dependent protein kinases, and calcineurin B-like proteins. Recently, members of the Ca^{2+}-activated CAMTA transcription factor family were shown to function in cold acclimation. Phosphatases and MAP (mitogen activated protein) kinases are other signaling intermediates that regulate cold acclimation signaling.

Gene activation pathways. A recent analysis indicates that about 6% of genes are low temperature responsive. Different low temperature regulated suites of genes are implicated in cold acclimation, but for most the signaling determinants are not extensively characterized. As stated previously, low temperature signaling pathways are categorized into ABA independent and ABA dependent cascades (Yamaguchi-Shinozaki and Shinozaki 2006). It is not established if ABA dependent signaling controls cold acclimation but ABA independent transcriptional pathways are necessary, but not sufficient, for cold acclimation, and chilling and freezing tolerances. The most characterized pathway is the CBF (C-repeat binding factor)/DREB1 (dehydration responsive element binding factor 1) signal transduction pathway that is named after the first identified transcription factor in the pathway.

A clear demonstration of cold acclimation function was in the report of Jaglo-Ottosen and colleagues from the Thomashow laboratory in which CBF1 over-expression activated cold acclimation responsive genes at 22°C, a non-acclimating temperature. It is now recognized that CBF1, 2, and 3 (DREB1B, DREB1C, and DREB1A, respectively) have redundant as well as unique functions in cold acclimation. CBFs/DREB1s bind to the core cis-element A/GCCGAC (C repeat, CRT) to activate expression of genes in the regulon. CBF1, 2, or 3 expression induces regulon expression and facilitates cold acclimation and freezing tolerance. Between 12 to 28% of low temperature responsive genes are in the CBF regulon.

Positive regulators of CBF expression include ICE1 (inducer of CBF expression), SIZ1, CAMTA (Yamaguchi-Shinozaki and Shinozaki 2006; Miura *et. al.* 2007; Zhu *et. al.* 2007; Doherty *et. al.* 2009). HOS1 (high expression of osmotically responsive genes) and MYB15 are negative regulators of CBF activity. The CBF transcriptional regulatory cascade is conserved in higher plants based on available genome sequence information, and has been implicated in cold acclimation of crops, most notably rice, wheat, and tomato. As mentioned previously, the CBF/DREB1 pathway is not the only regulatory cascade that is necessary for cold acclimation.

Chapter 27

Ethylene : The Gaseous Hormone

ETHYLENE IN THE ENVIRONMENT ARISES BIOTICALLY AND ABIOTICALLY

Even away from cities and industrial air pollutants, the environment is seldom free of ethylene. Ethylene can be formed either photochemically (*e.g.*, in the atmosphere and upper oceanic layers) or enzymatically by plants and micro-organisms. The production of ethylene in plants is highest in senescing tissues and ripening fruits (>1.0 nL g-fresh-weight^{-1} h^{-1}), but all organs of higher plants can synthesize ethylene. Ethylene is biologically active at very low concentrations—as low as 1 part per billion (1 nL L^{-1}). The internal ethylene concentration in a ripe apple has been reported to be as high as 2500 μL L^{-1}.

Young developing leaves produce more ethylene than do fully expanded leaves. In kidney bean (*Phaseolus vulgaris*), young leaves produce 0.4 nL g^{-1} h^{-1}, compared with 0.04 nL g^{-1} h^{-1} for older leaves. With few exceptions, nonsenescent tissues that are wounded or mechanically perturbed will temporarily increase their ethylene production several fold within 30 minutes. Ethylene levels later return to normal.

Gymnosperms and lower plants, including ferns, mosses, liverworts, algae, and certain cyanobacteria, have shown the ability to produce ethylene. Ethylene production by fungi and bacteria contributes significantly to the ethylene content of soil. Certain strains of the common enteric bacterium *Escherichia coli* and of yeast (a fungus) produce large amounts of ethylene from methionine.

There is no evidence that healthy mammalian tissues produce ethylene, nor does ethylene appear to be a metabolic product of invertebrates. However, both a marine sponge and cultured mammalian cells can respond to ethylene, raising the possibility that this gaseous molecule acts as a signaling molecule in animal cells.

ETIIYLENE READILY UNDERGOES OXIDATION

Ethylene is a flammable gas and readily undergoes oxidation to ethylene oxide :

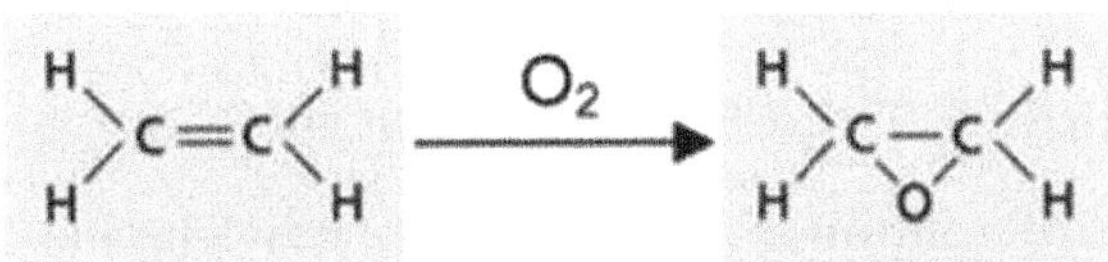

Ethylene **Ethylene oxide**

Ethylene oxide can then be hydrolyzed to ethylene glycol :

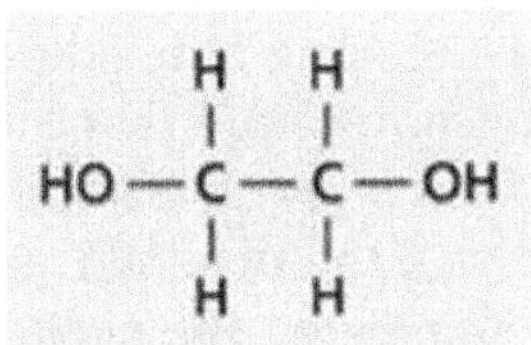

Ethylene glycol

In most plant tissues, ethylene can be completely oxidized to CO_2, in the following reaction :

Complete oxidation of ethylene

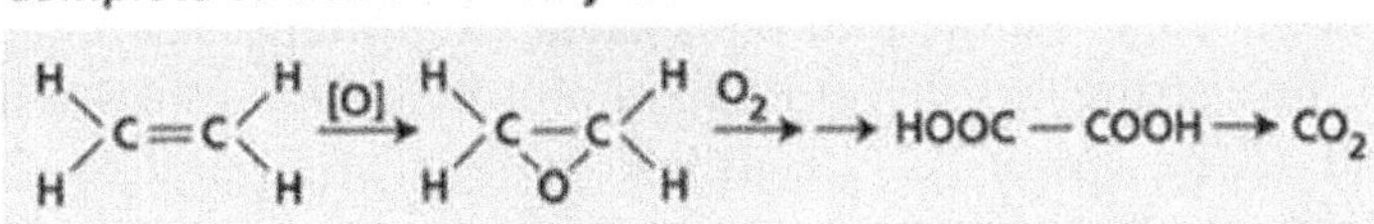

Ethylene **Ethylene oxide** **Oxalic acid** **Carbon dioxide**

ETHYLENE CAN BE MEASURED BY GAS CHROMATOGRAPHY

Historically, bioassays based on the seedling triple response were used to measure ethylene levels, but they have been replaced by **gas chromatography**. As little as 5 parts per billion (ppb) (5 pL per liter) of ethylene can be detected, and the analysis time is only 1 to 5 minutes.

Ethylene is released easily from the tissue and diffuses in the gas phase through the inter-cellular spaces and outside the tissue. At an ethylene concentration of 1 $\mu L\ L^{-1}$ in the gas phase at 25°C, the concentration of ethylene in water is 4.4×10^{-9} *M*. Ethylene concentrations are normally expressed in terms of their gas phase concentrations, which are easier to measure.

Usually the ethylene produced by a plant tissue is allowed to accumulate in a sealed vial, and a sample is withdrawn with a syringe. The sample is injected into a gas chromatograph column in which the different gases are separated and detected by a flame ionization detector. Quantification of ethylene by this method is very accurate. A more sensitive, but less common method to measure ethylene uses a laser-driven photoacoustic detector that can detect as little as 50 parts per trillion (50 ppt = 0.05 pL L^{-1}) ethylene.

CLONING OF THE GENE THAT ENCODES ACC SYNTHASE

A major hurdle in the understanding of ethylene bio-synthesis was the difficulty in purifying the low abundance, unstable protein and the cloning of the corresponding gene. Researchers overcame this obstacle by raising antisera against partly purified ACC synthase from zucchini fruit slices that were induced to produce large amounts of ethylene. The resulting antisera contained a mixture of antibodies, including antibodies to ACC. The investigators then removed the contaminating antibodies by allowing them to bind to proteins extracted from zucchini tissue that lacked the enzyme. The key to the success of this procedure was the availability of tissue largely lacking the enzyme with which to subtract the non-ACS antibodies from the polyclonal antibody mixture. This circumvented the need for highly purified ACC synthase protein. The purified ACC synthase antibodies were then used to isolate the gene encoding the enzyme.

Once the gene encoding ACS was isolated, it could be used to express the ACC synthase protein in the bacterium *E. coli* at high levels and isolate purified ACC synthase in large amounts. The cloned zucchini gene also allowed genes encoding ACC synthase to be isolated from numerous other plant species, using low stringency hybridization and degenerate PCR. The deduced amino acid sequences of ACC synthase genes were found to show a high degree of similarity to the *aminotransferase* superfamily of enzymes. These enzymes catalyze the transfer of amino groups among amino acids and require pyridoxal phosphate as a cofactor.

CLONING OF THE GENE THAT ENCODES ACC OXIDASE

The gene that encodes ACC oxidase was cloned by a circuitous route. Researchers isolated a set of genes that are highly induced during tomato fruit ripening. They then blocked the expression of these genes by introducing an antisense version of each into transgenic tomatoes. Blocking of one such gene, *pTOM13*, repressed fruit ripening and inhibited ethylene production. Expression of *pTOM13* in *Xenopus* oocytes as well as in yeast conferred the ability to convert ACC to ethylene, indicating that the gene encodes an ACC oxidase enzyme. Using this gene, researchers then isolated ACC oxidase clones from numerous plant species.

ETHYLENE BINDING TO ETR1 AND SEEDLING RESPONSE TO ETHYLENE

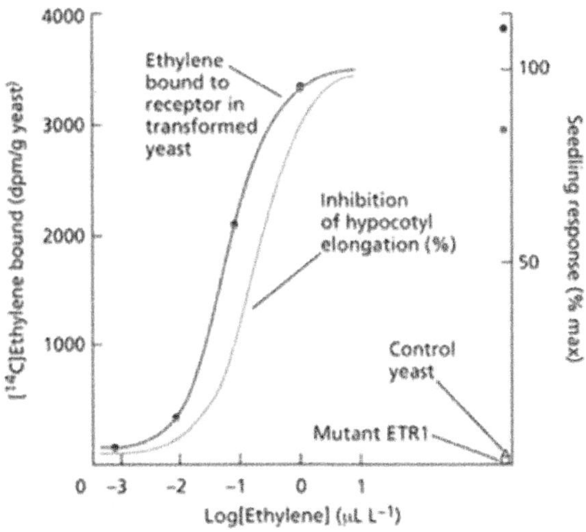

Figure 27.1 : Comparison of the response of ethylene binding to ETR1 with the response of seedlings to ethylene. The *ETR1* gene was expressed in yeast, and the ability of the transformed yeast cells to bind radio-active ethylene was tested. The continuous curve (in black) with the open and closed circles represents the amount of ethylene bound to duplicate samples of yeast cells at different ethylene concentrations. The open triangle and square represent ethylene binding by yeast expressing the vector alone or the mutant ETR1 protein. The green line represents the inhibition of hypocotyl elongation by ethylene. The similarity of the two curves strongly suggests that the *ETR1* gene encodes the ethylene receptor.

CONSERVATION OF ETHYLENE SIGNALING COMPONENTS IN OTHER PLANT SPECIES

Genes that are similar to the *Arabidopsis* ethylene signaling components have been found in other plant species. For example, the tomato genome, like that of *Arabidopsis*, encodes a family of ethylene receptor genes, including *Never-ripe*, which are differentially expressed throughout development. The tomato ethylene receptor genes are homologous to ETR1. A tobacco homolog of EIN3 has been found, and genes similar to all the *Arabidopsis* signaling components can be found in the rice genomic sequence. These and other studies suggest that the components of these proteins probably represent ubiquitous mediators of ethylene signaling in plants.

ACC SYNTHASE GENE EXPRESSION AND BIO-TECHNOLOGY

As we have seen, ACC synthase is a cytosolic enzyme that catalyzes the first committed step in ethylene bio-synthesis in higher plants. It is a key regulatory enzyme that catalyzes the production of the ethylene precursor ACC from AdoMet (*S*-adenosylmethionine). However, this enzyme is difficult to purify because it is labile and present in low abundance in plant tissues. The molecular cloning and functional expression of ACC synthase in *E. coli* and yeasts have facilitated bio-chemical and structural studies of this enzyme.

ACC synthase cDNAs and genomic sequences have been cloned from numerous plant species. The emerging picture from the study of these genes is that ACC synthase is encoded by members of a divergent multigene family in which each gene is differentially regulated by various environmental and developmental factors during the growth of a plant. *Arabidopsis*, for example, has twelve and tomato has at least nine gene members, which are differentially regulated by inducers such as auxin, fruit ripening, and wounding (Olson *et. al.* 1991; Rottmann *et. al.* 1991; Abel *et. al.* 1995).

The protein synthesis inhibitor cycloheximide is a potent inducer of many ACC synthase genes. Cycloheximide inducibility is a characteristic of *primary response genes* such as the early auxin-inducible genes. Two mechanisms have been proposed to explain this inducibility.

- First, the transcription of these genes may be under the control of a short-lived repressor protein whose synthesis is prevented by the addition of cycloheximide;
- Second, the mRNA transcripts of these genes may be short-lived and are stabilized when cycloheximide prevents the synthesis of a nuclease.

The induction of at least one of the genes for ACC synthase (*ACS2*) in *Arabidopsis* by cycloheximide appears to be due to transcriptional activation, and thus is likely regulated by a short-lived transcriptional repressor.

A clue to the signal transduction mechanism involved in the regulation of ACC synthase gene expression has come from experiments with lithium (Li^+). Li^+ is known to interfere with the phosphoinositide signaling pathway by inhibiting the activity of inositolphosphate phosphatase. Li^+ is also one of the strongest inducers of ACC synthase activity in plants. This finding suggests that the regulation of at least some ACC synthase genes may involve Ca^{2+} mobilization mediated by inositol trisphosphate (IP_3).

Because ethylene plays such an important role in fruit ripening, the cloning of the genes for ACC synthase and ACC oxidase has enabled scientists to use bio-technology to reduce ethylene production during fruit ripening, and thus to prolong the storage life of fruit.

In the case of tomato, two different strategies have been employed to control ethylene production.

- The first strategy was to use the Ti plasmid of *Agrobacterium tumefaciens* to transform tomato tissue with antisense DNA for ACC synthase or ACC oxidase. When the genes for the two enzymes are expressed in the reverse direction in the cell, the resulting "antisense transcript" binds to and inactivates the "sense transcript" produced by the wild-type gene, effectively shutting down expression of the wild-type gene. After transformation, intact plants can be regenerated in tissue culture.

- In the second strategy for reducing ethylene production, tomato tissue was transformed (using *A. tumefaciens*) with genes encoding enzymes that metabolize ACC or AdoMet. In this case, the ACC that is produced is rapidly degraded, and no ethylene can be synthesized.

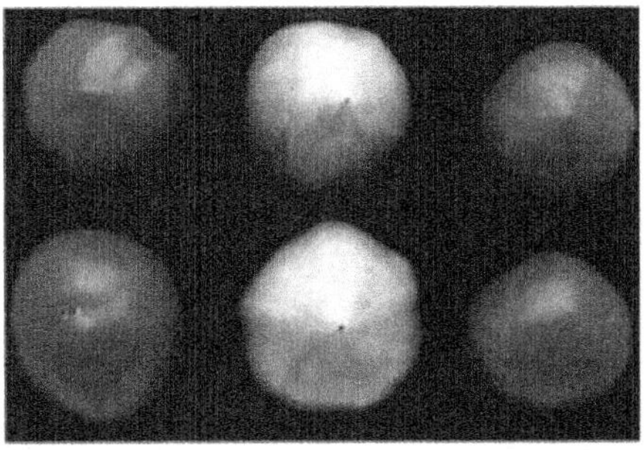

Figure 27.2 : Inhibition of tomato fruit ripening in detached tomato fruits by antisense ACC synthase RNA. Antisense tomato fruits (center) remain firm and green unless treated by ethylene. Ethylene-treated antisense fruits (left) are indistinguishable from naturally ripened fruits (right) in terms of color, flavor, aroma, and texture.

Regardless of the strategy used, all fruits from these genetically engineered plants show significant delays in ripening. Tomato fruits from plants transformed with the ACC synthase antisense DNA fail to ripen unless treated with ethylene.

Similar approaches are now being carried out in other agronomically important crops, such as melon and banana. In addition, these approaches are being used to retard floral senescence in carnations, thereby reducing post-harvest losses. The research on ACC synthase and other ethylene bio-synthesis genes thus provides an excellent example of how basic research can lead to improvements in agricultural and horticultural crops, the full benefits of which we are only beginning to realize.

THE *HOOKLESS* MUTATION ALTERS THE PATTERN OF AUXIN GENE EXPRESSION

A gene required for formation of the apical hook, ***HLS1*** (*HOOKLESS 1*) (so called because mutations in this gene result in seedlings lacking an apical hook), was identified in Arabidopsis. In addition to regulating apical hook formation in response to ethylene, *HLS1* also functions to repress hook opening in the dark. Disruption of this gene severely alters the pattern of expression of auxin-responsive genes. When the *HLS1* was over-expressed in Arabidopsis, it caused constitutive hook formation even in the absence of ethylene.

HLS1 encodes a putative *N*-acetyltransferase that may acetylate, and thereby destabilize, one of the auxin response factors (ARFs). Recall that ARFs are a large family of transcription factors that bind to auxin-response elements in the promoters of many auxin-regulated genes. These findings indicate that hook formation and opening are regulated by the interactions of light, ethylene, and auxin, and that *HLS1* plays a role in the integration of these signals.

ETHYLENE INHIBITS THE FORMATION OF NITROGEN-FIXING ROOT NODULES IN LEGUMES

Nitrogen-fixing rhizobacteria play an important the assimilation of nitrogen by leguminous plants. Through the process of nitrogen fixation, which occurs in specialized structures on the roots of legumes called nodules, these bacteria are able to convert atmospheric N_2 into organic forms that can be assimilated by plants. The bacteria obtain energy in the form of dicarboxylic acids from the plant in exchange for ammonia produced during nitrogen fixation, which is an energetically costly process.

The process of nodulation is highly regulated by a number of signaling pathways, including ethylene. The first evidence that ethylene was involved in nodulation came the observation that treatment with exogenous ethylene suppressed nodulation, and conversely treatment with inhibitors of ethylene action (*e.g.*, Ag^+ or AVG) can increase nodule numbers. The finding that the sickle mutant in *Medicago truncatula*, which displays hypernodulation, is ethylene-insensitive and encodes an ortholog of EIN2 provided genetic support for the role of ethylene in the regulation of rhizobial symbiosis. Ethylene acts at multiple stages of the legume–rhizobial interaction, including the initial response of root hairs upon exposure to Rhizobia bacteria and the progression of infection threads into the cortex.

ETHYLENE BIO-SYNTHESIS CAN BE BLOCKED WITH ANTI-SENSE DNA

Another very useful genetic approach that offers direct evidence for the function of specific genes is based on transgenic plants. Through genetic engineering technology, the roles of both ethylene and cytokinins in the regulation of leaf senescence have been confirmed.

One way to suppress the expression of a gene is to transform the plant with antisense DNA, which consists of the gene of interest in the reverse orientation with respect to the promoter. When the antisense gene is transcribed, the resulting antisense mRNA is complementary to the sense mRNA and will hybridize to it. Because double-stranded RNA is rapidly degraded in the cell, the effect of the antisense gene is to deplete the cell of the sense mRNA.

Transgenic plants expressing antisense versions of genes that encode enzymes involved in the ethylene bio-synthetic pathway, such as ACC synthase and ACC oxidase, can synthesize ethylene only at very low levels. Consistent with a role for ethylene in senescence, such antisense mutants have been shown to exhibit delayed leaf senescence, as well as delayed fruit ripening, in tomato.

ABSCISSION AND THE DAWN OF AGRICULTURE

Civilization as we know it is entirely dependent on an abundant food supply, which can be provided only by agriculture. Before the development of agriculture, people lived in small, nomadic groups whose migration patterns followed the seasonal availability of foods derived from wild plants and animals. The shift from hunting and gathering to cultivation and pastoralism led to the establishment of permanent settlements with ever-expanding populations. This transition, considered to be the most significant (and by some, the most disastrous!) cultural change in human history, was made independently at least seven times at various locations around the world.

Remains of some of the earliest settlements are found in the region of the Near East called the Fertile Crescent, a broad arching zone of grassland and woodland that begins at the eastern edge of the Mediterranean and curves around some 2000 km eastward to the Zagros Mountains. One of the oldest known agricultural settlements in the world is found here, in the ancient city of Jericho, located 14 miles north of the modern city of Jericho in the Jordan Valley. Here archaeologists have unearthed evidence for a transition from hunting and gathering to agriculture that occurred around 10,000 years ago.

The first step in the transition from hunting and gathering to agriculture was the domestication of plants and animals. The seven primary domesticates of the Fertile Crescent were barley, emmer wheat, einkorn wheat, sheep, goats, cattle, and pigs. The first of these to be domesticated were the cereals. The Mediterranean region is particularly well stocked with cereals suitable for domestication. Of the world's 56 species of large-seeded grasses, 32 grow wild in the Mediterranean area. Recent DNA analyses have pinpointed the area near the Karacadâg mountains in Turkey as the probable site of origin of einkorn wheat.

All the evidence suggests that wheat and barley were domesticated over a period of only 2 or 3 centuries, between 10,000 and 9700 B.P.. Considering the large number of morphological and physiological differences between even the most primitive of the domesticated forms of wheat and barley today and their wild progenitors, it is difficult to envision how so many changes would come about in such a brief period by natural selection alone. Clearly, some type of artificial selection had to be carried out (most likely by women) to accelerate the evolutionary process. Archaeologists and botanists have attempted to reconstruct this selection process.

Of all the anatomical and physiological differences between wild and domesticated cereals, two are regarded as most crucial. The first is the **loss of seed dormancy**—the failure of certain seeds to germinate when planted in moist soil unless given additional environmental signals, such as cold or light. The process of plant domestication probably began in small garden plots planted near temporary settlements. It is easy to imagine how the loss of seed dormancy might be selected for by a nomadic group, since seeds that germinated and grew rapidly would be more likely to have been gathered than those whose germination was delayed.

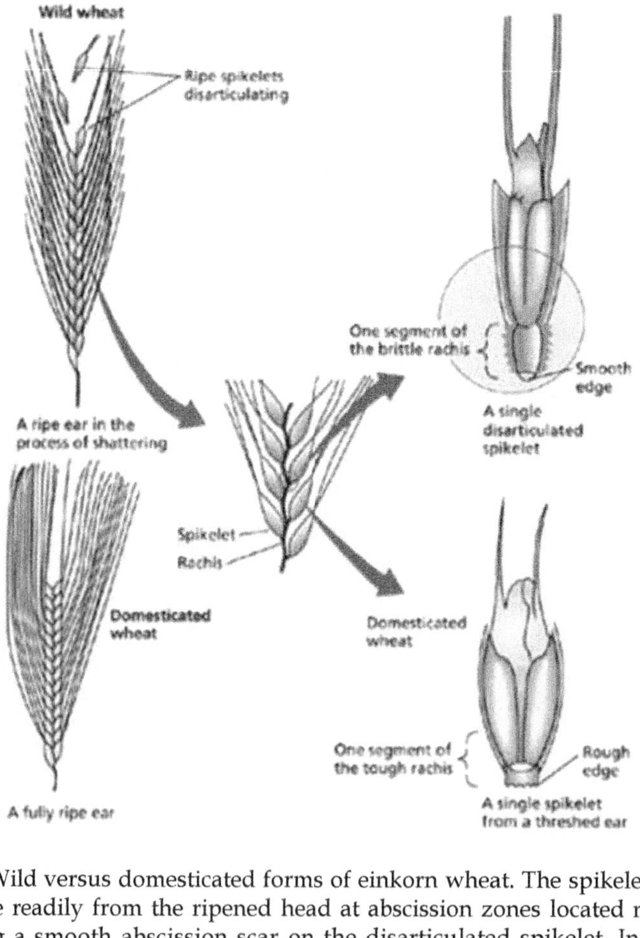

Figure 27.3 : Wild versus domesticated forms of einkorn wheat. The spikelets of the wild variety abscise readily from the ripened head at abscission zones located nearby on the rachis, leaving a smooth abscission scar on the disarticulated spikelet. In contrast, the abscission zones of the domesticated varieties that have tough rachises fail to develop properly, and the spikelets remain on the head until released by threshing. Note that the breakage of the rachis below the spikelet leaves a rough edge, which tends to impede penetration into the ground.

The second critical change that took place during cereal domestication was the development of a **tough rachis**. The **rachis** is the main axis of the inflorescence, or **spike**, of wheat and other cereals, to which the spikelets are attached. **Spikelets** are small inflorescences bearing one or more **florets**, or small flowers, along with a set of miniature bractlike leaves. After pollination, the fruits (**caryopses**) develop within the spikelets. When the caryopses of the wild-type wheat are fully ripened, a series of abscission layers forms that divides the rachis into dispersal units consisting of a single spikelet attached to a short segment of the rachis. This **disarticulation** of the rachis begins at the top of the inflorescence and works its way downward. A ripe head of wild wheat is thus easily shattered into dispersal units when touched or blown by the wind and is referred to by archaeologists as a *brittle rachis.*

A brittle rachis is an extremely undesirable trait from a human standpoint. Imagine the plight of a small nomadic group that carefully timed its peregrinations to coincide with the ripening of wild cereals. Because of the extreme fragility of the rachis, it would be crucial to arrive close to the time of ripening, but before disarticulation of the rachis. A slight miscalculation could spell disaster for the group. Arriving too late, they would be greeted by a field of barren stalks yielding only a handful of remnant grains. As a consolation, however, they might find the rare mutant spikes that remained intact because the abscission layers of the rachis failed to develop. This property is referred to by archaeologists as a *tough rachis*. If mutants with the tough-rachis phenotype were then gathered and planted in gardens, their numbers would, over time, increase.

A second scenario for the selection of a tough rachis involves the method of harvesting. The most efficient method for harvesting wheat is to cut it off at the base of the stalk, and evidence suggests that flint-bladed sickles were used for this purpose. However, harvesting wild wheat by cutting would have to be performed before formation of the abscission layers, or the fragile spikes would shatter, sending their spikelets flying to the ground. This method of harvesting would therefore favour the selection of mutations that delayed the abscission process. Hillman and Davies have demonstrated in field experiments that over many generations, the population would come to be dominated by the varieties with tough rachises. Once cut, the stalks of these varieties of wheat would then be transported to another location, where the grains would be removed by threshing.

Is the mutation rate in wild cereals rapid enough to account for their domestication in a period of only 200 to 300 years? Sharma and Waines have determined that two genetic loci are involved in the evolution of the tough-rachis character in einkorn wheat. Even if we assume that a mutation that converts a brittle rachis to a tough rachis occurs in only one out of every million wild-type wheat plants, Hillman and Davies have estimated that varieties with tough rachises would become the dominant forms in wild wheat populations after only 200 years of human management. Higher rates of mutation would shorten this time considerably, possibly to within 20 years.

Thus the experimental results and calculations are in good agreement with the archaeological evidence. Agriculture in the Fertile Crescent, which ultimately led to the founding of Greece and Rome and all of Western civilization, began with mutations in genes that control abscission.

SPECIFIC INHIBITORS OF ETHYLENE BIO-SYNTHESIS ARE USED COMMERCIALLY TO PRESERVE CUT FLOWERS

Specific inhibitors of ethylene bio-synthesis and action have proven useful in the post-harvest preservation of flowers. Silver (Ag^+) has been used extensively to increase the longevity of cut carnations and several other flowers. The potent inhibitor AVG retards fruit ripening and flower fading, but its commercial use has not yet been approved by regulatory agencies. The strong, offensive odor of *trans*-cyclooctene precludes its use in agriculture. On the other hand, the ethylene binding inhibitor 1-methylcyclopropene, which is marketed under the trade name EthylBloc, is nearly odorless and is used for a variety of post-harvest applications, most notably increasing the shelf life of cut flowers. It has also been commercialized as "SmartFresh^{SM}" to help maintain the freshness and quality of fruits such as apples and melons during storage and shipping. The near future may see a variety of agriculturally important species that have been genetically modified to manipulate the bio-synthesis of ethylene or its perception. The inhibition of ripening in tomato by expression of an antisense version of ACC synthase and ACC oxidase has already been mentioned. Another example of this technology is in petunia, in which ethylene bio-synthesis has been blocked by transformation of an antisense version of ACC oxidase. Senescence and petal wilting of cut flowers are delayed for weeks in these transgenic plants.

BIO-TECHNOLOGY FOR ABIOTIC STRESS TOLERANCE

While working on genetic diversity studies amongst the mangroves, it was felt that these unique systems offer an amazing potential in harnessing genetic combinations that could help susceptible crop species cope with increasing salinity in the soil due to global warming and sea level rise. Therefore, MSSRF undertook an anticipatory research programme for developing location specific crop varieties offering tolerance/ resistance to coastal salinity. Significant leads have been obtained in this area of research. With limited resources, and sustained funding by the Department of Bio-technology, the group has been able to isolate and characterize a number of promising genes, and promoters with implications on abiotic stress tolerance and has also mobilized some of these genes into transgenic systems. Limited field trials of the transgenic rice varieties containing mangrove genes have been undertaken under contained conditions. The group has already 36 full-length genes isolated from the mangrove species, *Avicennia marina*; as also a number of partial clones for abiotic stress conditions. Eight patent applications has been filed. **This research work has been rated as among the ten most important scientific work carried out during the last decade in India, which can change our lives for better.**

On a competitive bidding process, MSSRF has been supported under the *Centre for Excellence* programme of the Department of Bio-technology, Govt. of India.

Genes for Abiotic stress from *Avicennia marina, Prosopis juliflora,* and *Porteresia coarctata* : *Avicennia* is a monotypic pantropical mangrove genus with eight species, of which A. *marina* is widely distributed both latitudinally and longitudinally. The high salt tolerance of A. *marina* is a consequence of water use efficiency, which balances the relation between carbon gain, water loss, and ion uptake with the transpiration stream on a low but constant level. A. *marina* grows in coastal regions where the salt concentration can be as high as 9%. The regulation of inorganic ions occurs partially by exclusion at the roots and also by excretion via salt glands, the excretion rate for sodium and chloride ions being 0.41 mol m-2 S-I and 0.0461 mol m-2 s-1,respectively. It is thus an ideal candidate plant for mining genes for salt tolerance.

As a first step towards the characterization of genes that contribute to combating salinity stress, we have constructed a cDNA library from A. *marina.* Random expressed sequence tag (EST) sequencing of 1,841 clones produced 1,602 quality reads. These clones were classified into functional categories, and BLAST comparisons revealed that 113 clones were homologous to genes earlier implicated in stress responses, of which the dehydrins are the most predominant in this category. Of the ESTs analyzed, 30% showed no homology to previously uncharacterized genes in the public plant databases. Of these 30%,52 clones were selected for reverse Northern analysis : 26 were shown to be up-regulated and five shown to be down-regulated under stress. The results obtained by reverse Northern analysis were confirmed by Northern analysis.

Prosopis Jllliflora, a leguminous a phreatophytic perennial deciduous thorny small tree is now the dominant woody plant found in about 45 million ha of grazing lands worldwide from sea level to 1500 m. It grows in very hot and dry regions with temperatures as high as 48°C and annual precipitation of 150-750 mm. The roots penetrate to great depths in the soil and the tree can grow in a variety of areas including saline, alkaline, sandy and rocky soils. An introduced species in.' India, it has subsequently spread aggressively across the country. P. *Jllliflora* can withstand high leafto- air vapour pressure deficit (VPD) by reducing metabolic activity and by effective adjustments in the partitioning of electron flow between assimilation and non-assimilation processes, which in turn imposes a strong limitation on the potential carbon gain. P.*Jllliflora* is also heavy metal tolerant and due to the heavy metal accumulating property of P.*Jllliflora,* it has been suggested as a green solution for soils contaminated with cadmium, chromium and copper. The plant, therefore, is an ideal candidate for mining genes for abiotic stress tolerance including drought tolerance.

Random expressed sequence tag (EST) sequencing of 1750 clones produced 1467 highquality reads. These clones were classified into functional categories, and BLAST comparisons revealed that 114 clones were homologous to genes implicated in stress response(s) and included heat shock proteins, metallothioneins, lipid transfer proteins, and late embryogenesis abundant proteins. Of the ESTs analyzed, 26% showed no homology to previously pncharacterized genes in the databases. Fifty-two cl,ones from this category were selected for reverse Northern analysis : 21 were shown to be upregulated and 16 downregulated under stress. The results obtained by reverse Northern analysis were confirmed by Northern analysis. Two of the abundant genes coding for a non-specific lipid transfer protein and late embryogenesis abundant protein were sequenced completely. Northern analysis (after polyethylene glycol stress) of the 2 genes was carried out.

Antiporters

The ion pump, vacuolar H+-ATPase (V-ATPase) plays a major role in the maintenance of cellular pH, and influences the transport of cations into the vacuoles of plant cells. A cDNA clone (PcVHAc1) encoding the c sub-unit of V-ATPase was isolated from the salt-tolerant wild rice, *Porteresia coarctata*. The DNA sequence of PcVHA-c1 showed significant homology with V-ATPase sub-unit c of rice. The deduced amino acid sequence of PcVHAc1 and other reported c subunits were compared, and sequence relationships have been drawn to know their genetic relatedness. Southern analysis suggested the presence of multiple coding regions for sub-unit c in P. *coarctata*. Northern and Western analyses of salt-treated P. *coarctata* plants revealed that sub-unit c of V-ATPase was upregulated by NaCl treatment at both transcriptional and translational level.

Reactive Oxygen Species Genes

Antioxidant enzymes play an important role in conferring abiotic stress tolerance. Superoxide dismutase (SOD) is the first enzyme in the enzymatic antioxidative pathway. Halophytic plants like mangroves have been reported to have a high level of SOD activity, which plays a major role in defending the mangrove species against severe abiotic stresses. We have isolated of *Sod* 1, a cDNA encoding a cytosolic copper zinc super oxide dismutase from the mangrove plant *Avicennia marina* and studied its mRNA expression pattern during various oxidative and abiotic stresses. The cDNA was cloned into a binary vector (pCAMBIA 1300) and transformed into *indica* rice. Southern hybridization analysis of transgenic rice plants revealed stable integration of the Sodl trans gene in the rice genome. The mRNA transcript of Sodl was detected by Northern hybridisation in the transgenic rice plants. SOD isozyme assay of the transgenic rice plants revealed the stable expression of the transgenic Sod1 protein. The transgenic plants withstood salinity stress of **150 mM of NaCl** for a period of eight days while the untransformed control plants wilted at the end of the stress treatment in hydroponics. The transgenic plants also revealed better tolerance to drought stress in comparison to untransformed control plants.

To investigate the functions of antioxidative enzymes in a mangrove plant, we isolated three cDNAs encoding cytosolic Cu-Zn SOD (*Sod* 1), catalase (*Cat* 1) and ferritin (*Fer 1*) from *Avicennia marina* cDNA library. *Sod* 1, *Cat* 1and *Fer* 1 cDNA encoded full-length proteins with 152, 492 and 261 amino acids respectively. We studied the expression of these antioxidant genes in response to salt, iron, hydrogen peroxide, mannitol and light stress by mRNA expression analysis. *Cat* 1, *Fer* 1 showed short-term induction while Sodl transcript was found to be unaltered in response to NaCl stress. A decrease in mRNA levels was observed for *Sod* 1, *Cat* 1 while *Fer* 1 mRNA levels remained unaltered with osmotic stress treatment. *Sod* 1, *Cat* 1and *Fer* 1 mRNA levels were induced by iron, light stress and by direct HP2 stress treatment, thus confirming their role in oxidative stress response.

Ascorbate peroxidase has a key role in scavenging ROS and in protecting cells against their toxic effects in algae and higher plants. A cDNA encoding a peroxisomal ascorbate peroxidase, Am-pAPXl, was isolated from salt stressed leaves of *Avicennia marina* (Forsk.) Vierh. by EST library screening and its expression in the context of various environmental stresses was investigated. The peroxisomal localization of Am-pAPXl was confirmed by stable transformation of the GFP- (Ala)lO-Am-pAPXl fusion in tobacco. RNA blot analysis revealed that Am-pAPXl is expressed in response to salinity (NaCl) and oxidative stress (high intensity light, hydrogen peroxide application and excess iron).

Transcription Factors

Fifteen putative transcription factor cDNA clones were identified from the sequenced A. *marina* ESTs. These transcription factors were categorized into five major groups based on the domain it constitute. These transcription factors are MYB, NAC, AP2/EREBP, bZIP and Zinc finger and are known to play a vital role in abiotic and biotic stress tolerance. There is very limited report on role of MYB, bZIP and plant specific NAC transcription factors from mangroves with respect to stress tolerance. Northern analysis was done to further characterize two cDNAs PA272 and PD126 coding for MYB and NAC transcription factors respectively. Genomic clone of *Am-AmMYB 1 that is 1500bp* in length was isolated using AmMYBlcDNA forward and reverse VTR primers. AmMYBl cDNA coding for MYB transcription factor was cloned in pCAMBIA1301 binary vector under control of 35SCaMV promoter (Constitutive) and mobilized into *Agrobacterium tumefaciens* strain LBA4404 and was used to transform *Nicotiana tabacum* cv. Petit Havana by the leaf disc method. Preliminary selection of transformed plants were selected using Gus staining and PCR analysis with *AmMYBl* gene specific primers and further characterizations of this gene in contribution to different abiotic conditions are in progress.

NAC transcription factors are a family of functionally diverse proteins responsive to biotic and abiotic stresses. A full-length cDNA isolated from the salt stressed mangrove plant *Avicennia marina* showed high sequence identity to NAC proteins induced upon abiotic stress in tomato and potato. The predicted protein sequence had all the highly conserved sub domains characteristic of NAC domain containing proteins. Northern analysis for AmNACl expression under tolerable (250mM) concentration of NaCl revealed up regulation of the transcript after 48 h and higher transcript level after 10 days of treatment. Induction of AmNAC1 after 12 h of ABA treatment was similar to the treatment with stressful (500 mM) concentration of NaCl. The results suggest the involvement of AmNAC1 in early salt stress response and long-term adjustment to salt, besides a role for ABA in its expression under salt stress conditions.

Genes for Bioremediation

Prosopis juliflora is a tree species that grows well in heavy metal laden industrial sites and accumulates heavy metals. To understand the possible contribution of metallothioneins (MTs) in heavy metal accumulation in *P. juliflora,* we isolated and compared the metal binding ability of three different types of MTs (*PjMT1-3*). *PjMT1* and *PjMT2* are induced by copper and zinc respectively while *copper, zinc and cadmium induce PjMT3*. Variation in induction of *PjMTs* in response to metal exposure and their differential binding to metals suggests that each MT has a specific role in *P. juliflora*. Of the three MTs analyzed, PjMT1 shows maximum heavy metal sequestration and is thus a potential candidate for use in heavy metal phytoremediation.

Promoters

Promoter is a region of DNA that facilitates the transcription of a particular gene. Promoters are typically located upstream of the genes they regulate. For some traits, the expression pattern desired is constitutive, sometimes constitutive expression may be detrimental and expression limited to specific organs. Wide variety of promoters has been described from crop species, but relatively few have been tested for use in crop plants. Using â-glucuronidase (GUS) reporter gene fusions, we have characterized the expression patterns in transgenic tobacco of several nominally strong constitutive promoters, several number of gene promoters, inducible promoters from heterologous species such as *Avicennia marina, Porteresia coarctata* and *Prosopis juliflora*.

Promoter sequences upstream of PcNHX (Na+/H+ antiporter) from *P. coarctata*; PR244, a salt inducible gene from the mangrove *A. marina;* MYB1 transcription factor from *A. marina;* Ascorbate peroxidase from *A. marina;*

Lipid Transfer Proteins from *P. juliflora;* atypical LEA protein from *P. juliflora;* Photosystem II *PsbR* from *P. juliflora;* Metallothioneins from *P. juliflora* are being characterized.

Identification of Genes that are Uniquely Regulated during Oil Bio-synthesis in *Jatropha Curcas* Seeds

Fossil fuel resources are reaching finite limits and a sharp increase in oil prices all over the world has resulted in an unprecedented impetus for research on alternative sources of fuel, especially from that of plants. Many plants have the property of accumulating oil in the seeds. *Jatropha curcas* is one such oil yielding plant that has gained widespread attention in recent times as a source of bio-diesel. In addition to many factors that influence seed oil yield; regulation of genes that control the expression of proteins that bring about oil accumulation in seeds is one very basic factor that differentiates a high oil yielding plant from that of the average.

In order to identify genes that control active oil accumulation, *J. curcas* seeds were grouped into four maturation stages as follows : Early maturation stage, mid-maturation stage, late-maturation stage and full-maturation stage. Since protein expression will be more active when the oil accumulation is still progressing, mid- and late-maturation stages of seeds was used for the identification of genes. The experimental design essentially involved 'subtraction' of genes which are common to the stages mentioned above wherein genetic information pertaining to late stages of oil bio-synthetic pathway would only be left out as 'unique', for further analyses. In this regard, a cDNA subtraction library consisting of 'unique' information that relate to genes involved in seed oil bio-synthesis in *J. curcas* was constructed last year. To identify and characterize unique information, annotation of the same was performed through sequencing of a small portion of the 'expressed' genetic information called, Expressed Sequence Tag (EST). Information annotation of about 300 ESTs have been performed till date, of which, a few show homology to genes that have already been reported to be involved in the regulation of oil bio-synthesis in other plants.

Standardization of tissue culture methodologies for transforming *J. curcas* plants through gene transfer has been initiated using emerging cotyledons as starting plant material. Tissue culture methodology development will aid in transferring genes controlling active oil accumulation into *J. curcas* to obtain high and uniform oil yield.

Flower and Sex-specific Genes

In an effort to identify flowering as well as putative sex-specific genes, a cDNA library was constructed by isolating good quality, high concentration total RNA from a fresh flower of *Pandanus* using a modified LiCl method. A total of 979 flower-specific ESTs were thus obtained. CAP3 analysis performed on the above data set grouped them into 82 contigs (549 ESTs) and 430 singlets. Approximately 512 unigenes were identified. Important genes like Dehydrin, metallothionein, LEA, GST, Serine protease inhibitor, ATPases, chaperons, transcription factors, Zinc finger proteins, cellular transport proteins, ribosomal proteins, proteins involved in carbohydrate metabolism, lipid metabolism, proteolysis etc. have been isolated from the library. In addition to these several putative sex-specific genes like pollen specific protein, Arabinogalactan, MADS-box transcription factor, LIM transcription factor, profilin etc were also isolated. Expression studies (Northern) and Southern hybridizations are currently being performed to identify putative sex-specific genes in this plant.

Development of Transgenics

Salt and drought stress treatment : Seeds collected from the transgenic plants were germinated on MS medium and grown for a period of 2 weeks in 16 h light/8 h dark cycle. Then the plants were transferred to Yoshida medium maintained at pH 5.0 and grown for a period of three weeks in a hydroponic solution, in a green house with 12 h dark/light photoperiod. After three weeks of hydroponic growth the nutrient solution was replaced with fresh solution consisting of 150 mM NaCl to study the effect of salt stress on the transgenic plants.

For pot experiments both transgenic and control plants were used for the study. Plants of untransformed control and homozygous transgenic plants of both the lines were used for the pot experiments of salinity and drought stress experiments respectively. For pot experiments 40 day old plants, grown in pots were watered with salt water (100 mM NaCl) every four days. For drought stress treatment, the transgenic plants were grown in pots for a period of fifty days with normal watering everyday. Drought stress treatment was provided by withholding irrigation. Two days prior to the drought stress treatment the irrigation was stopped to ensure the soil was dry and no moisture content was present in the soil. The drought treatment was done for two cycles of 100 h of drought stress in each cycle and the plants were irrigated once between the two cycles.

Three limited field trials were also conducted with the approval of the regulatory authorities of Govt. of India. ***This is the first ever report on the increased tolerance to salinity stress and drought stress by over expressing a cytosolic SOD in rice.*** Our study reconfirms the importance of cytosolic Cu/ZnSOD isoform in conferring abiotic stress tolerance in plants SOD genes exist as multiple gene families in mangroves. Isolating the genes coding for different isoforms of SOD from *A. marina* and other mangroves and studying the gene regulation of these isoforms would provide more input on the role of SOD enzyme in providing protection to the mangrove plants against abiotic stresses. These studies would in turn throw light on how these genes can be utilized for transforming to crop plants for enhancing their abiotic stress tolerance. ***This happens to be the first report on the transfer of a gene isolated from a mangrove to rice, which is a crop plant.*** The study revealed that mangroves could be potential sources for isolating novel genes for abiotic stress tolerance given the fact that they are halophytes and is capable of tolerating a high degree of salinity and abiotic stress.

Introgression to Local varieties : Once the trangenics are generated they could be crossed using breeding technologies, introgressed into any related varieties to create a diverse background for the transgenics. From the population of 200 BC2F1 plants in each variety of rice, the positive plants (50%) which possess the transgene were identified initially by PCR analysis and among them those plants with a higher resemblance to the respective local variety were identified by phenotypic selection and tagged, and the presence of the gene was confirmed by Southern blot analysis. The expression of the gene was confirmed through the isozyme analysis in the case of *AmSOD*. Such plants were used as the female parent; pollen from the respective local variety was dusted and BC3F1 seeds were obtained. The BC3F1 plants were screened as per the same procedure followed for BC2F1 plants, crossing was done and BC4F1 seeds were obtained. BC4F1 plants are being screened through PCR analysis for identification and selection of positive plants to serve as female parents for crossing. In addition, a few panicles in the BC3F1 plants, used as female parents for crossing, were selfed and the seeds were collected. These seeds were sown to get the BC3F2 generation. Twenty plants are maintained in the BC3F2 generation and the expected Mendelian ratio of 3 : 1 was observed with respect to the corresponding transgene. Homozygous plants will be identified in the next generation and subjected to stringent screening for salinity and drought stress tolerance.

Co-expression of Active Oxygen Species (AOS) scavenging genes in rice and evaluation of the same for increased salinity tolerance : In *Agrobacterium* mediated transformation of pCAM *Am-APX* + *Am-MDAR* in rice, two lines regenerated under hygromycin selection. PCR analysis for Am-APX and Am-MDAR revealed that line DC1 was positive for both genes, the 1.1kb and 1.9kb fragment specific to Am-APX and Am-MDAR respectively but in DC2 only MDAR was amplified but not APX. This might be due to partial removal of T-DNA during infection. Genomic DNA was isolated from DC1, DC2 and untransformed control and digested with *Sac* I (linearises the T-DNA) to determine the copy number of integration of Am-APX and Am-MDAR in the rice genome.

Twenty-four seeds (T0 generation) were sown from the DC1 lines. DNA was isolated from all the lines and PCR analysis revealed that among the 24 lines, 18 lines were positive for both APX and MDAR. The results revealed that the segregation of DC1 in T1 generation was in 3 : 1 Mendelian ratio.

Chapter 28

Breeding for Salt Tolerance in Rice

SALT-EFFECTED SOILS AND THEIR EXTENT

In general, the term salinity includes all the problems due to salts present in the soil while in strict term, these soils are categorized into two types : sodic (or alkali) and saline, however, there is third type of salt-affected soils are also found and referred as saline-sodic soils. Sodic soils are dominated by excess sodium on exchange complex and a high concentration of carbonate / bicarbonate anions. Such soils have high pH (greater than 8.5 and sometimes up to 10.7) with a high sodium absorption ratio (SAR) and poor soil structure. Saline soils are again dominated by sodium cations with electrical conductivity (EC) more than 4 dSm^{-1}, but the dominant anions are usually soluble chloride and sulphate. pH and SAR values of these soils are much lower than in sodic soils. Saline-sodic soils, also called saline-alkali, have both high ESP and EC.

The extent of salt-affected land has long been uncertain and remains so. World wide the estimates range from 340 to 1,200 x 10^6 ha. Despite this much area either barren or with very low productivity, there are only few instances where salt tolerant cultivar have been developed. The reason being that it is a very complex trait and has many components, which are probably controlled by polygenes.

WHY DO WE NEED SALT TOLERANT CULTIVARS?

Worldwide, the research to overcome salt related problems is based on two approaches; (i) either change the growing environment (make it normal) suitable for the normal growth of plants; (ii) or select the crop and/or change genetic architecture of the plant so that it could be grown in such areas. The first approach involves major engineering and soil amelioration process which need lot of resources are often out of the reach of small and marginal farmers. The second approach *i.e.* breeding crop varieties with *in-built* salt tolerance is realized as the most promising, less resource consuming /economical and socially acceptable approach. So the ability of the plant to tolerate the salt stress upto an extent is of paramount importance to mange the resources optimally and this is the reason to develop the tailored crops with higher salt tolerance suited to salt stress environments.

There is third approach as well which can be termed as hybrid approach as it is the combination of both environment modifying and biological approach. It is highly productive, less resource consuming and economically viable approach. Nowadays major soil reclamation programmes in different states involve both biological and hybrid approaches to combat the salt problem.

PREREQUISITE FOR THE DEVELOPMENT OF SALT TOLERANT CULTIVARS

3.1 **Wide spectrum of variability in available germplasm** : Existence of genetic variability for salt tolerance within species is of paramount importance in crop improvement programme. Therefore choice of germplasm in breeding programme is most crucial as the success lies on it. Extensive germplasm collection provides a useful source of genetic diversity for the studied traits.

3.2 **Target environment / site characterization** : Before designing any ideal plant type, it is crucial to define the soil and agroclimatic conditions of the target areas for which they are to be develop. Genotypes which are suitable for coastal areas may or may not not be fit for sodic soils or inland saline soils and *vice-versa*. Therefore exact site characterization is an important aspect to meet the objective(s).

3.3 **Availability of the defining traits / selection criteria** : Ideally germplasm should differ as much as possible for the traits to be improved or introgressed. Other traits should not vary too much otherwise keeping all the desirable traits into one superior agronomic background become very difficult.

3.4 **Repeatable screening techniques** : Reliable and repeatable screening techniques are the mainstay of any successful breeding programme specifically for biotic or abiotic stress breeding. Though screening techniques vary with crop species, growth stage and type of stress imposed but ideally it should be rapid, reproducible, easy and affordable.

MANIFESTATION OF SALT STRESS ON PLANT

Degree of salt stress can affect the different crops differently. For rice, soil salinity beyond ECe ~ 4 dS/m is considered moderate salinity while more than 8 dS/m becomes high. Similarly pH 8.8 - 9.2 is considered as non-stress while 9.3 - 9.7 as moderate stress and ^{3} 9.8 as higher stress. Extremely high salt stress conditions kills the plant but the moderate to low salt stress affect the plant growth rate and thereby manifest symptoms which could be associated with morphological, physiological or bio-chemical alterations.

Morphological Effects

Most of the parameters like low tillering, spikelet sterility, less florets per panicle, low 1000 grain weight and leaf scorching, are affected uniformly under both sodicity and salinity, however it is not a thumb. Major symptoms are :

- White leaf tip followed by tip burning (salinity)
- Leaf browning & necrosis (sodicity)
- Stunted plant growth
- Low tillering
- Spikelet sterility
- Low harvest index
- Less florets per panicle
- Less 1000 grain weight
- Low grain yield
- Change in flowering duration
- Leaf rolling
- White leaf blotches
- Poor root growth
- Patchy growth in field.

Physiological and Bio-chemical Effects

It is observed that crop varieties and breeding lines do differ for their inherent capability to modify various physiological and bio-chemical processes in response to the salt stress. Though numerous physiological and bio-chemical changes take place under altered stress environment but only few of them change very significantly and also contribute a lot to the salt tolerance mechanism. These changes control the solute and water balance and their distribution on whole plant and tissue basis. Based on the studies it was observed that most of the crop plants and varieties encounters following physiological and bio-chemical manifestation under higher salt stress conditions.

- High Na^+ transport to shoot
- Preferential accumulation of Na in older leaves
- High Cl^- uptake
- Lower K^+ uptake
- Lower fresh and dry weight of shoot and roots
- Low P and Zn uptake

- ❖ Change in esterase isozyme pattern
- ❖ Increase of non-toxic organic compatible solutes
- ❖ Increase in Polyamine levels.

MODE OF TOLERANCE TO SALT STRESS IN PLANTS (GLYCOPHYTES)

Under salt stress conditions, the crop plants either try to avoid the stress, which is indeed not an actual tolerance mechanism or employ the following mechanisms to overcome the salt damage in a sequential adaptations :

1. Minimize the initial entry of salt from roots
2. Intra-cellular compartmentation
3. Plant level transport of salt and its compartmentation.

Avoidance

Most of the crop plants are relatively sensitive at early seedling and flowering stage. Rice, being transplanted crop, can alleviate the salt stresses at seedling stage by management *i.e.* transplanting of aged seedlings but cannot avoid stress at flowering stage. However, under coastal saline conditions salinity sometime increase near the terminal growth stage of the plant. In that case, plants mature fast to complete their life cycle. It is a typical case avoidance rather than tolerance but it works as far as the productivity is concerned. The false signals due to avoidance should be carefully separated out from the actual tolerance.

Initial Entry of Salts from Roots

Plants roots experience the salt stress when Na^+ and Cl^- along with other cations are present in the soils in varying concentration (1 to 150 mM - for glycophytes; more for halophytes). Ion uptake depends upon the plant growth stage, genotype, temperature, relative humidity and also light intensity. Excessive amount of salt in the rhizosphere retards the plant growth, limits yield and even cause the plant death. The toxic ions sneak into the plant along with the water stream which moves from soil to the vascular system of the root by different pathways like symplastic and apoplastic. In symplastic pathway, water goes in the roots through plasma membranes of epidermis and then moves cell to cell through plasmodesmeta until offloading to the xylem. In apoplastic pathway, water moves through intra-cellular spaces to unload the salt in xylem. Differential osmotic potential is the driving force of energy driven pathways *i.e.* symplastic, while apoplastic is a non-energy driven pathway. So based on osmotic potential, plant can check the toxic ions like Na^+ to enter into the cell through energy driven pathway. Na^+ and K^+ are mediated by different transporter which is clearly demonstrated by Garciadebleas *et. al.*. They used Ba^{2+} which inhibits Na^+ uptake but not K^+ uptake in rice roots to demonstrate the differential transport pathways.

Intra-cellular Compartmentation

A number of mechanisms are reported to affect the salt tolerance in plant based on cell level tolerance.

a) Ion Homeostasis Pathway

Ion homeostasis in cell is taken care of by the ions pumps like antiporters, symporters and carrier proteins on membranes (plasma membrane or tonoplast membrane). Salt Overly Sensitive (SOS) regulatory pathway is one good example of ion homeostasis. This pathway is activated after the receptor perceives the salt stress to alter protein activity and gene transcription by signaling intermediate compounds. One of its example is, three salt-overly sensitive mutants (*sos*) which are hypersensitive to specific salt NaCl (rather than osmotic effect as they are not sensitive to mannitol). *SOS1, SOS2* and *SOS3* mutants exhibits different phenotype with reference to Na^+ accumulation. *SOS3* is calcium binding protein which directly interacts and activate *SOS2,* serine / threonine protein kinase. *SOS3* inducts *SOS2* on the plasma membrane, where *SOS3-SOS2* complex protein kinase complex phosphorylates *SOS1* to stimulate the Na^+/H^+ antiporter activity of Arabidopsis thaliana.

Na^+ which enters leaf cells, is pumped into cytoplasm before it reaches to toxic level for enzymatic activities. This pumping activity is controlled by valuolar Na^+/H^+ antiporters. Addition of salt induce the Na^+/H^+ antiporter activity but it increases more in salt tolerant than salt sensitive species. This mechanism has been emphasized by certain experiments where over- expressing of vacuolar transporter (*NHX1*) has increased the salinity tolerance of Arabidopsis, Tomato, *Brassica napus* and rice. This increase uptake of Na^+ to short vacuoles could facilitate enhanced storage of Na^+ and ultimately conferring greater tolerance by reducing Na^+ in cytosol.

b) Synthesis of Osmoprotectants

Though osmoprotectant enable plants to tolerate more salinity but still a significant amount of Na^+ needs to be compartmentalized for better tolerance. Therefore, it is desirable that over-production of osmoprotectant is governed by the pleiotropic control of vacuolar Na^+/H^+ antiporter activity. Most of the plants, bacteria and many other organisms accumulate certain organic solutes such as sugar, alcohol proline, quarternary ammonium compounds in response of osmotic stress hence they are called osmoprotectants and also termed as compatible solutes because even in high concentration they do not interfere with enzymatic activities. These are localized in cytoplasm and the inorganic ions such as Na^+ and Cl^- are preferentially sequestered into vacuole, thus leads to the turgor maintenance for the cell under osmotic stress.

Trehlose, a non-reducing sugar, possess a unique feature of reversible water storage capacity to protect biological molecules from desiccation damages. Recently there has been growing interest of utilization of trehlose metabolism to ameliorate the effects of abiotic stresses. Garg *et. al.* have demonstrated the expression of trehlose bio-synthesis conferred the tolerance to multiple abiotic stress. The increase in trehlose levels in transgenic rice lines of Pusa basmati 1 using either tissue specific or stress dependent promoter, resulted into the higher capacity for photo-synthesis and concomitant decrease in the extent of photo-oxidative damage during salt drought and low temperature stresses.

c) Signal Pathway - Transcription Factors

Another mechanism act in response to stress is known as transcription factor. Indeed transcription factor bind to specific sequence of the promoter regions of target genes which needs to be activated collectively or sequentially in response of stress (drought, salinity or temperature). These promoter regions include dehydration-responsive elements (DRE's) and ABA responsive elements (ABRE's) which are involved in the plant responses to the osmotic effect. The transcription factor *DREB1A* specifically interacts with *DRE* and induces the expression of stress tolerance genes. Constitutive overexpression of source of the genes encoding for these protein can induce the constitutive expression of many genes resulting into increased tolerance but it associated with reduced growth even under unstressed condition. *DREB1A* activated at least 12 genes in Arabidopsis but caused dwarfism of the plants. Hence, such genes, when were used with stress inducible promoter (*rd29A*), plants looked normal and showed enhance stress tolerance.

d) Stress Activated Protein Pathway

Plants produce many kind of stress responsive proteins induced by various kind of stresses like heat, cold, salt or drought etc. Major one of them is like LEA and dehydrins etc.. Rice accumulates LEA family of proteins during stress. Chaurey *et. al.* observed six prominent shoot salt stress induced proteins (SSPs) of 18.5, 22, 23, 26, 40 and 46 kda upon 100mM NaCl stress to rice variety (Bhura rata) for 2 weeks. These six proteins are induced early and synthesized throughout the salt stress and accumulate to high level. However, there are many small proteins also which expressed transiently. Out of this four SSRs have identified as LEA protein. These are reported to play an important protective role during desiccation/salt stress in rice plants. Some of these proteins make amphiphic a-helic structure which readily binds to ions. They are reported to be analogous to the *HVA* gene product in barley.

e) Reactive Oxygen Species (ROS)

Salt stress in plants induce higher concentration of ROS/intermediate such as superoxide, H_2O_2 and hydroxy-radicals due to the impaired election transport processes in chloroplast, mitochondria and photo-respiration pathway. Under normal growth conditions, the production of ROS in cell in as low as

240μMS^{-1} superoxide and the steady state level of H_2O_2 in chloroplast is 0.5μM. However, under salinity, the level of ROS production reaches to as high as 720μMS^{-1} (3 fold) and H_2O_2 level as high as 15μM (30 fold). It is reported that H_2O_2 concentration of 10μM reduces the net photo-synthesis rate by 50%. Superoxide and H_2O_2 toxicity have been attributed to the cascade reactions that result into the production of hydroxyl radicals and other destructive species like lipid peroxidases. Indeed hydroxyl radicals are very reactive and can damage vital macromolecules by protein denaturation, mutation and peroxidation of lipids.

Plants have devised different system for scavenging of ROS by using the enzymes like SOD peroxidases, catalases and antioxidants like ascorbate and reduced glutathione. There is variability among rice genotypes for the enzymatic and non-enzymatic scavenging system hence it is possible to tag the genes coding for both enzymatic and non-enzymatic ROS scavenging agents and use them in engineering the desired plants of MAB.

Plant Level Transport of Salt and its Compartmentation

Plant level compartmentation is the most important mechanism conferring the salt tolerance. Plant, as an intelligent entity, transport the toxic ions to the older leaves and leaf sheaths which are ready to sacrifice for early senescence and/or death at the cost of saving young growing meristematic tissues. Ultimate aim of any crop plant is to undergo life cycle which completes with reproductive stage and maturity. Most of the crop plants are very sensitive to salt stresses specifically during reproductive stage, hence plant try to avoid transport the toxic ions to the flowering tissues. However there is difference in capacity to check the Na^+ to the reproductive tissues among the varieties. Salt tolerant restrain the transport of toxic ion better than the sensitive ones.

Screening Criteria

Reproducible differential manifestation in plants with respect to their morphological, physiological or bio-chemical parameters in response to salt stress qualifies for a reliable screening criterion. Reliable screening is an integral part of any successful breeding programme. Salt related problems seldom occur in isolation and are coupled with many associated problems. Complexity of the salt tolerance, soil heterogeneity and various interactions are the major hurdles for the repeatability of the results. Following parameters are considered screening :

Morphological Parameters

Though there is no single definite morphological marker available for salt tolerance or sensitivity in any crop, but a combination of criteria give a good indication toward the salt response of crop plants. Therefore, several parameters are used in combination for the effective and reproducible screening.

(a) **Germination studies :** Germination percentage, coleoptile and radicle length under varying degree of salt stress for different crops is a good salt tolerance indicator at initial stages. Higher salt concentration delays or reduces the germination..

(b) **Survival of the plant :** It is mainly limited to the seedling studies; however, in some of the adult plant studies it has also been considered. Under moderate stress, plant survival is not a problem but under higher stress, it is a good selection criterion.

(c) **Injury score :** Individual plant or group of genotypes are scored usually on 1 to 5 or 1 to 9 scale where lower score towards 1 states tolerant and higher score denotes sensitive genotypes. In rice IRRI's Standard Evaluation System for salinity, sodicity and Singh *et. al.* technique for sodicity on 1-9 scale is followed.

(d) **Phenotypic expression :** Excessive tip burning especially in younger leaves, spikelet sterility and stunted growth are considered for the overall phenotypic expression of the genotype under stress environment.

(e) **Grain Yield :** In the absence of any simple and reliable selection criteria, the 50% reduction in grain yield of the genotypes under salt stress in relation to the normal (non-stress) yield has been considered as critical limit for selection/rejection of the genotypes.

(f) **Stability of traits over environments :** The genotype with high mean, near to unit regression value (bi) and uniformity of regression (Sdi^2) under multiple stress environments are adjudged as suitable, stable and adaptable genotypes for sustainable productivity in problem soils. Selection for high mean yield is a fundamental selection criterion for all the varieties while selection based on unit regression value and least deviation from regression indicates the stable performance across a range of environments.

(g) **Mean tolerance index (MTI) :** It is a product of Stress Resistance Index (SRI) and Response Index (RI) of the genotype. Here SRI is the performance of a genotype for that trait averaged over all the stress environment in conjunction with performance under corresponding non-stress environment and RI is the mean of the genotype for the trait under all the stress environments divided by the mean performance of all the genotypes over all the stress locations. Beside yield, it has been used for germination, seedling screening, grains per panicle, ear bearing tillers /m^2 and many more traits for the screening of the genotypes. Similar to MTI, another criteria Stress tolerance index (STI) and Susceptibility indices were used using seed yield, main shoot length, siliqua/main shoot and primary branches / plant in Indian mustard.

Physiological Parameters

Extremely high salt concentration kills the plant but the moderate salt stress exhibits the growth differences among the crop varieties. The tolerance to salt stresses is complex phenomena because it may require the combination of different independent and/or interdependent mechanisms and pathways. A tolerant genotype can be expected to have more than one adaptation.

(a) **Na and K Uptake :** In general, tolerance of a crop variety was found to be associated with its ability to restrict potentially toxic ion uptake like Na^+ and associated with preferential uptake of the balancing ion like K^+. It is like an adaptation for the survival of plants so that the vital metabolic activities are not hampered. There are larger differences in ion (Na^+ and K^+) uptake between the species in comparison to the genotypic differences within a crop species. These are most studied parameter for the salt tolerance in crop plants.

The tolerant varieties maintained lower Na concentrations besides maintaining K concentrations under high sodicity. On the other hand, the sensitive rice varieties were unable to effectively prevent accumulation of Na^+ as well as the depletion of K^+. The success of the tolerant varieties in gaining higher fresh and dry weights at all the stages widens the differences in their Na concentrations still further by dilution. Tolerant genotypes like CSR 1 showed regulation over distribution and accumulation of Na taken up by the plants *i.e.* the delicate and vital organs like young and photosynthetically active leaves as well as the reproductive organs like panicles are kept relatively free of Na, beside having an assured supply of K even under higher salt concentration.

No such regulation of Na and K seems to operate in case of salt sensitive genotype, which further aggravates the stress situation. Thus the differences in the distribution of ions in individual plant organs and with the age and position of these organs on plant are more important indicators of its tolerance potential than the difference in the average salt content of the plant.

(b) **Na/K ratio :** Although the uptake of both Na and K is entirely independent, but lower Na/K ratio is considered as desirable trait as it maintains the ion balance. Na is transported to shoot usually through apoplastic pathways (passive transport) while K transport takes place through symplastic pathway *i.e.* through membrane / plasmalemma (active transport). Younger leaves relatively have lower Na than K as compared to the older leaves, which in turn resulted in higher Na/K ratio in case of the older leaves.

Thus Na / K ratio increased steeply with salt concentration and leaf age, and the two rice varieties behaved differently. Consequently the tolerant variety keeps their leaves relatively free of the toxic ions besides having assured K supply. This factor along with the higher number of the leaves and higher leaf area probably, contributes to its success under high salt concentration. However this parameter is also not a universal phenomena hence could be taken as a selection criteria but with cautiously.

(c) **Tissue tolerance :** Tissue tolerance is measured in terms of LC_{50} which is an analogue of LD_{50}. Here LC_{50} is the concentration of sodium (in mmol g^{-1}

ethanol-insoluble dry wt.) in the leaf tissue which causes a 50% loss of chlorophyll. It is taken as an indicator of metabolic damage to the tissue due to the salt load. LC_{50} values differs for different genotypes and not all the varieties in high tissue tolerant category were highly salinity tolerant and vice-versa. For example, GR11 and IR2153-26-3-5-2 are highly tissue tolerant varieties but phenotypically they are not highly salt tolerant varieties. Similarly not all the low tissue tolerant varieties like CSR 10 having low tissue tolerance are highly salt tolerant variety.

SCREENING METHODOLOGY

***In-situ* Field Evaluation :** Field screening is the best way to identify the most suitable genotypes because salt tolerance is very complex phenomena but spatial variability in the field sometimes give the false positive results due to escape. Therefore extensive field testing for soil salinity gradient and blocking of the same is the best way to minimize the error chances. Long rows of 10 to 30 meters with space planting depending upon the type and generation of material minimise the effect of spatial variability. It allows maximum possible exposure of all genotypes to varying soil sodicity conditions. The layout for such a test is generally incomplete block design or augmented design in which a set of check varieties is replicated many times. It increases the potential number of test varieties allowing screening of large number of genotypes at the same time. It is further possible to cut across the long plots in several parts as replicate to obtain varietal performance at varying levels of soil stresses. The involvement of more number of checks have been found better in data processing and finalizing the tolerant lines.

Screening in Microplots : Soil heterogeneity and spatial variability hinders the reliability of the response of genotypes in true and dependable way. This attracted the development of mini-field environments with varying levels of controlled salinity and sodicity environments. These devices are series of dug-out cavity structures made of brick-mortar-concrete materials and filled with artificially prepared or natural transported problem soil. It is possible to create and maintain desired levels of sodicity and salinity in these microplots in a manner very much comparable to field conditions minus the soil heterogeneity. The plot size in these pits are kept very small (single row) to 2-3 rows but because of a good control over micro-environment, it is highly representative of the genotypic performance. The microplots are being used to screen mostly early segregating and also stabilised populations besides genetic studies.

Screening in Pots : For more precise study of the individual plant response under a constant stress, round porcelain pots of 20 or 30 cm dia, with a capacity of 8 or 16 kg soil with a provision to allow or plug off leaching from bottom, are used. Pot studies also allow for periodic uprooting of plants at a given growth stage for specific studies associated with age of plant. The edaphic environment of the pots is more or less uniform throughout the growth period of plant in respect of degree of stress. The genetics for salt tolerance is being studied in pots. Sand and soil cultures are being used in the pots for the controlled studies.

Screening in Solution Culture : A separate hand handout for this has already been provided.

BREEDING METHODOLOGY

Conventional Breeding

Almost all the conventional breeding methods have been followed for the development of the salt tolerant materials *i.e.* introduction, selection, hybridization, mutation and shuttle breeding approach. Most of the initial salt tolerant rice varieties Damodar (CSR 1), Dasal (CSR 2), and Getu (CSR 3) were the pure line selections from the local traditional cultivars prevailing in the Sunderban areas in West Bengal, India. Later on other salt tolerant varietal series like CSR 10, CSR13 CSr23, CSR27, CSR30 and CSR36 were developed through recombination breeding following methods like :

- Pedigree method
- Modified bulk pedigree method : Individual F_2 plant harvest is bulked upto F_4 generation followed by panicle selection and handling the population as in pedigree method.
- Shuttle Breeding.

The segregating materials are put in the long rows under salt stress with space planting in F_2 particularly. Selection pressure is gradually increased with the generation advancement simultaneously in moderate stress and high stress of sodicity and salinity.

Recombination Strategy

Grouping of the genotypes based on the inherent physiological mechanism responsible for salinity tolerance, inter-mating of the genotypes with high degree of expression of the contrasting salinity tolerance mechanism and identifying / screening of the recombinants for pooling of the mechanisms is being followed to enhance the further level of salt tolerance. The genotypes are grouped into different categories based on the physiological mechanism for salt tolerance. Crosses are made between the parents/donors possessing contrasting physiological traits like tissue tolerance, Na^+ exclusion, K^+ uptake and Cl^- exclusion to pyramid the genes governing or contributing for salinity tolerance into one agronomically superior background. The ideal high yielding salinity tolerant variety should posses following traits :

- Able to withstand high amount of Na^+ (tissue tolerant)
- Per day uptake of Na^+ is minimum (takes more number of days for LC_{50} stage)

- High uptake of K^+ per day
- Good initial vigour
- Agronomically superior with high yield potential

Recent Strategies to Supplement the Conventional Breeding Programs

Marker Assisted Selection (MAS)

DNA MAS has been seen as a means of improving the speed and efficiency of plant breeding programs because it is growth stage independent, unaffected by environment; no dominance effect and efficient to use in early generations. Most widespread use of MAS to date has done in the marker assisted back-crossing (MAB) of major genes to into already established varieties, mega-varieties (which occupies a large area within the country on across the countries) or elite cultivars. These markers could reduce the linkage may around the target gene, and also recover the recurrent parent background within less number of generation in comparison to conventional breeding.

(a) Marker-Assisted Selection in Early Generation

MAS in early generation is most useful for relatively less number of genes but which are affecting the important traits and difficult to phenotype. Two important factor need to be satisfied for effective MAS strategy, first : the markers are tightly linked (1-2 cM) to loci with large effects on trait which are difficult or costly or appear lately (maturity) for accurate phenotyping. Second, specific marker alleles are associated with desired alleles at target loci consistently across the different breeding populations. But unfortunately both of these two situations are not applicable for most traits and most populations. MAS for soybean *cystx* nematode is a good example where maker alleles (2 flanking marks) are consistent with desired allele (*rhg1*) across the populations and it is difficult to phenotype.

(b) Marker Assisted Back-crossing

Most wide use of markers in conventional breeding have been in back crossing purpose where previously evolved varieties or elite material through conventional breeding is augmented with selected alleles with major effects for which they are lacking. Young and Tanksley, demonstrated that large amount of DNA from the donor can remain around the target gene even after many generation of back-crossing. This surrounding material contributes toward "linkage drag" especially if the donor parent is a wild relative. So markers are used to select the same progeny in which recombination near the target gene have as little chromosome segment as possible. This is called "fore-ground selection".

(c) MAS for Polygenic Traits

Most of the traits of economic importance like yield and stresses are controlled by polygenes and considerably influenced by environment and g x e interaction for their expression. These traits are most difficult to breed conventionally and using non-conventional techniques like MAS as well.

DNA markers could be of great importance to plant breeding if they are used to aid selection for quantitative traits. Now a days micro-satellites or simple sequence repeats (SSR) markers are the first choice of molecules biologist while single nucleotide polymorphism (SNPs) are going to be the most preferred markers of the future. After the completing the draft sequence of rice genome with >99% accuracy, 18,838 new class I SSR markers are reported on its physical map.

Major difficulty in the employing MAS for polygenic traits is the limited phenotyping accuracy of the quantitative trait loci (QTL) because QTLs do not have direct phenotypic variation hence their chromosomal location is typically inferred by calculating the LOD (likelihood of odds) value. The chromosomal location with maximum LOD value has the likelihood for the QTLs of the trait but there is always a good possibility that the QTL is not located precisely at the maximum likelihood position. This precision could be increased by increasing size of mapping population..

(d) Quantitative Trait Loci (QTL) and Candidate Genes for Salt Stress

Identification of molecular markers tightly linked to salt tolerant genes can serve as land marks for the physical localization of such genes facilitating marker assisted selection (MAS). In rice a major gene for salt tolerance has been mapped to chromosome #7 using RFLP markers. Random amplified polymorphic DNA markers were also shown to be linked to salt tolerance using the same population. Sahi *et. al.* found six novel clones which showed insignificant homology to any of the existing expressed sequenced tagged (EST) database and are differently regulated in salt tolerant and sensitive (Pusa Basmati 1) rice varieties under salt stress. Several QTLs for salinity tolerance have also been identified. Lang *et. al.*, 2001 detected one micro-satellite marker, RM 223 associated with salt tolerance at vegetative stage. Based on RFLP markers, Koyama *et. al.* have identified the QTLs for Na and K uptake and Na : K ratio. Major QTLs were found on chromosome 6, 4, 1 and 9. IRRI also reported major gene for salinity tolerance on chromosome 1 which could be similar to the *Sal*T gene of the rice Cornell map. Glenn *et. al.* also tried to construct an AFLP map with the help of 206 markers and found major QTLs for high K absorption, low Na uptake and low Na : K ratio on 5 chromosomes (No. 4, 12, 3, 1 and 10) including the big affect QTL *i.e. Saltol* on chromosome 1. Molecular markers based approach (MAS) has enormous potential to be used as reliable tool to the breeders for the effective, fool proof and early generation screening.

Ren *et. al.* mapped eight QTLs responsible for the K^+ and Na^+ content of F_2 RILs of cross Nonabokra and Koshikari in rice. *SKC1* was mapped as a major QTL for shoot K^+ content on chromosome 1. They used BC_2F_2 populations of 2,973 plants to generate high resolution map and also to find out the Koshikari background with clone of *SKC1* from salt tolerant Non-abokra. Under salt stress, the shoot K^+ content in the NIL (*SKC1*) containing small Non-abokra region between caps markers K159 and K 061 was substantially higher than that in the isogenic control Koshikari, hence regulating the K^+/Na^+ homeostasis under salt stress.

(e) Practical Application of MAS

Major research program on MAS at International Rice Research Institute (IRRI) are concentrated on introgressing *Sub1* and *Saltol* QTLs for submergence and salinity tolerance respectively into the improved germplasm or adapted varieties. The major QTL for salinity tolerance *i.e. Saltol* is also being transferred to the improved background as well as adapted rice varieties. *Saltol* is located on chromosome 1 and linked with SSR markers like RM 8094, RM 493 and RM 3412. However, in both the cases, these markers are population specific which limits the scope of the MAS application across the diverse material. Flowers *et. al.*, while reporting the putative AFLP markers for ion transport and selectivity for salinity tolerance from a custom made mapping population of rice, also cautioned against any expectation of general applicability of markers for these physiological traits.

GENETICS OF SALT TOLERANCE

The Inheritance of sodicity tolerance was inferred from the phenotypes observed in F_1 and bulked F_3 generation of 3 crosses between two tolerant parent and two sensitive parents (Basmati 370 and Pakistani basmati) under artificially created sodic soils (pH 9.8) in lysimeters. All the F3 plants, scored based on IRRI scoring method at adult stage, showed continuous variation suggesting that sodicity tolerance is controlled by polygenic trait acting additively in most of the cases along with interactions between the alleles at some loci. A similar inheritance study for salinity was conducted involving same populations. Results indicated the role of few major genes alongwith numerous minor genes involved for salinity tolerance. It was also inferred that salinity tolerance trait is polygenic in nature and lacks maternal influence.

IMPACT OF THE PROGRAMME

A long chain of introduction, selection and recombination processes led to the development of improved salt tolerant materials which directly benefited the farmers by increasing their harvest in salt affected lands. Increased food production in fields with low or zero productivity not only improved economy and well being of the poor farmers / owners of the land but also increased the employment avenues to the local people during the crop season in addition to the improved status of the state's and country's granaries.

There are many rice varieties which have been released as salt tolerant varieties in many countries. Philippines has released many IRRI developed materials as variety like IRRI-112 as PSBRc48 (Hagonoy); IRRI 113 as PSBRc50 (Bicol); IRRI 124 as PSBRc84 (Sipocot); IRRI 125 as PSBRc86 (Matnog); IRRI 126 as PSBRc88 (Naga) and IRRI 128 as NSICRc106. In other countries also many salt tolerant rice varieties have been released for commercial cultivation like CSR10, CSR13, CSR23, CSR27, CSR30, CSR36 and Lunishree, Vytilla 1, Vytilla 2, Vytilla 3, Vytilla 4, Panvel 1, Panvel 2, Sumati, Usar dhan 1, 2 & 3 (from India); BRRI dhan 40, BRRI dhan 41 (from Bangladesh); and OM2717, OM2517, OM3242 (from Vietnam).

Chapter 29

Solute Transport

RELATING THE MEMBRANE POTENTIAL TO THE DISTRIBUTION OF SEVERAL IONS ACROSS THE MEMBRANE : THE GOLDMAN EQUATION

Many different ions permeate the membranes of living cells simultaneously; therefore, the Nernst potential for any single ionic species seldom describes the membrane's diffusion potential. Instead, the contributions of all the ions to the diffusion potential across the membrane are described by the Goldman equation :

$$\Delta E = \frac{RT}{F}\left(\ln \frac{P_{K^+}C_{K^+}{}^{o} + P_{Na^+}C_{Na^+}{}^{o} + P_{Cl^-}C_{Cl^-}{}^{i}}{P_{K^+}C_{K^+}{}^{i} + P_{Na^+}C_{Na^+}{}^{i} + P_{Cl^-}C_{Cl^-}{}^{o}} \right)$$

This equation relates the standing ion gradients across a membrane to the diffusion potential that develops. $P_K{}^+$, $P_{Na}{}^+$, and $P_{Cl}{}^-$ represent the membrane permeabilities for K^+, Na^+, and Cl^-, respectively. Although the equation should include terms for all ions passing through the membrane, K^+, Na^+, and Cl^- have the largest membrane permeabilities and highest concentrations in plant cells and therefore dominate the equation.

The relationship between the Goldman equation and the Nernst equation can be visualized if one imagines a membrane that is permeable to only one ion, say K^+, so that both $P_{Cl}{}^-$ and $P_{Na}{}^+$ equal zero. Under these conditions, the Goldman equation reduces to the Nernst equation for K^+. Although biological membranes are never permeable to only a single ionic species, artificial membranes can approximate this situation and are used in the manufacture of pH electrodes and other ion-selective electrodes.

PATCH CLAMP STUDIES IN PLANT CELLS

The use of the patch clamp method in plant physiology is generating information about the properties of plant membranes that is beyond the scope of other techniques. Conventional electro-physiological studies with intra-cellular electrodes are valuable for the measurement of membrane potentials and other electrical properties of plant cells, but these methods have limitations for the characterization of ion pumps and single ion channels. Patch clamping, on the other hand, is very well suited for that purpose.

During conventional intra-cellular recording of cellular electric potentials, cells are impaled with a glass micro-pipette electrode, and the electrode tip is located inside the cell, usually in the vacuole. For patch clamp experiments, the cell walls are first digested away by treatment with the appropriate cell wall enzymes, resulting in naked protoplasts. The tip of a glass micro-pipette is then brought into contact with the plasma membrane of the protoplast, and gentle suction is applied to facilitate the formation of a tight seal between the micro-pipette and the membrane. The unusual tightness of the seal that forms between the patch pipette and the membrane reduces the leakage-caused background electrical noise typical of conventional intra-cellular recording sufficiently to allow high-resolution recordings of the currents through single ion channels.

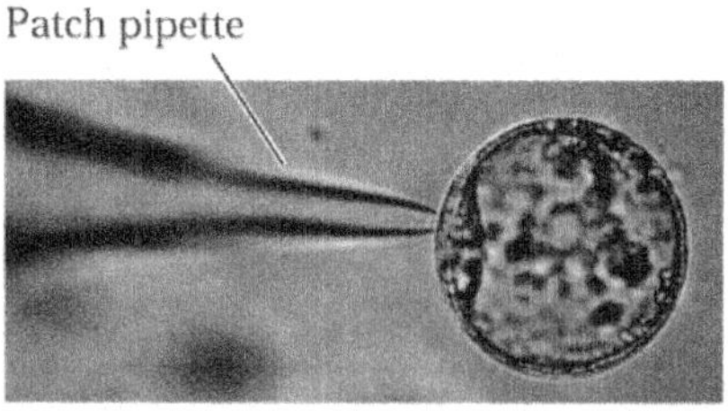

Figure 29.1 : Micrograph of a patch pipette attached to the surface of a barley aleurone cell protoplast.

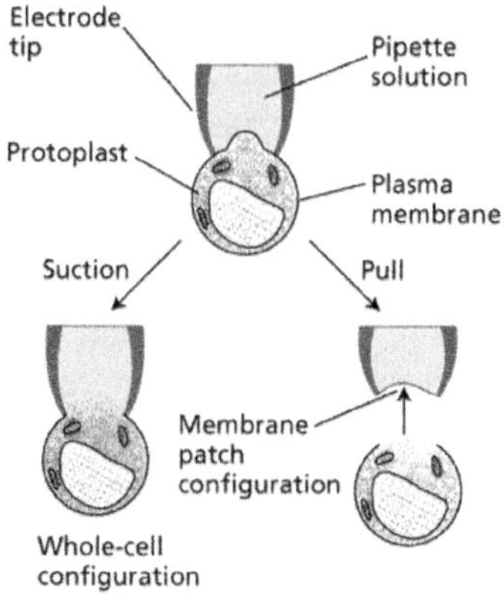

Figure 29.2 : Diagram of the whole-cell and membrane patch configuration.

When a tight seal is attained, several options become available to the investigator. Further suction can remove the portion of the membrane delimited by the opening in the pipette tip, exposing the interior of the cell to the micro-pipette solution (thus different solutions can be introduced into the cell through the micro-pipette). In this configuration, measured electric currents reflect the sum total of all electric charges carried by active and passive ion fluxes across the plasma membrane of the whole cell.

Pulling the electrode away from the cell produces a membrane patch that becomes exposed to both the pipette solution and the bathing medium. Because the membrane patches contain only a few channels, this configuration makes it possible to study the abrupt changes in current caused by the opening and closing of single ion channels. Remarkably, this technique is sensitive enough to detect the change in conformation of a single protein molecule.

Because protoplasts are spherical, their volume is easy to measure, and their lack of electrical connections with other cells facilitates calculations of charge fluxes per unit area of membrane. Other advantages of patch clamping are the ability to distinguish between electrical events at the plasma membrane and the tonoplast and the possibility of controlling the composition of both the cytosol and the bathing medium.

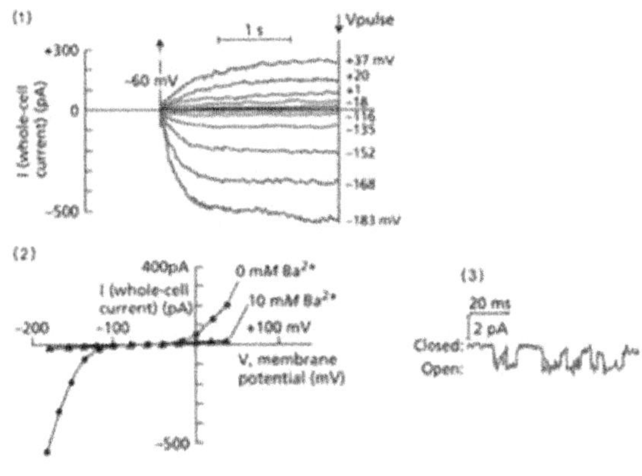

Figure 29.3 : Recordings of K^+ currents in whole cells and in single K^+-selective channels of guard cell protoplasts. (1) K^+ currents recorded in the whole-cell configuration when the membrane potential is clamped at different values. Upward deflections show outward currents; downward deflections show inward currents. K^+ concentrations were 105 m*M* in the pipette and cytoplasm, and 11 m*M* in the bathing solution outside the cell. (2) Data from the same experiment plotted as an I/V (current/voltage) curve, before and after the addition of Ba^{2+}, an ion that blocks K^+ channels. (3) Inward K^+ current through a single channel in a membrane patch.

Many of the patch clamp studies on plant membranes have been performed with guard cells. Whole-cell currents, using solutions in which K^+ is the main ion that can permeate the membrane. Initially, the membrane potential is held at a value close to the K^+ equilibrium potential. Then voltage pulses (in this case about 3 s long) are applied to the protoplast, and the electric current (in picoamperes) is followed with time (an upward deflection of the recording instrument indicates positive charges moving out of the cell). In this case, the current does not change instantaneously to a new stable value after a voltage change.

The gradual increase in current after a change in voltage reflects the opening of increasing numbers of channels in response to the membrane potential. Channels responding in this way are termed voltage-gated channels. In general, the gating of a channel is controlled independently of its ion selectivity, which is generally constant for a particular kind of channel protein. Again the current does not depend in a simple way on the driving force for K^+ diffusion (that is, the deviation of the potential from the K^+ equilibrium potential). An outward current is activated only at potentials more positive than -40 mV, and an inward current is activated only at potentials more negative than about -100 mV. This activation is a reflection of the gating of individual channel proteins with different properties.

The inward and outward currents can be shown to have different single-channel conductances (the ability of current to cross the membrane) and varying response times, and they are differentially affected by various activators and inhibitors. It is now known that the inward and outward currents are carried by two different kinds of K^+ channels, as demonstrated by the cloning and expression of the respective genes. The outward-rectifying channels open only when the membrane potential favours outward diffusion of K^+, and the inward-rectifying channels open only when the membrane potential favours the inward diffusion of K^+. Since each type of channel has its own regulatory mechanisms, the cell can independently control the rates of inward and outward diffusion.

Additional information about K^+ channels can be obtained from studies of currents in isolated patches of membrane in which single channels can be visualized. When held at a constant voltage, these patches show abrupt jumps in the level of current, as single K^+ channels open and close. This behaviour is seen only in membrane channels, never in ion pumps or carriers, because a single conformational change of a channel can allow hundreds of thousands of ions to cross the membrane, whereas a conformational change of a pump or a carrier can transport only one, or a few, molecules or ions.

Patch clamp studies have also shown that gated channels can be distinguished on the basis of how long the gates remain open in response to a prolonged stimulus. Two types of anion channels have been characterized at the guard cell plasma membrane : Rapidly activated (R-type) and slowly activated (S-type). *R-type channels* open and close very rapidly in response to a voltage stimulus; *S-type channels* remain open for the duration of the stimulus. Rapid and slow channels have also been identified in vacuolar membranes, and they are referred to as *fast vacuolar (FV)* or *slow vacuolar (SV)* channels.

CHEMIOSMOSIS IN ACTION

The British nobel laureate Peter Mitchell coined the term chemiosmosis to explain the coupling between gradient of electro-chemical potentials of protons across selectively permeable membranes and the performance of cellular work. In plants, proton gradients play a role in transport across the plasma membrane and the tonoplast and drive both transport and ATP synthesis on the inner membranes of chloroplasts and mitochondria. The passage of protons along their electric and chemical gradients can be coupled to cellular work because the change in free energy is negative. This process can be represented by the following equation :

$$\Delta\tilde{\mu}_{H^+} = F\Delta E + 2.3RT\left(\log\frac{[H^+]^i}{[H^+]^o}\right)$$

where $\Delta\mu_H^+$ is the proton gradient of electro-chemical potential, F is the Faraday constant, ΔE is the membrane potential, R is the gas constant, T is the absolute temperature, and the superscripts "i" and "o" refer to the inside and outside, respectively, of the cell or other membrane-enclosed compartment.

Mitchell introduced the term "proton motive force" (Δp) for the difference in electro-chemical potential between protons inside and outside a cellular compartment, μ_H^+. It is convenient to express Δp in units of electric potential, which we can do by dividing both sides of the foregoing equation by the Faraday constant, F, which changes the units to millivolts :

$$\Delta p = \frac{\Delta\tilde{\mu}_{H^+}}{F} = \Delta E + \frac{2.3RT}{F}\left(\log\frac{[H^+]^i}{[H^+]^o}\right)$$

Since pH = -log[H^+], the term log([H^+]^i/[H^+]^o) simplifies to -(pH^i - pH^o). If ΔpH is defined as pH^i - pH^o, the general expression for proton motive force results :

$$\frac{\Delta\tilde{\mu}_{H^+}}{F} = \Delta p = \Delta E - \frac{2.3RT}{F}\Delta pH$$

At 25°C, $2.3RT/F$ = 59 mV, and substituting this value into the foregoing expression results in the most commonly used equation for proton motive force :

$$\Delta p = \Delta E - 59\text{m } V\Delta pH$$

where Δp is expressed in millivolts.

Let's consider the example of a cell bathed in a solution of 1 mM K Cl and 1 mM sucrose. Proton pumping by an H^+-ATPase results in a membrane potential of -120 mV and a pH difference between the inside and the outside of the cell of 2 pH units. Thus, from the previous equation,

$$\Delta p = -120\text{mV} - 59(2)\text{mV} = -238\text{mV}$$

How much K^+ can the cell take up by using this Δp? Because potassium is positively charged and the membrane potential differece across the cell membrane is inside negative, potassium will be taken up through ion channels by the electrical component of Δp, which equals -120 mV. Using the Nernst equation, we can calculate that an external K^+ concentration of 1 mM and a ΔE value of -120 mV will equilibrate with an internal K^+ concentration of 100 mM. Thus, if the electrical component of Δp is used, the cell can generate a 100-fold K^+ gradient across the membrane.

The proton motive force can also be used to take up sucrose against a concentration gradient, usually via a proton–sucrose symporter. Because sucrose is cotransported with a proton, both components of the gradient of electro-chemical potential (-238 mV in our example) can be used for sucrose uptake. At equilibrium, $\Delta\mu_H{}^+$ will be equal to $F\Delta p$; thus we can calculate that an external sucrose concentration of 1 mM and a Δp of -238 mV will equilibrate with an internal sucrose concentration of 10 M. In real life, however, such concentration gradients would not exist : Sucrose would diffuse back out of the cell, and regulatory mechanisms at the membrane would repress the function of the symporter after certain critical concentrations were attained.

KINETIC ANALYSIS OF MULTIPLE TRANSPORTER SYSTEMS

Studies of K_m values have shown a feedback inhibition of carrier-mediated transport that allows the cell to maintain stable cytoplasmic ion concentrations over a wide range of external concentrations. For example, the apparent K_m value of the high-affinity K^+ carrier in barley roots changes with the plant's demand for K^+. Barley plants starved for K^+ have high-affinity K_m values in the range of 0.02 to 0.03 mM, whereas well-fertilized plants have K_m values that are four- to fivefold greater. An increase in K_m reflects a decrease in the carrier's affinity for K^+.

The change in apparent K_m probably reflects the expression of different members of a gene family, with their gene products thereby matching cellular needs to environmental ionic conditions. If the plants' adjustments were merely a matter of adding more transporters to the plasma membrane, there would be change in V_{max} without a change in K_m. This is not what is observed for K^+. However, plants that have been starved and then resupplied with sulfate or phosphate often respond by an increase in V_{max}. As the internal sulfate and phosphate levels recover, the V_{max} values for their uptake drop to the usual values as excess carriers are removed from the membrane.

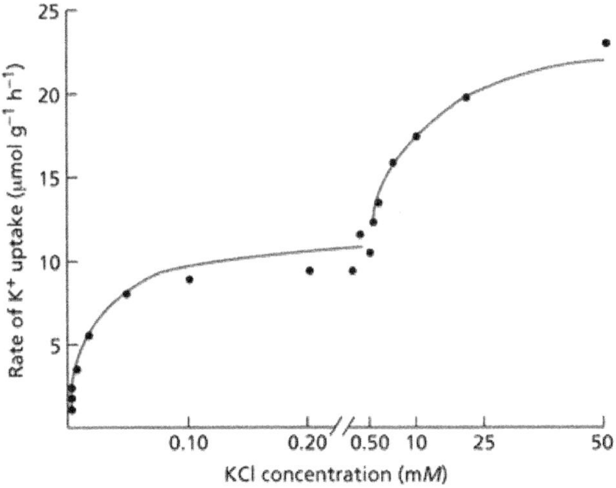

Figure 29.4 : The transport of potassium into barley roots shows two different phases. The biphasic kinetics of potassium uptake, accentuated in this figure by the change of scale at around 1 m*M*, suggests the presence of different types of transport systems for potassium. The high-affinity transport system, having a K_m value of 0.02 to 0.03 m*M*, is attributed to active transport by symporters; the low-affinity system (which may or may not show saturation) is attributed to diffusion through K^+ channels.

ABC TRANSPORTERS IN PLANTS

The ATP-binding cassette (ABC) transporters are members of a large family of active transport proteins energized directly by ATP hydrolysis. Originally identified in microbial and animal cells, ABC transporters are able to use the energy of ATP hydrolysis directly to pump organic molecules (especially large anionic molecules) across a membrane. Like the P-type H^+-ATP synthases, ABC transporters form a phosphorylated intermediate during catalysis, and they are therefore inhibited by vanadate.

A number of genes for ABC transporters have been cloned from various plants. Genes for ABC proteins are recognizable by a predicted protein structure consisting of one or two copies each of two structural units, a transmembrane domain (crossing the membrane several times) and a nucleotide-binding fold, which contains a characteristic signature sequence. The transmembrane domain provides the pathway for solute transport and determines the molecular specificity of the transporter.

An ABC transporter has been found to secrete anti-fungal terpenes across the plasma membrane of tobacco cells. More commonly, ABC transporters are found at the tonoplast, where they are sometimes referred to as glutathione conjugate pumps, or GS-X pumps, since they often transport molecules that have been covalently attached to glutathione. Glutathione (GS) is a tripeptide (Glu-Cys-Gly) that functions as an important cellular antioxidant molecule. The vacuolar GS-X pumps of plant cells function in herbicide detoxification, protection against oxidative damage, pigment accumulation, and the storage of antimicrobial compounds.

A family of enzymes called glutathione transferases (GSTs) is responsible for attaching glutathione to the molecule to be transported. The common chemical determinant for GS conjugation is a carbon–carbon double bond adjacent to an electron-withdrawing group (CH_2 = CH-Z). Compounds that contain this determinant include the anthocyanins, IAA (indoleacetic acid, or auxin), and various phenolic compounds. Metal-binding polypeptides called phytochelatins, which are synthesized from glutathione, are also transported into plant vacuoles along with their chelated heavy metals via GS-X pumps.

TRANSPORT STUDIES WITH ISOLATED VACUOLES AND MEMBRANE VESICLES

Just as studies with isolated thylakoid membranes led to the elucidation of the electron transfer reactions of photo-synthesis, the development of techniques for isolating functionally competent tonoplasts and plasma membranes has enabled scientists to study the transport reactions of these membranes in vitro. The use of isolated membrane vesicles to study transport avoids interference from metabolism. Since the contents of the vesicles can be fairly readily exchanged, the concentrations of solutes on each side of the membrane can be better defined. Two types of preparations have been used : intact vacuoles, and isolated plasma membrane or tonoplast vesicles.

Isolating Intact Vacuoles from Protoplasts

When protoplasts prepared by enzymatic digestion of the cell walls are maintained in the plasmolyzed state by incubation in medium containing a suitable osmotic solute, such as mannitol, the plasma membranes can be carefully lysed without breaking the tonoplast by a gentle osmotic shock. The intact vacuoles that are released by this treatment can then be purified by centrifugation through a sucrose density gradient. Intact vacuoles (along with contaminating cellular debris) are centrifuged in a tube containing a concentration gradient of sucrose (or another suitable solute). Because intact vacuoles tend to exclude the solute, they sediment to a lighter density on the density gradient than the other organelles and membranes do. In this way preparations of highly purified intact vacuoles can be obtained.

Isolating Tonoplast Membrane from Cell Homogenates

Alternatively, tonoplasts and plasma membranes can be purified as small membrane vesicles. When tissues are ground in a mortar and pestle or homogenized in a blender, broken pieces of various membranes tend to seal up to form small vesicles less than 1 μm in diameter. Because of a low ratio of protein to lipid in the tonoplast membrane, tonoplast vesicles usually have a relatively low density, and they can be separated from other cellular membranes by density-gradient centrifugation.

Isolating Plasma Membrane from Cell Homogenates

Plasma membrane vesicles can be purified by phase partition in a mixture of dextran and polyethylene glycol solutions. Such a mixture separates into two phases, and because of their surface properties, plasma membranes tend to accumulate in the upper phase, while intra-cellular membranes tend to remain in the lower phase.

Tissue homogenization often produces a mixture of vesicles of different orientation (right-side-out and inside-out vesicles). In the case of plasma membrane vesicles, phase partition selects for right-side-out vesicles, which therefore show little reactivity to ATP because the catalytic sites of the enzymes are facing inside the vesicle. If a detergent is added to such a preparation, ATP penetrates the vesicles and is hydrolyzed, but since the vesicles are no longer sealed, no transport is observed.

One kind of detergent (Brij 58) in low doses does not make the membranes leaky but, it is claimed, causes them to reverse their orientation. Apparently, this detergent accumulates in the inner half of the lipid bilayer, causing it to expand and flip the vesicles inside out. Since the catalytic site of the enzyme is now facing the external solution, in the presence of this detergent the vesicles respond to ATP by creating a gradient of pH and membrane potential.

Measuring Transport in Membrane Vesicles

Radioisotopes can be used to monitor the transport of various solutes into membrane vesicles. The vesicles are incubated for various times with the labelled solute, then separated by rapid filtration, and their radio-activity is measured in a scintillation counter. Other methods have been developed to measure the gradients of pH and membrane potential created by proton pumps. One such method uses fluorescent weak bases such as 9-aminoacridine (9-AA) or quinacrine, which are lipid-soluble in their uncharged form and readily diffuse across the membrane and enter the vesicle. Within the acidified vesicle, the amino group becomes protonated, trapping the charged form of the molecule inside. As the dye molecules accumulate inside the vesicles, they form dimers and other aggregates that result in a reduction in their fluorescence, a phenomenon termed fluorescence quenching. One can thus study ATP-driven proton transport across the vesicle membrane by measuring the rate of fluorescence quenching in an instrument called a fluorescence spectrophotometer.

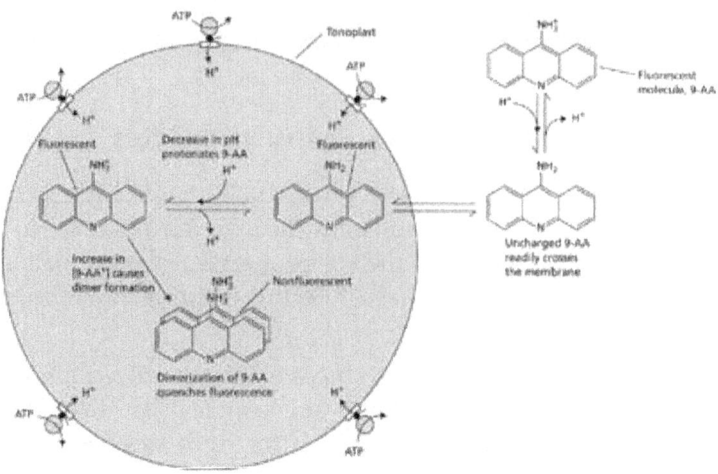

Figure 29.5 : Measuring H^+ pumping by the fluorescence-quenching method. Diagram of the reaction. In the presence of an ATP-driven pH gradient, the protonated (charged) form of a fluorescent weak base, such as 9-AA, accumulates inside the vesicle. The accumulation of 9-AA inside the vesicle results in a decrease in the total fluorescence of the suspension as a result of self-quenching caused by the dimerization of 9-AA.

Addition of ATP to membrane vesicles isolated from the plasma membrane or tonoplast leads to a rapid decrease in the fluorescence (quenching). A steady state is reached when the rate of leakage of protons out of the vesicle equals the rate of transport of protons into the vesicle. Gramicidin, a compound that induces non-specific holes in the membrane, rapidly dissipates the pH gradient, as indicated by the recovery of fluorescence. Other molecular probes, such as oxonol dyes, which associate with lipid membranes and show changes in fluorescence as the electric potential changes, can be used in a similar fashion to study ATP-driven membrane potentials.

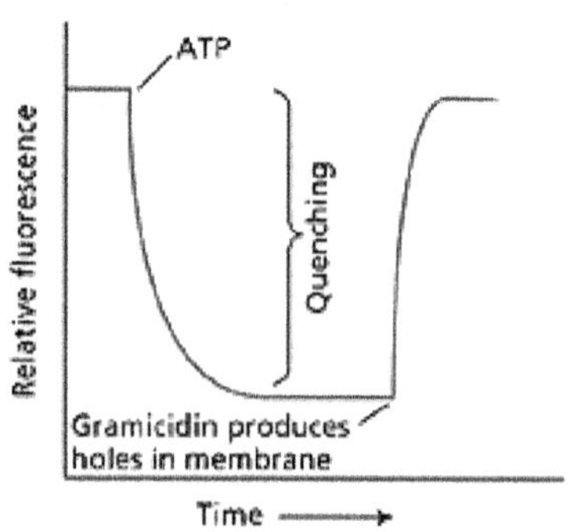

Figure 29.6 : Results of a typical fluorescence-quenching experiment, showing the decrease in fluorescence upon the addition of ATP, and the recovery of fluorescence when the DpH is dissipated using the channel-forming ionophore gramicidin, which produces nonspecific pores in the membrane.

Although it has proven difficult to make absolute measurements of the pH and potential gradients in membrane vesicles, this type of study has been very useful for relative measurements, and for defining the energy sources and activator and inhibitor sensitivities of various transport systems (Sze 1985). In combination with other experimental techniques, studies with membrane vesicles have provided many insights into transport in plant cells.

Chapter 30

Stress Resistance Breeding

INTRODUCTION

Stress refers to adverse conditions for crop growth and production imposed by either environmental factors or biological factors, thus stress is of two types, *viz.* (1) Biotic and (2) Abiotic. The stress that is caused by biological agents or factors, such as disease, insects and parasitic weeds, is known as biotic stress. When the stress is caused by environmental factors or non-biological factors, it is referred to as abiotic stress. Abiotic stress is generally caused by factors like deficiency or excess of nutrition, moisture, temperature and light; presence of harmful gases or toxicants; and abnormal soil conditions such as salinity, alkalinity and acidity. All crop plants suffer from both biotic and abiotic stresses to varying degrees. This chapter deals with breeding for resistance to disease.

PLANT BREEDING FOR DROUGHT RESISTANCE

Introduction

There are several environmental factors that have adverse effects on normal growth ad development of crop plants. Such factors include deficiency or toxicity of minerals, moisture deficit and low temperature, soil salinity and alkalinity. Soil acidity and environmental pollutants. These factors are known as abiotic or no biotic factors. The stress or adverse condition caused by such factors for growth and development of crop plants are referred to as abiotic stress or environmental stress. Crop plants often suffer from abiotic stresses resulting I significant reduction in both yield and quality. In some areas, the soil and weather conditions are so much unfavourable that cultivation of certain crops becomes very difficult and non-profitable. Under such situation, it is essential either to develop stress resistant cultivars or to modify the environment is the most practical way of solving such problems. This chapter deals with breeding for drought and salt tolerance.

Drought Resistance

In India, problems of drought, salinity and alkalinity are of major concern among environmental stresses. Drought refers to the condition of soil moisture deficiency or water scarcity. Soil drought is more common in the arid and semi-arid tropics and in the areas of steep slope. Thus desert areas are more prone to drought conditions. The ability of crop plants to grow, develop and reproduce normally under moisture deficit conditions is referred to as drought resistance. In other words, drought resistance refers to survival of plants under water deficit or scarcity conditions without injury. Improvement in the drought tolerance ability of a plant is known as drought hardening. Main features of drought are given below :

1. Drought is characterized with soil moisture deficit or low soil moisture.
2. Arid and semi-arid areas are more prone to drought than humid zones. About 36% of the land area constitute arid and semi-arid zones.
3. Drought leads to reduction in both yield and quality of economic product in crop plants. It has adverse effects on plant growth and development.
4. Drought damages chloroplasts and lowers output of the photo-synthetic apparatus.
5. There is an increase in proline level in the leaves of plants which are subjected to drought, Proline level can be used as an indicator of water stress, but not as a measure of drought resistance.
6. The occurrence of drought depends on the amount and distribution pattern of rainfall. If the rainfall is adequate and well distributed over the crop season, there are less chance of drought. On the other hand, when rainfall is erratic there are more chances of drought. Soil type and topography also affect drought.
7. Drought resistance is a genetically controlled physiological property of plant species. Resistance to drought is associated with various morphological and physiological features of the plant.
8. Xerophytic plants are more resistant to drought than mesophytes.
9. There is an increase in abscisic acid content in leaves of barley, and in ethylene level in cotton and wheat under drought conditions.

Mechanisms of Drought Resistance

There are four different mechanisms which help in survival of plants under moisture deficit conditions. There are : (1) Drought escape, (2) Drought avoidance, (3) Drought tolerance, and (4) Drought resistance. The last one refers to true drought resistance. These are briefly described below :

1. Drought Escape

The simplest way of survival under drought conditions is to escape drought. Generally, drought occurs either in the mid or late-crop season. Drought escape is most common in case of plants grown in desert regions. They complete their life cycles in 4 to 6 weeks. Drought escape also plays an important role in some crop plants. For Example, yields of early varieties of wheat, sorghum, maize, and rice are less affected by severe drought than late maturing ones. All these crops have determinate growth habit. In spring wheat, late maturing varieties give higher yield than early types especially when drought occurs early in the season and is over before anthesis.

2. Drought Avoidance

Drought avoidance refers to ability of the plant to maintain a favourable internal water balance under moisture stress. In other words, plants which avoid drought retain high water contents in their tissues. Drought avoidance can permit a longer growth period in the crop through reduced water use or increased water uptake. However, drought avoidance leads to reduction in photo-synthesis and thereby reduction in the growth of aerial parts. It leads to increase in root development and therefore, is more important than drought tolerance. In cereals, drought avoidance operates during vegetative phase, while tolerance operates during reproductive phase. Drought avoidance mechanisms are of two types. First those which reduce water loss through transpiration. Such features include stomatal characteristics and shape, size and orientation of leaves. The second, those which maintain water uptake during drought period.

3. Drought Tolerance

The ability of crop plants to withstand low tissue water content is referred to as drought tolerance. Drought tolerance is more desirable because the crop can produce more yield at lower water potential. In cereals, drought tolerance generally operates during reproductive phase. Tolerant cultivars exhibit better germination, seedling growth and photo-synthesis. In Sorghum, a drought resistant line exhibited higher photo-synthetic rate at leaf water potential than a less drought resistant line. Drought tolerance differs from drought avoidance in several aspects.

4. Drought Resistance

Drought resistance is the sum of drought avoidance and drought tolerance. In other words, drought resistance refers to the ability of crop plants to give good yield under moisture deficit conditions. Drought resistance is measured in terms of various mechanisms associated with drought tolerance and yield under soil moisture deficit. In winter wheat, both avoidance and tolerance features are important for drought resistance.

Table 30.1 : Difference between Drought Avoidance and Drought Tolerance.

Sr. No	Drought Avoidance	Drought Tolerance
1	Plants maintain favourable tissues water content.	Plants do not maintain favourable tissue water content.
2	It reduces photo-synthesis and increase root development.	Better seed germination, seedling growth and photo-synthesis.
3	In cereals, it operates during vegetative phase.	In cereals, it operates during reproductive phase.
4	Plants can not withstand low tissue water content.	Plants can withstand low tissue water content.
5	It involves various morphological and anatomical features of plant which reduce water loss through transpiration.	It generally involves those characters which support for better photo-synthesis under drought conditions.

Basis of Drought Resistance

Drought resistance is associated with various morphological and physiological features or factors of crop plants. Morphological characters which are associated with drought resistance included earliness, shape, size and structure of stomato, size, number and orientation of leaves; presence of cuticle; Waxiness on leaf lamina and stem, rooting pattern, growth habit etc. Various physiological characters which are related to drought resistance are photo-synthetic rate, transpiration rate, osmotic concentration etc. These factors are briefly discussed below :

Morphological Characters

1. Earliness

Earliness is a desirable character which leads to drought escape in many crops. For example, in wheat, sorghum, maize, and rice yield of early maturing varieties is less affected by severe drought than late maturing varieties.

2. Stomatal Features

Sunken, small size and less number of stomato are associated with drought resistance. Control of stomatal aperture is important in drought resistance. The rapid closing of stomata during development of drought helps in maintaining higher water potential in the tissues by reducing transpiration rate and thus resulting in drought avoidance.

The stomatal aperature is measured with the help of porometers. Drought resistant genotypes have rapid closing habit of stomata. Porometers are of two types, *viz.* (1) Viscous flow porometers which measure out flow rate of air through the leaf, and (2) diffusive flow porometers which measures the rate of diffusion of water vapour out of the leaf. Now infrared thermometers are used to measure leaf water status and stomatal activity. Leaves with closed stomato will exhibit higher temperature than those with open stomata. Leaves with open stomato have cooling effect due to water loss through transpiration.

3. Leaf Characters

Cuticular thickness and Waxiness of leaf surface help in reducing transpiration. These characters are genetically controlled. Leaf rolling is an indicator of stress. It can also serve as drought avoidance mechanisms. Leaf rolling reduced transpiration from 46 to 63% in some grasses of Mediterranean region. In cotton, small and thick leaves are associated with drought resistance. Leaf hairiness lowers the leaf temperature and thus reduces transpiration. In barley, light green and golden leaves reflect more light than dark green leaves and thus remain cooler. The genotypes which reflect more light have more cooling effect resulting in reduction of transpiration.

4. Rooting Patterns

Increase in depth, width and branching of root systems leads to decrease in plant water stress. Generally, deep rooted plants exhibit greater drought avoidance than shallow rooted ones. Wheat cultivars that produce greatest root mass under drought conditions are important in breeding for drought resistance if the depth is more than 60 cm. Breeding for root patterns associated with drought resistance has been successful in several crops such as soft and hard wheat, barley, corn and sunflower. Two drought resistant cultivars of durum wheat have been developed in Canada through the use of extensive root system. The new varieties combine deep root system, good grain quality and high yield. seedling root growth is an indication of root growth at maturity.

5. Growth Habit

In upland cotton, interminate genotypes yielded more than determinate genotypes in a semiarid environment. Interminate plants produce flowers throughout the growing seasons whenever sufficient moisture is availability. This is not possible in case of determinate genotype.

6. Awns

In wheat ad barley, presence of awns appears to be associated with high yield under drought conditions. The increase in yield from awns results due to increase in seed size. Awns play important role in growth and development of seeds through increase in photo-synthetic surface of spike.

Physiological Factors

Resistant genotypes maintain high photo-synthesis under moisture stress conditions by restricting transpiration water loss. Hence photo-synthesis is a reliable parameter to select drought resistant genotype in segregating materials in breeding programmes. In maize, photo-synthesis was reduced by stress and translocation within leaves was also reduced. In spring wheat, dehydration hardiness is associated with small cell size, high concentration of cell sap, increased protoplasmic permeability and high viscosity of cytoplasm.

Source of Drought Resistance in Plant Breeding

There are three main sources of drought resistance in crop plants : (1) Cultivated varieties, (2) Germplasm collections, and (3) Wild relatives and wild species. Transfer of drought resistance is easy from cultivated variety and germplasm of cultivated species, because such material can be easily used in the breeding programmes. Moreover, there is no problem of cross incompatibility. When the source of drought resistance is a wild species, the transfer of resistance poses several problems such as cross incompatibility, hybrid inviability, hybrid sterility and linkage of several undesirable genes with desirable ones. Wild sources of drought resistance have been reported in wheat, sugarcane, tomato, and several other crops.

Table 30.2 : Wild Sources of Resistance to Drought and Salinity in Some Crop Plants.

Name of Crop	Name of Wild Species	Resistant Available for
Wheat	Aegliops kotsehyi	Drought
	Ae.variabills	Drought
	Ae.speltoider	Drought
	Ae. Umbellulata	Drought
	Ae.squarrosa	Drought
	Agropyron seirpea	Salinity
	A.pontica	Salinity
	A.junceiformis	Salinity
	A.cliac	Salinity
Sugarcane	Saccharum spontaneum	Drought and Salinity
	Lycopersicon cheesmanii	Salinity

Plant Breeding Methods for Drought Resistance

Breeding methods for drought resistance are the same as for yield and other economic characters. Breeding for drought resistance refers to breeding for yield under soil moisture stress condition. In other words, it refers to yield improvement in environment represented b water deficit. Four breeding methods, *viz.* (1) Introduction, (2) Selection, (3) Hybridization, (4) Mutations are commonly used for development of drought resistant crop cultivars.

In self-pollinated crops, introduction, pure line selection, mass selection, hybridization (Pedigree method and backcrossing) and mutation breeding are used. When drought resistance is available in an exotic variety, such variety can be introduced and after through testing, if found suitable, can be released in the new area. When the drought resistant genotypes are available in the land races or mixed populations, either pure line selection or mass selection is adopted. When the resistant genes are available in the germplasm or wild species, the breeder has to resort to hybridization. Mutation is used when the desired genes are not available in the germplasm. Sufficiently large segregating populations have to be grown to select for drought resistance. In maize, drought tolerance is controlled by no less than genes pairs. Simple screening was effective in selection of drought resistant characters in one genotype, screening of material under drought conditions, use of large population in yield test and testing at several locations will help in selection of superior lines.

In cross-pollinated crops, the most commonly used methods are mass selection, backcross, hybridization of inbreds to develop hybrid cultivars, recurrent selection and formation of synthetic cultivars. Mass selection is partially effective because there is no control of pollination. The back-cross method is equally successful in self and cross-pollinated species. Heterosis is used in both species. However, hybrid vigour is more successful in cross pollinated species than in self pollinated ones.

Drought Hardening in Plant Breeding

Improvement in the drought tolerance ability of a genotype through various treatments is referred to as drought hardening. Two types of treatment are in common use for drought hardening. These are pre-sowing treatment and post sowing treatment.

Pre-sowing Treatment

Seeds are treated before sowing to induce drought hardiness. Seeds are soaked in water for 24 hours and then sundried. The treated seeds are sown in the field. Plants developed from such seeds exhibit improved drought resistance or tolerance. The hardened plants exhibit higher yield under drought, higher water contents, increased viscosity of protoplasm, more bound water, higher metabolic rate and stronger roots. The same seed treatment, in sunflower, resulted in more stable metabolism of nitrogen ad phosphorous besides higher yield under drought conditions than unrelated plants. In barley, seed treatment was effective in improving drought tolerance of some genotypes, while it was ineffective for others. This indicated genetic differences for drought hardening. Seed treatment also improved drought tolerance in some genotypes of wheat. Various modifications of seed treatment method have been suggested by various workers.

Post-sowing Treatment

The post-sowing treatment is given to the young seedlings. Mild drought stress in early stages of growth increase resistance to water stress under subsequent and more severe drought. In wheat and oats, such treatment was effective in improving drought resistance. However, cultivars differed in the hardening capacity in these crops. Several cycles of stress helped in identification of the most drought resistant and the most susceptible genotypes by measuring their survival.

Measurement of Drought Resistance in Plant Breeding

Various procedures are used for measuring drought resistance in crop plants. The most commonly used procedures include : (1) Leaf water retention, (2) Photo-synthesis, (3) Yield performance, and (4) Root lengths of seedlings. These techniques can be used for large scale screening of segregating populations in breeding programme.

1. Leaf Water Retention

In this method leaves are excited from the plant and are allowed to dry. The slower drying genotypes are considered as drought tolerant. In other words, the high water retainer genotypes are considered as drought tolerant. The cut leaf method was used in wheat, barley and oats. In wheat, two varieties Pelissier and Pitic 62 were significantly better water retainers when leaves were excised from three week old plants. These two varieties are highly resistant to drought. Some investigators use tissue water potential as an index of water stress under drought conditions. The tissue water potential is measured with the help of thermocouple psychrometer. The portable field psychrometer is widely use for measuring drought resistance in segregating populations.

2. Rate of Photo-synthesis

The rate of photo-synthesis during and after moisture stress is an important index of drought resistance. In wheat, Pitic 62 (a drought resistant variety) exhibited high photo-synthetic rate under drought conditions in several tests. Now portable non-destructive photo-synthesis analyzers are available which can be used in the fields for large scale screening of germplasm as well as segregating populations in standing crops. The genotypes which have high photo-synthetic rate under moisture stress are considered as drought resistant, because such genotypes give higher yield than those having low photo-synthetic rate. A simple portable photo-synthesis analyser made it possible to measure photo-synthesis of many plants within a short time. Now photo-synthesis is used as a criterion to select for drought resistance.

3. Yield Performance

Superior yield performance under moisture stress conditions is an important and reliable index of drought tolerance. The yield tests should be conducted in drought prone areas at several locations or for several years. These will help in identification of genotypes with drought resistance and also in the elimination of drought susceptible lines. The yield test should be conducted under both field as well as glass house conditions. Moreover, large number of populations should be grown. This will enhance chances of obtaining superior drought resistant genotype.

4. Root Length of Seedlings

The root length during seedling stage is also used a measure of drought resistance. In wheat during root length of 5-7 days old seedlings grown in sand was related to root mass at maturity. In a more recent study it was observed that root mass after 30 days was reliable index of root mass at maturity. Thus those genotypes which have longest root during seedlings stage also exhibit extensive root system at maturity. This is a simple and quick method of measuring drought resistance in each crop season. Moreover, after screening superior plants can be replanted and grown to maturity. Some workers use hydroponics tank to measure the root growth of seedlings.

In rice, the force required to pull out the seedlings from paddy soil is used as a measures of drought tolerance. The force required is correlated with root weight, branching and number. This technique may be used for large scale screening of rice populations for drought resistance.

Limitation of Drought Resistance Plant Breeding

Breeding for drought resistance has several advantages. It helps in extending cultivation of several field crops in drought prone areas. Drought resistant varieties are more stable to environmental fluctuations and are expected to stabilize crop production under conditions of water stress. Thus, drought resistance breeding is useful in many ways. However, there are some limitations of drought resistance breeding which are given below :

1. Breeding for resistance to various types of stress generally leads to reduction in yield due to strict selection pressure for resistant genes. Breeders are fortunate if any drought resistant line is equal to the best parent in yield. Sometimes, simply inherited characters associated with stress resistance have adverse effect on yield potential.
2. In Spring wheat, a cultivar which gave the highest yield under normal moisture level gave lower yield than many cultivars where grown under drought conditions. Thus selection for wide adaptability is a selection for medium or even low yield.

3. Genes for drought resistance are mostly found in the germplasm lines which are generally unadapted types. Utilization of unadapted lines poses problems of linkage between desirable and undesirable characters.
4. In some cases, the drought resistant genes are available in the wild species. Transfer of resistant genes from wild sources poses several problems such as hybrid inviability, hybrid sterility and linkage between desirable and undesirable genes.

   This greatly reduces the chances of obtaining desirable combination of favourable genes.
5. Breeding for drought resistance requires cooperation among geneticist, breeder and crop physiologist.

Plant Breeding for Salt Resistance

Salt tolerance refers to the ability of plants to prevent reduce or overcome the injurious effect of soluble salts present in their root zone. The problem of salinity is of global significance because saline and alkaline soils are found in almost all countries of the world. In India, more than seven million hectares land is estimated to be salt affected. Higher concentration of salt in the soil adversely affect the growth and development of plants by disturbing various physiological processes.

There are two ways to overcome the problem of salinity. One is reclaim the salt affected soil and the other is to develop salt tolerant genotypes/ cultivars. The first method is very costly, time consuming and short method, on the other hand, is long lasting, more effective and less costly. Genetic differences exist among cultivars for their salt tolerance capacity, moreover, plants adapt themselves to adverse environments for their survival. High salt tolerant plants in crop are found in the salt affected areas.

Breeding Approaches

In India, breeding work for salt resistance is carried out at Central Soil Salinity Research Institute (CSSRI), kernel and its various regional centres. In India, two types of approaches are followed for salt tolerance breeding. The two approaches are :

1. Improving yield level of salt tolerant cultivars, and (2) Transfer of salt tolerant genes to high yielding cultivars. In the first traditional cultivars of salt affected areas are improved for their productivity without affecting their salt tolerance ability. In the second approach, salt tolerances gene from locally adapted (Salt tolerant) cultivars are transferred to high yielding cultivars through hybridization and selections.

Screening Techniques

Various methods are used for screening segregating material for salt resistance. The commonly used methods are : (1) Lysimeter micro plots, (2) Sand culture, and (3) Solution culture tanks. The Lysimeter micro plots of 6 m X 3 m X 1 I size are used at CSSRI, Karnal. Several levels of salinity are used and replicated experiments are conducted to screen for salt resistance. Some workers prefer to use sand culture technique using nutrient solution. Solutions of various salinity levels are applied in the sand. Another technique is use of solution culture tanks with several levels of salinity. Replicated experiments are conducted over seasons to get more reliable result. Genotypes which survive under salinity conditions are considered as tolerant and screened further.

The salt tolerance capacity differs from species to species. For example, crops like barley, cotton, sugarcane, oilseeds and grasses are known to be more salt tolerant than other crops. The salt tolerance capacity also differs within a crop based on ploidy level. For instance, in wheat, hexaploid bread wheat is more tolerant to salt than durum (tetraploid) and einkorn (diploid) wheat. In Brassica, tetraploid species are more tolerant than diploids. In rice, late maturing, tall and coarse grain varieties exhibit maximum tolerance to salinity conditions. In crested wheat grass, selection for tolerance under artificial salinity conditions improved the salt tolerance level of selected lines. In sugarcane, vast differences for salt tolerance are observed among strains. In Hawaii, salt shock treatment is used for improving salinity tolerance. In sugarcane, moreover, land races are more tolerant to salinity than high yielding cultivars.

Breeders require close co-operation of geneticist, physiologist, bio-chemist, and soil scientist in developing salinity resistant varieties. The new approaches such as tissue culture and genetic engineering may be rewarding in developing more salt tolerant genotypes. The main limitations in the breeding for salt tolerance are : (1) Lack of efficient selection criteria, (2) Lack of inter disciplinary cooperation, and (3) paucity of funds, etc.

Practical Achievements in Drought and Salt Resistant Breeding

Breeding for drought resistance is an important objective of plant breeding programmes in many crops. In India about 73.5% the agriculture is represented by dry land farming. In the rainfed areas, crops generally suffer from drought at one or the other stage. Hence, breeding work on drought resistance is carried out in almost all the International and National Crop Research Institutes. In India, drought and salinity resistant or tolerant varieties have been developed in several crops.

Drought resistant varieties have been developed in several crops in many countries throughout the world. In wheat, drought resistant varieties have been developed in China and USSR. All these varieties have been developed through induced mutations. In Lathyrus saliva, a drought resistant variety, has been developed by induced mutations in USSE. In India, drought resistant varieties have been evolved in cotton. Rice, sugarcane, (Co258, Co312, Co740, Co997, Co1148 and COLK 8001) and several other crops. In cotton, drought resistant varieties have been developed through induced mutations.

In India, breeding work for salinity resistance is mainly carried out at Central Soil salinity Research Institute, Karnal (Haryana). Several salt resistant varieties of rice, barley, wheat and sugarcane (CO453 and CO62175) have been released. Soil salinity is a serious problem throughout the world in some areas. Hence breeding work on salt resistance is carried out in all important International and some National Crop Research Institutes. Salt resistant variety of rice has been released from Indonesia and that of wheat from china. In India, salinity resistant varieties have been developed in sugarcane, okra, rice, onion, and barley.

ABIOTIC STRESS AND BREEDING FOR DROUGHT RESISTANCE

Introduction

Stress is an altered physiological condition caused by factors that tend to disrupt the equilibrium. Strain is any physical and chemical change produced by a stress. The term stress is used with various meanings, the physiological definition and appropriate term as responses in different situations. The flexibility of normal metabolism allows the response initiation to the environmental changes, which fluctuate regularly and are predictable over daily and seasonal cycles. Thus every deviation of a factor from its optimum does not necessarily result in stress. Stress being a constraint or highly unpredictable fluctuations imposed on regular metabolic patterns cause injury, disease or aberrant physiology. Plants are frequently exposed to many stresses such as drought, low temperature, salt, flooding, heat, oxidative stress and heavy metal toxicity, while growing in nature.

Drought is a meteorological term and is commonly defined as a period without significant rainfall. Generally drought stress occurs when the available water in the soil is reduced and atmospheric conditions cause continuous loss of water by transpiration or evaporation. Drought stress tolerance is seen in almost all plants but its extent varies from species to species and even within species. Water deficit and salt stresses are global issues to ensure survival of agricultural crops and sustainable food production. Conventional plant breeding attempts have changed over to use physiological selection criteria since they are time consuming and rely on present genetic variability. Tolerance to abiotic stresses is very complex, due to the intricate of interactions between stress factors and various molecular, bio-chemical and physiological phenomena affecting plant growth and development. High yield potential under drought stress is the target of crop breeding. In many cases, high yield potential can contribute to yield in moderate stress environment. Drought stress is considered to be a moderate loss of water, which leads to stomata closure and limitation of gas exchange. Desiccation is much more extensive loss of water, which can potentially lead to gross disruption of metabolism and cell structure and eventually to the cessation of enzyme catalyzed reactions.

Drought stress is characterized by reduction of water content, diminished leaf water potential and turgor loss, closure of stomata and decrease in cell enlargement and growth. Severe water stress may result in the arrest of photo-synthesis, disturbance of metabolism and finally the death of plant. Water stress inhibits cell enlargement more than cell division. It reduces plant growth by affecting various physiological and bio-chemical processes, such as photo-synthesis, respiration, translocation, ion uptake, carbohydrates, nutrient metabolism and growth promoters. In plants, better understanding of the morpho-anatomical and physiological basis of changes in water stress resistance could be used to select or create new varieties of crops to obtain a better productivity under water stress conditions.

The reactions of plants to water stress differ significantly at various organizational levels depending upon intensity and duration of stress as well as plant species and its stage of growth. Understanding plant responses to drought is of great importance and also a fundamental part for making the crops stress tolerant.

Breeding for Drought Resistance

Drought seems to be rather difficult to define more difficult to quantify. For example, the common criteria used in the various definitions are precipitation, air temperature, relative humidity, evaporation from free water surface, transpirations from plant, wind, air flow, soil moisture and plants conditions. A working definition of drought may be "the inadequacy of water availability, including precipitation and soil moisture storage capacity, in the quantity and distribution during the life cycle of a crop to resist expression of its full genetic yield potential". Therefore, under conditions of drought, water stress develops in the plants as the demand exceeds supply of water; this may occur due to atmospheric or soil conditions, and is reflected in a gradient of water potentials developed between the soil/soil-root interference and leaf, the transpiring organ. Thus moisture stress may be defined as the inability of plant to meet the evapotranspirational demand.

Moisture stress is likely to develop to a different rate in different plant organs along this gradient. Phenotype is the result of genotype and environmental interaction. Therefore, assessment of desired genotypes is highly dependent on proper environmental conditions. Abiotic stresses (particularly drought, high temperature, salinity and others) generally reduce crop productivity. These stresses are location-specific, exhibiting variation in frequency, intensity and duration. Stresses can occur at any stage of plant growth and development, thus illustrating the dynamic nature of crop plants and their productivity.

Drought is the primary abiotic stress causing not only differences between the mean yield and potential yield but also causing variation from year to year, resulting in yield instability. Although selection for genotypes with increased productivity in drought-prone environments has been an important aspect of many plant breeding programs, the biological basis for drought tolerance is still poorly understood. Also, drought stress is highly heterogeneous in time, space, degree of stress, growth stage and time of stress exposure and is unpredictable. Due to their secondary mode of life, plants resort to many adaptive strategies in response to different abiotic stresses such as high salt, dehydration, cold and heat, which ultimately affect the plant growth and productivity.

Against these stresses, plants adapt themselves by different mechanisms including change in morphological and developmental pattern as well as physiological and bio-chemical responses. Drought tolerance comprises drought escape (the ability of a plant to escape periods of drought, especially during the most sensitive periods of its development), drought avoidance (the ability of a plant to withstand a dry period by maintaining a favourable internal water balance under drought) and drought tolerance mechanisms (the ability of a plant to recover from a dry period by producing new leaves from buds that were able to survive the dry spell.

Effect of Drought on Plant Growth and Development

Water stress has marked effect on cellular processes, plant growth, development and economic yield. Water stress is usually measured as leaf - water potential since leaves are directly involved with the production of assimilates for growth and yield. As water potential declines, pressure or torgor potential also declines; the decline in the turgor potential is much more when there is no osmoregulation in response to water stress. Osmoregulation or osmotic adjustment refers to the active accumulation of solutes in the cell during the period in which water stress develops. The solutes accumulated by different plants are considerably different : they ranges from photo-synthetic products like sugar,fructans etc through important. At the cellular level, it affects the flowering structures processes :

1. Structures and membranes and organelles.
2. Hydration and structures of macro molecules like proteins and nucleic acids.
3. Pressure differential across the membranes-cell wall complex, which in turns affect cell expansion.

Stomatal and Non-stomatal Limitation on Photo-synthesis of Droughted Plants

The rate of CO_2 assimilation in the leaves is depressed at moderate leaf water deficits or even before leaf water status is changed in response to a drop in air humidity or in soil water potential. The relative part of stomatal limitation of photo-synthesis depends on the severity of water deficit. Under mild stress it is a primary event, which is then followed by adequate changes of photo-synthetic reactions. Stomatal control of water loss has been identified as an early event in plant response to WD under field conditions leading to limitation of carbon uptake by the leaves. Stomata close in response either to a decline in leaf turgor and/or water potential, or to a low-humidity atmosphere. As a rule, stomatal responses are more closely linked to soil moisture content than to leaf water status. This suggests that stomata are responding to chemical signals (*e.g.* ABA) produced by dehydrating roots.

A clear time dependency in stomatal responsiveness to air humidity and water status was also found, suggesting that some of diurnal changes in stomatal function may result from metabolic processes with a circadian rhythm. Changes in cell carbon metabolism are also likely to occur early in the dehydration process as shown by Lawlor. The drought-tolerant species control stomatal function to allow some carbon fixation at stress, thus improving water use efficiency, or open stomata rapidly when water deficit is relieved. Although stomatal closure generally occurs when plants are exposed to drought, in some cases (severe stress) photo-synthesis may be more controlled by the chloroplast's capacity to fix CO_2 than by the increased diffusive resistance. Non-stomatal responses of carbon fixation such as PS2 energy conversion and the dark reaction of Rubisco carbon fixation are resistant to WD. In addition, stomatal closure occurs before inhibition of photo-synthesis and restricts CO_2 availability at the assimilation site in chloroplasts.

When WS is imposed slowly as is generally the case under field conditions a reduction in the bio-chemical capacity for carbon (C) assimilation and utilization may occur along with restriction in gaseous diffusion. For example, in grapevines grown in the field, CO_2 assimilation was limited to a great extent due to stomatal closure as summer drought progress, but there was also proportional reduction in the activity of various enzymes of the reductive Calvin cycle.The tight correlation between mesophyll photo-synthesis and stomatal aperture may reflect a down-regulation of photo-synthetic apparatus by low C availability. According to Ort *et. al.* the response of photo-Plant responses to drought and stress tolerance190 synthesis to internal cell CO_2 (Ci) indicates that the bio-chemical demand for CO_2 was downregulated in response to declining CO_2 availability.

Drought Stress and PS2 Activity

Photo-synthetic carbon reduction and carbon oxidation cycles are the main electron sink for PS2 activity during mild drought. It was shown that PS2 functioning and its regulation were not quantitatively changed during desiccation. The CO_2 molar fraction in the chloroplasts declines as stomata close in drying leaves. As a consequence, in C3 plants RuBP oxygenation increases and becomes the main sink for photo-synthetic electrons. Depending on the prevailing photon flux density (PFD), the O_2 through photorespiratory activity can entirely replace CO_2 as an electron acceptor or not. Havaux has investigated the impact of various environmental stresses (drought, heat, strong light) applied separately or in combination on the PS2 activity.The existence of a marked antagonism between physicochemical stresses (*e.g.* between water deficit and HT) was established, with a water deficit enhancing the resistance of PS2 to constraints as heat, strong light.Similar results were obtained on bean plants. The data show that quantum yield of PS2, as related to Calvin cycle metabolism, is reduced only under drastic water deficit.

Long-term drought reduction in water content led to considerable depletion of pea PS2 core. The remaining PS2 complex appeared to be functional and reorganized with a unit size (LHCP/PS2 core) two-fold greater than that of well irrigated plants, and enhanced degradation of CP43 and Dl proteins. In addition,PS2 complexes are able to change their location and structure as in PS2-β centers and state-transitions. The decline in PS2 efficiency is regulatory, serving a photoprotective role. Increased levels of energy dissipation which decrease Ψ PS2 may help to protect PS2 from over-excitation and photo damage studied the metabolic consumption of photo-synthetic electrons and dissipation of excess light energy in tomato plants under WS. They established that O_2 evolution, O_2 uptake, net CO_2 uptake and CO_2 evolution declined. It was concluded that PS2, the Calvin cycle and mitochondrial respiration are down regulated under WS. The same authors calculated the percentages of photo-synthetic electrons dissipated by CO_2 assimilation, photorespiration and the Mehler reaction in control leaves more than 50% of the electrons were consumed in CO_2 assimilation, 23% in photorespiration and 13% in Mehler reaction. Under severe stress the % of electrons dissipated by CO_2 assimilation and the Mehler reaction declined while the % of electrons used in photorespiration doubled. The consumption of electrons in photorespiration may reduce the likelihood of damage during WS.

Noctor *et. al.* provided quantitative estimation of the relative contributions of the chloroplast electron transport chain and the glycolate oxidase load placed on the photo-synthetic leaf cell.

Assuming a 10% allocation of photo-synthetic electron flow to the Mehler reaction, photo respiratory H_2O_2 production would account for about 70% of total H_2O_2 formed. When chloroplastic CO_2 concentration rates are decreased photorespiration becomes even more predominant in H_2O_2 generation. At the increased flux through photorespiration observed at lower ambient CO_2 the Mehler reaction would have to account more than 35% of the total photosynthetic electron flow in order to match the rate of peroxisomal H_2O_2 production. According to the authors, the interaction between oxidants, antioxidants and redox changes in draughted plants can modify gene expression and photorespiratory H_2O_2 can play role in signaling and acclimation.

Rubisco, Specific Proteins and Drought Stress

The mechanism by which Rubisco may be down regulated in the light due to tight binding inhibitors could be pivotal for tolerance and recovery from stress and may be central to integrating the midday depression of photosynthesis. Additionally, enhanced rates of oxygenase activity and photorespiration maintain the ET rate in response to drought and are quantitatively much more important than the Mehler reaction found a close relationship between Rubisco content and maximal O_2 evolution rate measured at high photo-synthetic photon flux density (PPFD) during leaf dehydration. It was established that below -2.0 MPa inhibition of photo-synthesis in two maize cvs is in part attributed to stomatal conductance but mostly to the decreased activities of carbonic anhydrase, phosphoenol pyruvate carboxylase and Rubisco. As mentioned above, the primary site of limitation of maximal O_2 evolution rate, measured at high PPFD, seemed related to significantly reduced RuBP content, not to the amount of Chl or Rubisco. But as mentioned above, Rubisco is not a prime target of water deficit and is not limiting net CO_2 assimilation of leaves submitted to desiccation. Decreased supply of CO_2 to Rubisco under both mild and severe water deficit is primarily responsible for the decrease in CO_2 fixation.

Specific proteins display particular structural features such as the highly conserved domain predicted to be involved in hydrophobic interaction leading to macromolecular stabilization. The majority of new proteins belong to dehydrin-like proteins,which are abundantly induced during embryo maturation of many higher plants as well as in water stressed seedlings. Dehydrins are synthesized by the cell in response to any environmental influence that has a dehydration component,such as drought, salinity or extra-cellular freezing. Dehydrins may stabilize macro-molecules through detergent and chaperone like properties and may act synergistically with compatible solutes. The steady state levels of major PS2 proteins, including the Dl and D2 proteins in the PS2 reaction center, declined with increasing water deficit possibly as a result of increased degradation. The effects of WD on PS2 protein metabolism, especially on the reaction center proteins may account for the damage to PS2 photo-chemistry.

Drought Stress and Lipids

Along with proteins, lipids are the most abundant component of membranes and they play a role in the resistance of plant cells to environmental stresses. Strong water deficit leads to a disturbance of the association between membrane lipids and proteins as well as to a decrease in the enzyme activity and transport capacity of the bilayer established that for *Arabidopsis*, polyunsaturated trienoic fatty acids may be an important determinant of responses of photo-synthesis and stomatal conductance to environmental stresses such as vapour pressure deficit. When *Vigna unguiculata* plants were submitted to drought the enzymatic degradation of galacto- and phospholipids increased. The stimulation of lipolytic activities was greater in the drought-sensitive than in drought-tolerant cvs. Drought stress provoked considerable changes in lipid metabolism in rape *(Brasica napris)* plants. The decline in leaf polar lipid was mainly due to a decrease in MGDG content. Determination of molecular species in phosphatidylcholine (PC) and MGDG indicated that the procaryotic molecular species of MGDG (C18/C16) decreased after DS while eukaryotic molecular species (C18/C18) remain stable.

It was suggested that the prokaryotic pathway leading to MGDG synthesis was strongly affected by DS while the eukaryotic pathway was not. Strong WD results in a profound overall drop in MGDG, the major leaf glycolipid. In drought sensitive seedlings of *Lotus corniculatus* the ratio of MGDG/DGDG declined 3-fold, while the relative part of MGDG was 12-fold lower. The lipid composition of desiccated *Ramonda* leaves is profoundly modified : the ratio of phospholipids (PLs) to galactolipids (GLs) increased and the relative proportion of MGDG to DGDG drastically decreased. An increase in the PLs relative to GLs in leaves indicate a preferential degradation of the chloroplast membranes.

Oxidative Stress and Antioxidant Defense Systems

It was established a link between tolerance to oxidative stress induced by WD and rise in antioxidant concentration in photo-synthetic plants. This shows that plants are well endowed with antioxidant molecules and scavenging systems. Enzymatic free radical processing systems include SOD, catalysing the dismutation of superoxide (O_2-) into H_2O_2 and O_2 and those involved in the detoxification of H_2O_2 - catalase, peroxidase, glutathione reductase (GR-ase). In optimal conditions leaves are rich in antioxidant enzymes and metabolites and can cope with activated O_2, thus minimizing oxidative damage. Antioxidant metabolites as ascorbate and glutathione are present in chloropiasts in very high concentrations and apart from their obvious role as enzyme substrates, they can react chemically with almost all forms of activated O_2.

The hydrophilic antioxidants ascorbate and glutathione are effective chemical scavengers of oxygen radicals. Enzymatic detoxification systems either quench toxic compounds or regenerate antioxidants with the help of reducing power provided by. Foyer *et. al.* showed that overexpression of GRase in chloropiasts doubled the concentrations of ascorbate and glutathione (GSH) in leaves and conferred increased resistance to oxidative stress. According to their results drought caused a decrease in the content of reduced glutathione and an increase in that of vitamin E. Carotenoids and vitamin E are the main lipid soluble antioxidants of plant cells. In stressed leaves vitamin E increased significantly.

Types of Drought Environment

The breeding methodology as well as the resistance mechanism that should be developed will depend, to a large extent, on the type of drought environment to which the crop will be subjected to. In general following three types of environments can be associated with drought : (a) stored moisture environment, (b) variable moisture environment and (c) optimal moisture environment. It may, however, be pointed out that numerous combinations of these environment occur in reality.

(a) **Stored moisture environment :** In this type of environment, the crop completes it life cycle on the moisture stored in the soil during a prior wet or rainy season. As a result the level of moisture stress will depend upon the amount of moisture stored in the soil, the duration of the crop and the rate of evapotranspiration. In such environments crops become subjected to moisture stress during their terminal phase of their growth and development. The likelihood of success of breeding for drought resistance is rather high, and a spectrum of traits can be exploited for this purpose.

(b) **Variable moisture environment :** This type of drought environment is characterized by alternate drying and wet periods of varying lengths. Plant grow in such environments can be able to take the advantages of the periodic rainfall and also to survive, with minimum detrimental effects, the periods of the water stress. The periodic and variable nature of water stress is likely to reduce the chance of breeding programmed for drought resistance.

(c) **Optimal moisture environment :** The crop grown with adequate moisture during of its life cycle; drought occurs occasionally at highly unpredictable stages of growth and development. The period of drought may be limited to a part of one day when evapotranspiration greatly exceeds the root uptake. But ordinarily it is associated with a period less than normal precipitation. The effect of drought in such environment are likely to be rather serve in view of the inadequate time available for the plants to become adjusted to water stress. Breeding for drought resistance for such environment would be extremely difficult.

Drought Resistance : Drought resistance may be defined as the mechanism causing minimum loss of yield in drought environment relative to the maximum yield in a constraints free environment for the crop. However it does not exist as a unique heritable plants attribute. The various mechanism by which a crop can minimize the loss in the yield during to drought are grouped into the following three categories (1) Drought escape (2) dehydration avoidance (3) dehydration tolerance; these are briefly disused below.

1. Drought Escape

Drought escape describes the situation where a drought susceptible variety performs well in a drought environment simply by avoiding the period of drought. Early maturity is an important attribute of drought escape, and is suitable for the environments subjected to late-season drought stress. Early varieties generally have lower leaf area index, lower evapotranspirational and lower yield potential. Therefore, this attributes is not suitable for variable moisture and optimal moisture drought environments. Seasonal length for maize under rainfed conditions is often defined as that time when precipitation is equal to or exceeds 50% of potential evapotranspiration, as determined by radiation, wind, and temperature. A major goal of breeding is to develop cultivars that can escape drought by being sufficiently early in maturity as to complete their life cycle within a given season length. In the lowland tropics, the lower limit of average seasonal rainfall for successful maize cultivation (> 1 t/ha) is around 400-500 mm; in mid altitude areas the minimum is about 350-450 mm; in the highlands it is around 300-400 mm.

Because WUE is lower in the warmer lowlands, maize requires more rain fall than in the highlands. Selection for earliness matches the phenology of the crop to the pattern of water availability. Since the time from sowing to flowering or physiological maturity is a highly heritable trait, selection for earliness can easily be accomplished. However, earliness carries a yield "penalty" when rainfall is higher than average. Under those circumstances, the yield of an early maturing cultivar is limited by the amount of radiation the cultivar can capture—normally less than that for a later maturing cultivar.

2. Dehydration Avoidance

Dehydration avoidance is the ability of a plant to retain a relatively high level of hydration under conditions of soil or atmospheric water stress. The results of various physiological, biochemicals and metabolic process of plants those are involved in the growth and yield not being internally exposed to stress, and there, by they are protected from water stress. The common measures of dehydration aviodence are tissue water status as expressed by water turgor potential under condition of water stress. This can be achieved by reducing the transpiration rate or increased by water uptake. Wild species are readily classifiable as "water savers" and water spender but crops plants ordinarily exhibits a combination of both features, probably as a result of selection by man.

- **Reduced Transpiration :** Water saving mechanism is common in xerophytes, which have evolved for survival under extreme water stress conditions; ordinarily, they show poor biomass production. Water saving species reduces transpiration mostly by closure of their stomata in response to water deficit well before wilting. Stomata are responsible for the bulk of transpiration, and also for gas exchange in respiration and photo-synthesis. Therefore, stomatal closure is likely responsible for the interference with photo-synthesis, and drought resistance mechanism based on the stomatal sensitivity and reduced transpiration are generally opposed to maintainance of a higher yield potential.in water stress plant, stomata may open early in the morning hours and close during the day time as solar radiation increases.

- **Osmotic adjustment :** It is also important mechanism responsible for the drought avoidance. Osmo-regulation is positively associated with higher yield under water stress conditions, as it allows growth and result in delayed leaf death by maintaining torgour pressure and possibly, some other unknown mechanism, but this mechanism of dehydration avoidance may reduce photo-synthesis upon recovery and could lower potential yield if it is associated with smaller cell size. The role of different mechanism may change with the stage of plant development. For example in sorghum, stomatal sensitivity to water stress seems to be the main mechanism during vegetative phase, while after flowering osmo-regulation and torgour maintenance were important.
- **Abscisic Acid (ABA) :** ABA is known as stress hormone as its concentration increases in response to stress, including water stress. Water deficit is sensed by roots. This begins to synthesis ABA within one hour of the onset of water stress. ABA is transported via xylem from roots to leaves within minutes to hours, its half life in leaf being 30 minutes. Xylem ABA concentration decrease sharply and stomata open in less one day after watering of the stressed plants. ABA plays a major role in water stress avoidance by effecting stomata closure reduction in leaf expansion and promotion of root growth. As a result mutants partly deficient in ABA bio-synthesis are more stressed at the cellular level then are normal plants, when both are subjected to the same level of water stress. In some crops, ABA accumulation was positively associated with yield under stress, while in several other the association was not clear.
- **Cuticular wax :** Transpiration also occurs through cuticle; the amount of transpiration depends mainly on the wax deposited within and over the cuticle. The genotypic potential for wax deposition is best in evaluated in plants subjected to water stress. But the effect of Cuticular wax on transpiration is small and, for a given plant, increase wax in load beyond a given threshold would not reduce transpiration. The shape and angle of wax deposition may affect leaf reflectance within the spectrum range of 400 to 700nm, which in turn may affect net radiation and leaf temperature. For example, increase in glaucousness in wheat and sorghum reduced net radiance and leaf temperature, which improves their yield under water stress.
- **Increased water uptake :** water uptake depends mainly on the characteristics of root system, which may be described and measured in various ways, *e.g.*, root- length density, root axial resistance, root radial resistance etc. some broad generalization about root system and its possible role in water stress resistance are as follows : When soil moisture is unlimited in deeper soil horizon, a deep root system is a distinct and effective component of drought resistance. Root distribution pattern is affected by water status of soil. In a situation of transient soil drying and wetting, a dense root system and or a low root resistance is important in the maintenance of higher leaf water potential.

3. Drought Tolerance

Dehydration tolerance describes the ability of plants to continue metabolizing at low leaf water potential and to maintain growth despite dehydration of the tissue or to recover after release from stress conditions. According to Hsiao, and Boyer,, translocation is one of the more dehydration-tolerant processes in plants. It would proceed at levels of water deficit sufficient to inhibit photo-synthesis. Ample information has been accumulated in the cereals to show that grain growth is partially supported by translocated plant reserves stored mainly in the stem during the pre-anthesis growth stages. When water stress occurs and the current photo-synthetic source is inhibited, the role of stem reserves as a source for grain filling increases, both in relative and absolute terms. Stem reserves may therefore be considered as a powerful resource for grain filling in stress affected plants during the grain filling stage.

Genetics of Drought Tolerance

Existence of different variation for drought resistance has been demonstrated in many crops. Drought resistance was estimated as yield stability (*e.g.* in wheat, rice, maize, barley, sorghum), lead water potential (sorghum, wheat, rice, soya bean, cotton), leaf rolling (rice), root growth (sorghum, rice, oat, wheat, maize,) root xylem diameter (wheat), osmotic adjustment (wheat, sorghum), stomatal conductance, ABA accumulation, canopy temperature, seedling establishment and growth, seeding recovery after stress, growth under stress, resistance to flower shedding and sustained pod formation under stress and prolie accumulation (barley and Brassica sp.).

The genetic control of these traits ranges from polygenic to oligogenic. Generally, leaf character like waxy bloom, glossy trait, and glaucousness, glabrous leaves are oligogenic control. Some other traits like ABA accumulation in wheat, constitutive proline accumulation in barley mutant and resistance to flower abscission and ability to support pod formation in rajma seem to be determined by oliogenes.

Sources of drought resistance : Drought resistance may be available in cultivated varieties, landraces, related wild species, or may be introduced by genetic engineering in different plants called transgenic.

Selection criteria : A good selection criteria should have following attributes; it should be easy to estimate\score, it should be have high heritability, a large genetic variability should exists for the trait, it should exhibit a significant association with yield under stress. A major factor that prevented progress for improving yield in water limited environment is the lack of knowledge of the critical traits that should be selected for achieving the goal. The various selection criteria used in breeding for drought resistance crops are written as follows :

1. Dehydration Avoidance

Leaf rolling : leaf rolling is visually scored from 0-5 in rice either in the morning or at mid day. It is extensively used as selection criteria at IRRI, Philippines. It is likely that leaf rolling will predict leaf water potential in species of low osmotic adjustment in rice.

A combination of leaf rolling and leaf firing is being extensively used in maize and sorghum. Leaf firing is the drying of leaves due to insufficient transpiration cooling. In maize leaf senescence under stress was negatively correlated with yield.

Canopy temperature is readily measured with infrared thermometer. In maize it was negatively correlated with yield under water stress and in rice it was negatively correlated with spikelet fertility. A sufficient level of water stress is a major pre-requisite for applying the method to selection work. The crop must fairly dense and free of skips measurements should made immediately after noon, and windy conditions should be avoided. Measurement should be repeated 2-3 times a week as stress progress.

Leaf attributes like dense pubescence, glaucousness etc. are scored visually. Epicuticular wax load can also measure relatively rapidly. But these traits are unreliable indicators of drought resistance; it is desirable that be incorporated into an integrated selection index with greater weight given to the visual symptoms of wilting/leaf rolling/canopy temperature.

Leaf water retention may be useful in some materials, and its usefulness as the sole criterion of selection is opened to question.

Root characteristic are very difficult to evaluate in masses. In any cases, root attributes should be reflected in canopy response to stress, which are far easier to evaluate.

2. Dehydration Tolerance

Seedling growth under PEG stress is a useful selection criterion. In alfalfa, a successful selection program for drought tolerance had delayed wilting in a PEG-containing solution as one of the components. Depending on the species, the extent of genetic variation and PEG concentration, a visual ranking of the response to PEG may be sufficient for selection.

Growth under stress in the field may be used as a selection index provided the measure is sufficiently simple and rapid.

Plant phenology may be used as a an index of stress tolerance as drought stress delays or accelerates flowering depending on the growth stage at which stress occurs and on stress intensity, delayed heading under stress has been used a selection index in rice. The time interval between pollen shedding and silking in maize under drought stress is shorter in selections with higher yield under stress, and can be used as selection index.

Grain filling by translocated stem reserve is an important attribute of resistance of cereals to drought after anthesis. In wheat, it may be used as follows : plants are sprayed to complete wetting by a solution of magnesium/sodium chlorate at 14 days after anthesis. The entire plant surface is bleached without killing the plant. Grain filling is now proceeding only due to reserve materials. The 1000-kernals weight of the treated plots is compared with that of the corresponding untreated plots. A smaller difference between the treated and untreated plots indicates a greater reserve translocation. This approach is used experimentally in Israel and Australia.

Difficulties in Breeding for Drought Resistance

1. The moisture regime or drought environment prevailing in the region for which the variety is to be developed must be clearly defined. This is essential since it would often determine the combination of drought resistance traits that should be incorporated in the new variety.
2. Selection for drought resistance has to be performed under moisture stress. Creation of controlled moisture stress in the field/greenhouse is usually difficult, and greenhouse results have to e confirmed in the field.
3 Drought resistance in plants usually is the consequence of a combination of characters. As a consequence, no single character can be used for selection index may developed.
4. Measurement of many drought resistance traits is difficult and problematic.
5. Many drought resistance traits reduce yield; *e.g.*, earliness, stomatal sensitivity, etc. incorporation of such traits may reduce yields of the varieties. In any case, it would at least require additional breeding effort to enhance the yield of the varieties in such cases where the drought resistance trait is not positively associated with yield.
6. Selection for yield for yield has to be performed under optimal moisture, while that for drought resistance must be done under moisture stress. This makes it necessary to develop a suitable and elaborate breeding scheme to develop a drought resistant variety with high potential.
7. The use of wild relatives as sources of drought resistance is problematic and their value for improving the drought resistance of crops is questionable.
8. Several transgenes holds promise for the development of drought resistant varieties. The results from such work are quite encouraging but so far a commercial example is not available.

Conclusion

Drought seems to be rather difficult to define and more difficult to quantify. For example, various definition used the term *i.e.*, precipitation, air temperature, relative humidity, winds, air flow, soil moisture condition, transpiration, soil moisture storage capacity and plant condition. A working definition may be drought is inadequacy of supply of water to the plant to meet their physiological activity for the growth and development. Increasing the drought area is increasing day by day due to many cause *e.g.* climate change, global warming. Against these stresses, plants adapt themselves by different mechanisms including change in morphological and developmental pattern as well as physiological and bio-chemical responses. Drought tolerance comprises drought escape *i.e.*, the ability of a plant to escape periods of drought, especially during the most sensitive periods of its development, drought avoidance *i.e.*, the ability of a plant to withstand a dry period by maintaining a favorable internal water balance under drought, and drought tolerance mechanisms the ability of a plant to recover from a dry period by producing new leaves from buds that were able to survive the dry spell.

Chapter 31

Abiotic Stress Responses in Tea

ABSTRACT

Abiotic stress is limiting to tea productivity. The severity and duration of the abiotic stress are critical, as it has multifarious effects on crop plants. The tea plant is commonly grown in rain-fed ecosystems and thus it encounters seasonal water deficit conditions that induce loss in crop yield. In this review, we have highlighted the effects of abiotic stress on the growth, mineral nutrition, reactive oxygen species (ROS), and antioxidant metabolism in tea plants. This article also highlights the mechanism of drought stress amelioration in tea plants on a physiological and bio-chemical basis. Drought stress caused increased water loss rate (WLR) and decrease in relative water content (RWC), dry mass, chlorophyll, carotenoid, and total phenolic contents of leaf and antioxidants like ascorbate and glutathione in tea. Leaf antioxidant enzymes (*e.g.*, SOD, CAT, GR, etc.) showed differential activities whereas there was increase in ROS and lipid peroxidation with decreased POX activities with progressive stress. Drought stress altered antioxidative response with apparent decrease in mineral nutrient (Zn, Ca, Na, Fe, Mg, and K) contents of leaves suggesting that mineral deficiency mediated drought stress induced oxidative damage in tea. Tea plants exposed to heavy metals (*e.g.*, Cd, Cu, Al) also showed reduction in growth and antioxidative responses. Further, a post-drought recovery study in tea also reveals that drought induced bio-chemical damages are not permanent, as the plant recovers on rehydration. Mineral nutrients play an important role in post-drought recovery in tea. The process of recovery was significantly influenced by foliar spray of K, Ca, Zn etc., leading to improved antioxidant potential. Thus, drought tolerance and post-drought recovery can be improved by application of nutrients like K, Ca, Zn, B etc. However, molecular mechanisms of abiotic stress amelioration and post stress recovery mediated by these nutrients need to be explored in future.

INTRODUCTION

Tea is second only to water as the most consumed drink in the world. It has been used medicinally for centuries in India and China. Green tea is comparatively healthier than black and Oolong tea. Literature also suggests polyphenols as potential indicators of drought tolerance in Camellia sinensis because of variation of shoot epicatechin and epigallocatechin contents in response to water stress. Tea plants are grown in a wide range of latitudes in the world ranging from 45°N (Russia) to 30°S and longitude from 150°E (New Guinea) to 60°W(Argentina). In India, it is grown in Barak/Brahmaputra valley/Dooars and other hilly terrains of India (*i.e.*, Darjeeling, Himachal, Nilgiri, and Uttaranchal). In general, the plant is kept as an evergreen shrub by pruning. The first two leaves and a bud are plucked for tea processing. In tropical countries, tealeaves are harvested all year round, whereas in temperate countries harvesting is seasonal. Tea is an evergreen, perennial, cross-pollinated, C3 plant and in nature, the tea tree can attain a height of 20–30 m. However, under cultivated condition, the bush height of 60–100 cm is maintained for harvesting the tender leaves, which continues flashing even more than 100 years.

Abiotic stresses cause considerable loss to agricultural production worldwide.

Abiotic stress conditions such as drought, salinity, heat, or heavy metal, etc., have been the subject of intense researches. In plants, osmotic stress generated by either drought and/or salinity or heavy metal stress represents the most common environmental hazard for the plant's growth and productivity. In addition, it has been reported that when combined, multiple abiotic stresses may interfere with nutrient accumulation, thereby further contributing to growth inhibition. Consequently, it seems a prerequisite for plant species selected for their use in these areas to have adaptations that confer abiotic stress resistance through notably the optimized utilization of water and nutrients. However, in nature, tea and other crop plants are routinely subjected to a combination of different abiotic stresses.Recent studies have pointed out that drought may exacerbate the advers eeffects of abiotic stress on the plant nutrient status, but only a few attempts have been made to quantify the combined effect of drought, mineral nutrients, and other abiotic stress. *Camellia sinensis* (L.) O. Kuntze is a perennial plant growing in rain-fed ecosystems and has several medicinal values attributed to high polyphenol contents having tremendous impact on human health. Growth and productivity of tea plants largely depends on their capacity to adapt to abiotic stress, namely temperature, drought, metal, and nutritional disturbances. Although the effects of some individual environmental factors on the tea plant are well documented (Handique and Manivel, 1990; Chakraborty *et. al.*, 2002; Panda *et. al.*, 2003; Upadhyaya and Panda, 2004a; Jeyaramraja *et. al.*, 2005; Sharma and Kumar, 2005; Cheruiyot *et. al.*, 2007; Cheruiyot *et. al.*, 2008; Upadhyaya *et. al.*, 2008; Yadavand Mohanpuria, 2009; Upadhyaya *et. al.*, 2011; Upadhyaya *et. al.*, 2012; Das *et. al.*, 2012; Gupta *et. al.*, 2012), the effects of interacting abiotic factors remain poorly investigated. Therefore, the objective of this review is to highlight the effects of drought, heavy metal stress, and their interaction on growth, water relationships, nutrient status, and antioxidative responses during abiotic stress and its recovery in the tea plant.

ABIOTIC STRESS AND THE TEA PLANT

Plants grow in almost any part of the world under diverse climatic conditions differing in temperature, light quantity, and quality and availability of water. Plants that grow in a specific environment are adapted to these conditions, and can also cope with changes that might be adverse for their growth and development. Adaptation is required because plants cannot escape unfavourable conditions due to their sessile growth habit. This implies that species differ genetically in their adaptation and resistance to abiotic stresses. Abiotic stress is in fact the principal cause of the loss of crop productivity worldwide, reducing average yields for various crops by more than 50%. Abiotic stresses cause losses worth hundreds of millions of dollars each year due to reduction in crop productivity and crop failure.

Any external constraint that reduces the ability of a plant to develop to its genetically predetermined level is called stress. Plant species are highly variable in their optimum environments and tolerance to abiotic stress conditions (*e.g.*, drought, salinity, metals, temperature, etc.) Principal abiotic stresses that can affect growth and productivity of the tea plant include the following :

- i. Cold (chilling and frost)
- ii. Heat (high temperature)
- iii. Salinity (salt)
- iv. Drought (water deficit condition)
- v. Excess water (flooding)
- vi. Radiations (high intensity of ultraviolet and visible light)
- vii. Chemicals and pollutants (heavy metals, pesticides, and aerosols)
- viii. Oxidative stress (reactive oxygen species, ozone)
- ix. Wind (sand and dust particles in wind)
- x. Nutrient deprivation in soil

Abiotic stresses include salinity, drought, nutrient deficiency, floods, metal, temperature, high light intensity, pollutants etc. Crops including tea and their varieties differ in their responses to various abiotic and biotic stresses due to differences in their inherent antioxidant potential. Among the abiotic stresses, temperature, heavy metal, and drought stresses are of special significance. As most of the plant processes are influenced by these stresses, their effect depends on the dose, duration of stress, and developmental stages of plants. During abiotic stress, signaling stress is first perceived by the receptors present on the membrane of the plant cells; the signal is then transduced downstream resulting the generation of second messengers including calcium, reactive oxygen species (ROS), and inositol phosphates. These second messengers, such as inositol phosphates further modulate the intra-cellular calcium level. This perturbation in cytosolic Ca^{2+} level is sensed by calcium binding proteins, also known as sensory proteins, which change their conformation in a calcium dependent manner and then interact, initiating a phosphorylation cascade regulating stress responsive genes (*e.g.*, drought responsive genes) or the transcription factors controlling these genes. The products of these stress genes ultimately lead to plant adaptation and help the plant to confer abiotic stress tolerance. Thus, the plant responds to stresses as individual cells and synergistically as a whole organism. Stress induced changes in gene expression in turn may participate in the generation of hormones like abscisic acid (ABA), salicylic acid, and ethylene. These molecules may amplify the initial signal and initiate a second round of signaling that may follow the same pathway or use altogether different components of the signaling pathway. The most common features of abiotic stress represent the regeneration of ROS in tea plants resulting in a condition of imbalance between oxidants and antioxidants called oxidative stress. Tea being perennial, encounters occasional drought, temperature change, various metals, etc. that affect the growth and productivity of crops growing in an agro-ecosystem.

RESPONSES OF DROUGHT STRESS IN TEA

Drought denotes a period without rains during which soil water content is reduced and plants suffer from water deficit. Drought affects the morphology, anatomy, and physiology of the plants. Physiological, bio-chemical, and anatomical responses, however, occur much earlier than the usual symptoms of wilting, which may be permanent or temporary depending on the availability of soil moisture. Drought stress results in stomatal closure and reduced transpiration rate, a decrease in water potential of plant tissues, decrease in photo-synthesis and growth inhibition, accumulation of ABA, proline, mannitol, sorbitol, formation of radical scavenging compounds (ascorbate, glutathione, tocopherol, etc.), and synthesis of protein. Drought stress caused decreased mineral uptake, stomatal conductance, and photo-synthetic rate and ultimately reduced the growth rate of tea plants. There are reports of various physiological and bio-chemical responses of various tea clones to drought. However, there is a dearth of information on the effect of drought on mineral nutrition in tea. Drought caused osmotic stress and affects mineral nutrition. Drought stress may be perceived by plants through various osmosensors, and signal transduction involving various cascades activates stress responsive genes responsible for detoxification, osmoprotection, chaperone function, and water and ion movements in plants. Nutrient imbalance resulting from drought or other abiotic stress may cause oxidative stress in plants. We studied the impact of drought stress changes in the nutrient status of the tea plant and the result reveals that drought stress caused reduction in leaf mineral content, which varies in different clones. Drought induces an increase in ROS production and inactivation of the antioxidant system, resulting in various degrees of oxidative damage in different clonal varieties of tea. Drought induced decrease in mineral nutrients modulates oxidative stress in tea.

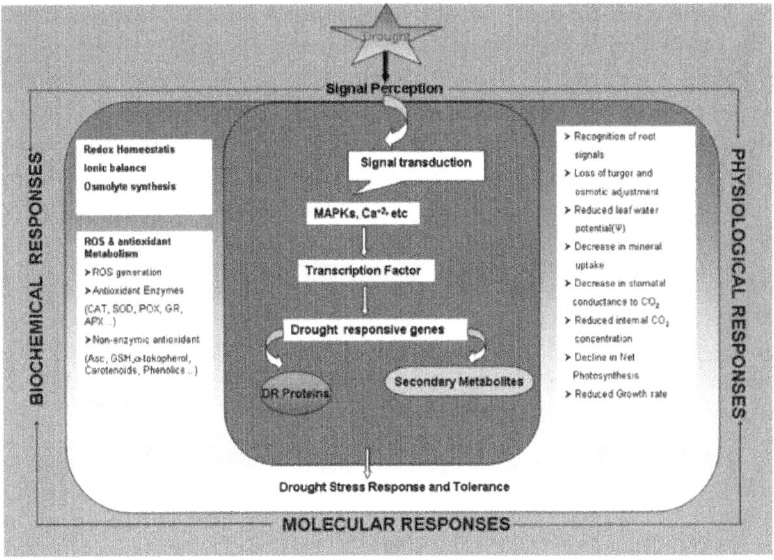

Figure 31.1 : Morphological, physiological, molecular and bio-chemical basis of drought stress responses in tea. Reactive oxygen species (ROS), catalase (CAT), superoxide dismutase (SOD), peroxidase (POX), glutathione (GR), ascorbate peroxidase (APX), ascorbate (Asc), glutathione (GSH), mitogen activated protein kinases (MAPKs) and drought responsive protein (DR protein).

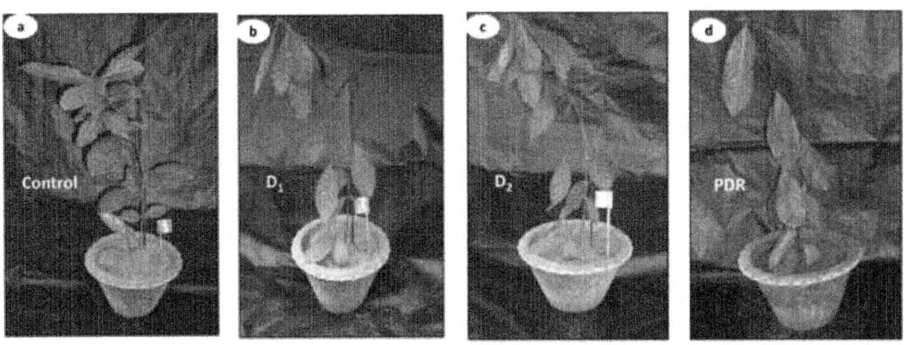

Figure 31.2 : Morphological changes in *Camellia sinensis* in response to drought and rehydration. Control (well irrigated), 7 days drought (D_1) and 14 days drought (D_2) conditions and 15 days of post drought rehydration (PDR).

Growth and yield of tea is greatly influenced by the seasonal fluctuations in weather variables such as rainfall, temperature, drought, and mineral nutrients etc. Drought stress is an important abiotic stress, which induces oxidative damage in the tea plant and affects antioxidant systems, altering different physiological and bio-chemical processes and causes significant crop losses. Even after 7 and 14 days of water withholding RWC of leaves significantly decreased, which is correlated with decreased leaf dry mass in a stressed plant with respect to control. Drought induced degradation of both total chlorophyll and carotenoid in tea plants. Decrease in chlorophyll and carotenoid in response to drought stress in tea was reported earlier. Such damage to chlorophyll pigments may eventually decrease photo-synthetic efficiency in plants, which might be one of the potent causes of reduction in growth of the tea plant. However, some studies also provide evidence for a significant positive relationship between transpiration efficiency and leaf chlorophyll concentration in plants.

Our previous experiment suggests that proline content increased in leaves of stressed tea plants. Proline acts as an osmoprotectant and accumulation of proline suggested genotypic tolerance of tea to water deficit stress, as this helps in maintaining water relations, prevents membrane distortion, and acts as a hydroxyl radical scavenger.

The data obtained from the previous study suggest that there is a close relationship between leaf nutritional status and drought stress induced oxidative damage in the growing seedlings of tea. Zinc content of leaves in stressed tea plants also decreased. Zinc is an essential nutrient that plays important roles in numerous physiological processes in plants, serving as a cofactor for many enzymes and as the key structural motif in transcriptional regulatory proteins. The level of zinc nutrition may affect plant water relations and alter stomatal conductance. Stomatal conductance and transpiration rates also declined under zinc deficiency. Gas exchange data presented by Hu and Sparks and Sharma *et. al.* indicated that zinc deficiency reduces transpiration efficiency of leaves. However, it has been reported that zinc fertilization and water stress affects plant water relations. Leaf Fe^{2+}content also decreased with progressive dehydration stress. Na^+ content of leaf decreased in tea under drought stress. Osmotic adjustment has been considered as beneficial to the drought tolerant mechanism in field crop species. Na^+ and K^+ contents of leaf affect and regulate the osmotic potential of the plant during stress conditions and hence the Na^+ /K^+ ratio should be adequate for stress acclimatization by plants through osmotic adjustment. Both Mg and K content decreased in leaves of drought induced tea seedlings. Mg nutrition is essential for the function of many enzymes, including adenosine triphosphatases (ATPases), RNA polymerases, phosphatases, protein kinases, glutathione synthase, and carboxylases. In higher plants, Mg is essential for chloroplasts, in the development of the chlorophyll molecule, and a bridging element for the aggregation of ribosome sub-units necessary for protein synthesis. Although K concentration of leaf is significantly correlated with increased proline content and its RWC, proline might not be sufficient to maintain leaf water status with concomitant decreased K content under drought stress. Potassium is one of the major nutrients, essential for plant growth and development. Literature suggests that improving K status of plants can greatly lower ROS production, imparting beneficial effects when plants are subjected to abiotic stress. Potassium is one important mineral regulating leaf osmotic potential, affecting the water use efficiency in plants.

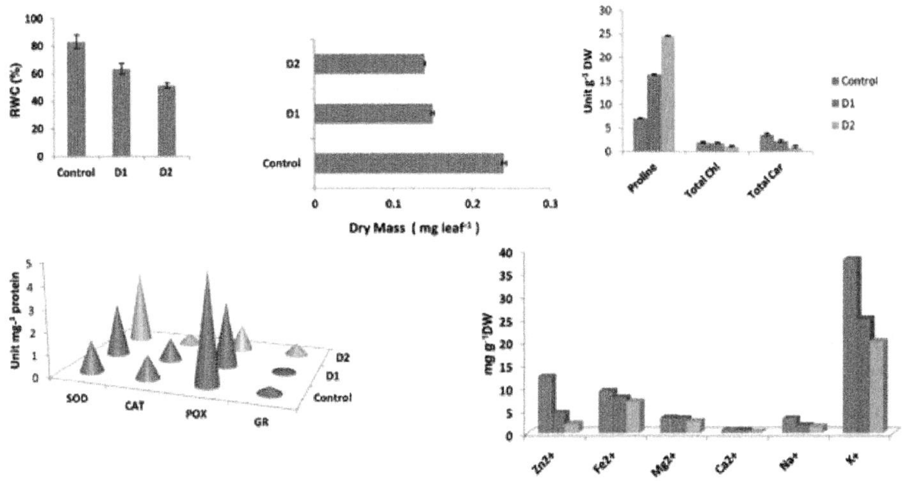

Figure 31.3 : Bio-chemical changes in *Camellia sinensis* in response to drought. Control (well irrigated), 7 days drought (D_1)and 14 days drought (D_2) conditions. Data presented are mean of three independent repeats.

Osmotic adjustment is an essential stress-acclimatizing feature of stressed plants. The effects of potassium on drought resistance of plants have also been reported. Calcium is another important nutrient that is well established as a secondary messenger and is also known to ameliorate water stress effects. In the present study, leaf Ca^{2+} content of growing tea seedlings decreased under drought stress. Calcium is known to be involved in oxidative stress signaling, participating in linking H_2O_2 perception and induction of antioxidant genes in plants. Calcium is a player in most cellular signaling processes and interacts strongly with ROS. Decreasing water availability under drought stress generally results in reduced total nutrient uptake and frequently leads to reduced concentrations of mineral nutrients in crop plants as reported by Baligar *et. al.*. The most important effects of water deficits are on the transport of nutrients to the root and on root growth and extension. Reduced absorption of inorganic nutrients results from interference with nutrient uptake, unloading mechanisms, and reduced transpiration flow.

Imposition of drought stress may increase generation of ROS, such as superoxide radicals (O_2^-), hydrogen peroxide (H_2O_2), hydroxyl radical ($\cdot OH$), alkoxyl radical (RO), and singlet oxygen ($\cdot^1O_2$) within the cell, particularly within the chloroplasts of water stressed leaves. Such ROS generation leads to lipid peroxidation, protein degradation, and enzyme inactivation and affects nucleic acid and almost every component of cell, leading to cell death.

Drought stress caused reduction in ascorbate and glutathione contents with differential responses of enzymic antioxidants in various clones of Camellia sinensis resulting in an oxidative stress situation. Drought induced oxidative stress was also evident from increased MDA, O_2^-, and H_2O_2 content in drought imposed plants. TV-1 clones showed less decrease in antioxidants; higher activities of POX, GR, CAT with higher phenolic contents suggested their better drought tolerance potential. On the other hand, higher ROS and membrane oxidation with decrease in phenolic content during stress in S3A3 and TV-29 proved their drought sensitivity. We reported variation in antioxidant efficiency and bio-chemical tolerance in response to drought stress in various clones of Camellia sinensis.

RESPONSES OF HEAVY METAL STRESS IN TEA

Heavy metal stress is a major threat to crop production. Heavy metal stress reduces crop yield and adversely affects consumers who are using that crop as food. Tea is grown in hilly areas like the north-western region of the Himalayas and weathering of rocks contributes a lot toward enhancing levels of metals such as cadmium. Plant exposure to high levels of Cd causes reductions in photo-synthesis, water uptake, and nutrient uptake. Plants grown in soil containing high levels of Cd show visible symptoms of injury reflected in terms of chlorosis, growth inhibition, browning of root tips, and finally death. Unlike Cd, Cu is an essential micro-nutrient required by the tea plant for various physiological functions, but excess Cu may lead to oxidative stress in the tea plant. Any imbalance in cellular redox homeostasis can be defined as oxidative stress, and imbalance in which the redox steady state of the cell is altered in the direction of pro-oxidants can result in the potentially dangerous ROS and is very common in tea plants subjected to heavy metal stress. ROS such as O_2^-, H_2O_2, and OH are commonly generated under stress conditions and act as strong oxidants that can damage biomolecules. In fact, these oxygen species represent intermediates emerging during the successive reduction of O_2^- to H_2O_2. Plants ' exposure to certain heavy metal ions shifts the balance of free radical metabolism toward an accumulation of H_2O_2. In the presence of redox active transition metals such as Cu and Fe^{2+}, H_2O_2 can be converted to the highly reactive $^{\cdot}OH$ molecule in a metal-catalyzed reaction via the Fenton reaction. The oxidized metal ions undergo a re-reduction in a subsequent reaction with superoxide radicals (O_2^-). An alternative mechanism of $^{\cdot}OH$ formation directly from H_2O_2 and O_2^- is the metal-independent Haber Weiss reaction.

Fenton reaction :

$$H_2O_2 + Fe^{2+}/Cu^{+} \rightarrow .OH + OH^{-} + Fe^{3+}/Cu^{2+}$$

$$\cdot O^{2-} + Fe^{3}/Cu^{2+} \rightarrow Fe^{2}/Cu^{+} + O_2$$

Haber Weiss reaction :

$$H_2O_2 + O_2^- \rightarrow {}^{\cdot}OH + OH^- + O_2$$

The .OH molecule is one of the most reactive with its ability to initiate radical chain reactions and is responsible for irreversible damage of various cellular components. Another ROS that might be involved mainly in lipid peroxidation is the protonated form of.O_2^-, the hydroperoxyl radical ($O_2{}^{\cdot}H$). These species exist in equilibrium. The metal ions inhibit the activities of antioxidative enzymes, especially of glutathione reductase, and also raise a transient depletion of GSH leading to ROS accumulation. Heavy metals like Cu, Zn, Cd, Pb, Al, etc. under toxic concentration inactivate the enzymic antioxidant defense system in tea plants resulting in increased ROS signaling, generally leading to death of a plant. Metals like Cd, Cu, etc. induce decrease in total chlorophyll and carotenoid, consequently reducing photo-synthetic rate and increase in proline content of tealeaf. Cd is more injurious than Cu as Cd induces decrease in leaf dry mass. However, antioxidant enzymes like SOD, CAT, POX, and GR showed differential activities in response to Cd and Cu. In tea and other plants, enzymatic antioxidant cellular machinery comprises a diverse array of enzymes including superoxide dismutase, catalase, peroxidases, and enzymes of ascorbate glutathione cycle, while nonenzymatic antioxidants include low molecular weight non-protein thiols like cysteine and glutathione that are involved in detoxifying ROS.

Al is a major limiting factor for crop production and quality in acid soil, resulting in ROS generation and enhancing lipid peroxidation in crop plants. It is known that tea-leaves can accumulate large amounts of Al from 8,700 to 23,000 mg kg^{-1}, and even up to 30,000 mg kg^{-1} without experiencing Al toxicity. Li *et. al.* demonstrate the adaptation mechanisms in the leaves of the tea plant with an increasing capacity to overcome ROS via antioxidative enzyme systems (SOD, CAT, GPX, APX, and GR) at low Al concentration with CAT and GPX activities of tea-leaves playing a more important role. At 0.53 mM Al concentration, the balance between the formation and detoxification of ROS is lost, leading to accumulation of H_2O_2, which is responsible for destruction of cell ultra-structure. Exposure of tea to high contents of Cu, Cd, Hg, and Ni causes decrease in chlorophyll and increase in proline content. Reduction in the chlorophyll content of tea has also been reported during Cd stress. Thus, heavy metal stress caused oxidative stress drastically affecting growth and productivity of the tea plant.

EFFECT OF MINERAL NUTRITION IN ABIOTIC STRESS RESPONSES IN TEA

Mineral deficiency reduced crop productivity under various environmental conditions, as limited mineral nutrition greatly affects plants 'performance and their response to abiotic stress. Mineral deficiencies exert unpredictable responses in plants' growth, chemical composition, and functioning of the antioxidative system, particularly abiotic stress tolerance of plant. Thus mineral deficiencies result in increase of intensity of abiotic stress induced damages in plants. Several studies suggest a significant effect of mineral nutrients on plant growth and productivity under optimum environmental conditions and thus mineral nutrients alleviate abiotic stress induced injuries. Tea, being an evergreen perennial shrub, is subjected to various abiotic stresses, particularly drought and metal stress.

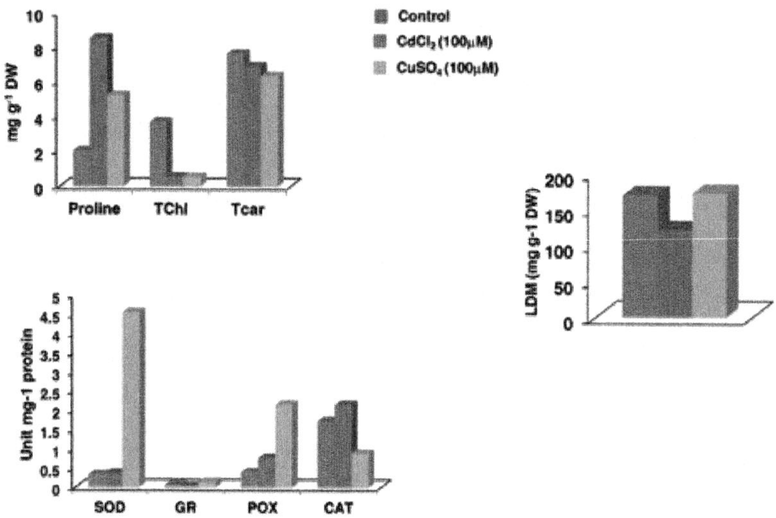

Figure 31.4 : Bio-chemical changes in Camellia sinensis leaf in response to 7 days of $CuSO_4$ (100μM) and $CdCl_2$ (100μM) treatment. Control plants are not treated with Cu or Cd. Data presented are mean of three independent repeats.

ABIOTIC STRESS RECOVERY RESPONSES IN TEA : AN INSIGHT

Studies on post-drought stress recovery on rehydration reveal that tea plants have the potential to retrieve growth and antioxidant function when subjected to normal growth conditions and the recovery potential varies in different clones of tea. During post-drought recovery, the tea plant showed differential response in activating and enhancing the co-ordinated antioxidant defense system to recover and resume growth after rehydration. In the drought acclimatization as well as during the recovery process several antioxidant enzymes play an important role in resuming the normal growth of the tea plant. Such studies help us understand the drought tolerance potential of various clones of the tea plant better. In this process, some of them can be recommended for growing in drought prone areas and in particular for the benefit of the tea industry at large. In the process of abiotic stress, particularly drought stress recovery, the role of mineral nutrients, particularly K, Ca, Mn, B, Zn, etc., has been emphasized. Decrease in RWC, dry mass of leaf, and antioxidants such as ascorbate and glutathione in tea has been reported as a consequence of drought but the damage was not permanent. Increase in phenolic content with decreasein O_2^-, H_2O_2, and lipid peroxidation was an indication of the recovery of stress induced oxidative damage following the post stress rehydration. Various bio-chemical changes occurred during post stress rehydration. After 15 days of rehydration (PDR), RWC of leaf increased to 89% relative to a stressed plant with 52%. Post-drought rehydration studies also showed revival of mineral uptake (*e.g.*, Zn, Mg, Ca, K, etc.) as evidenced by increased micro and macro nutrient levels in the leaf of a rehydrated plant relative to a drought stressed plant. Concurrently, rehydration improves antioxidant activity with respect to CAT, POX, SOD, GR etc., in the tea plant along with increased photo-synthetic pigments (chlorophyll and carotenoid) and leaf dry mass. Further, the post-drought recovery was enhanced by foliar spray of K, Ca, Mn, and B. Differential activities of enzymes like SOD, CAT, POX, GR, and PPO in response to foliar spray of nutrients in rehydrated plant improved the recovery process. We also observed that foliar spray of Zn and its combination with K, Ca, and B improves growth and antioxidative responses, resulting in enhancement of post-drought recovery potential in the tea plant. Such studies reveal that the mineral nutrients play a significant role in influencing growth and antioxidative responses during post stress recovery in the tea plant. However, the detailed mechanism of abiotic stress amelioration and post stress recovery at molecular level needs to be worked out in future.

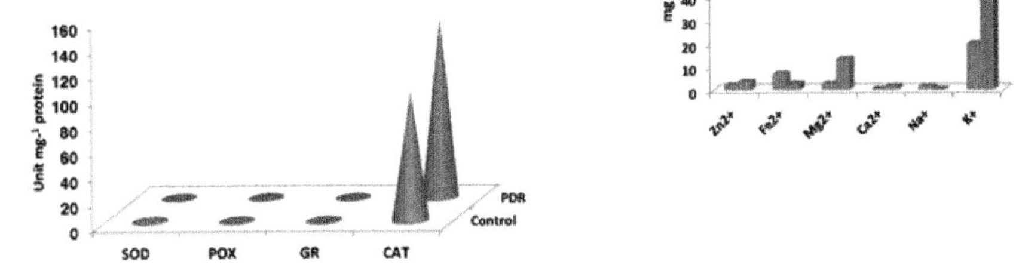

Figure 31.5 : Bio-chemical changes in Camellia sinensis in response to post drought and rehydration. Control (14G days drought) and 15 days of post drought rehydration (PDR). Data presented are mean of three independent repeats.

MOLECULAR PHYSIOLOGY OF ABIOTIC STRESS RESPONSES IN TEA

The molecular physiology of abiotic stress is well understood in various model plants. However, the molecular mechanism of abiotic stress, particularly drought and heavy metal in tea, is poorly investigated. Wang *et. al.* use a suppression subtractive hybridization analysis approach to identify genes involved in cold acclimation of tea [*Camellia sinensis* (L.) O. Kuntze]. A total of 10 complementary DNA (cDNA) clones, induced by low temperature in tealeaves, were identified. Das *et. al.* identified a total of 123 putative drought responsive genes that include candidate genes of ubiquitin-proteasome, glutathione metabolism, and sugar metabolism pathways and several transcription factors. In order to determine the possible expression, 10 genes associated with drought responsive pathways were further analyzed by reverse transcription polymerase chain reaction. Their study provides a basis for studying the drought tolerance mechanism of the tea plant, which will also be a valuable resource for the functional genomics study of woody plants in future. Rana *et. al.* reported that expression of tea cytosolic glutamine synthetase in tissue specific glutamine synthetase transcript by both salinity and Cd indicates that abiotic stress can induce the expression of this gene; thus, it contains stress responsive elements in its promoter region. Glutamine synthetase is an enzyme that plays an important role in nitrogen assimilation, growth, and development of plants. Expression of glutathione bio-synthetic enzymes γ-glutamyl cysteinyl synthetase (γ-ECS) and glutathione synthetase (GSHS) was reported to be elevated under Cu and Al stress. Phytochelatin synthase (PCS) was up-regulated at its transcript level more in the Chinary than in the Assamica variety of tea, suggesting that Chinary could be more tolerant than Assamica. Mohanpuria *et. al.* also reported that Cd induced oxidative stress influence on glutathione metabolic genes of *Camellia sinensis* (L.) O. Kuntze. The transcript levels of glutathione bio-synthetic genes, namely γ-glutamyl cysteine synthetase (γ-ECS) and glutathione synthetase (GSHS) increased upon Cd exposure. Furthermore, transcript levels of glutathione reductase (GR), an enzyme involved in reduction of oxidized glutathione (GSSG) to reduced glutathione (GSH), also showed up-regulation under Cd stress. Such studies suggest that Cd exposure induces oxidative stress in *Camellia sinensis* and the up-regulation in transcript levels of glutathione metabolic genes except GSH indicated the role of these enzymes in the protection of plants from Cd stress. Although there are few reports on molecular aspects of abiotic stress in tea, there is a dearth of information on the molecular physiological aspect of abiotic stress amelioration and post stress recovery in tea, which needs to be taken up in future.

CONCLUSION AND FUTURE PERSPECTIVES

The tea plant is commonly grown in rain-fed ecosystems and thus it encounters abiotic stress conditions that induce loss in crop yield. Abiotic stress affects the growth, mineral nutrition, ROS, and antioxidant metabolism in tea plants. Drought stress caused changes in leaf antioxidant enzymes (*e.g.*, SOD, CAT, GR, etc.) and there was increase in ROS, and lipid peroxidation with decreased POX activities with progressive stress. Hence, drought stress altered antioxidative response with apparent decrease in mineral nutrient (Zn, Ca, Na, Fe, Mg, and K) contents of leaves, suggesting that mineral deficiency mediated drought stress induced oxidative damage in tea. Heavy metal (*e.g.*, Cd, Cu, Al) stress also caused reduction in growth and antioxidative responses. Further, post stress recovery study in tea also revealed that the tea plant recovers on post stress rehydration and the post stress recovery potential varies in various clones of tea. However, the molecular mechanism of abiotic stress amelioration and post stress recovery mediated by various nutrients needs to be explored by applying genomics, proteomics, and metabolomics tools in future.

Chapter 32

Phytochrome and Light Control of Plant Development

MOUGEOTIA : A CHLOROPLAST WITH A TWIST

Mougeotia is a filamentous green alga whose cells contain a single ribbon-like chloroplast nestled between two large vacuoles and surrounded by a layer of cytoplasm. The chloroplast can rotate about its long axis and can respond to light by orienting *perpendicularly* to the direction of light. The perpendicular orientation is referred to as "face position."

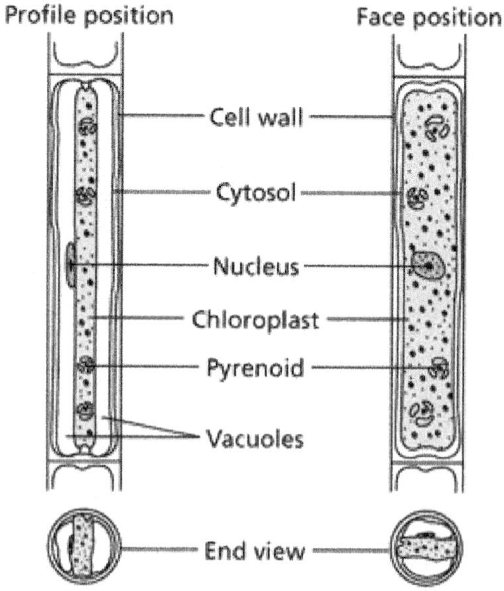

Figure 32.1.

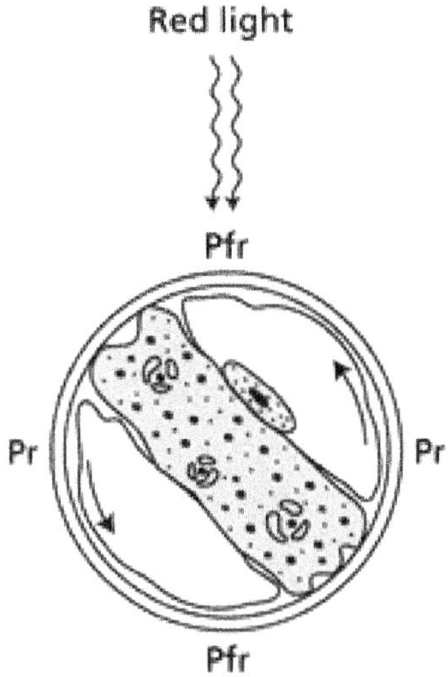

Figure 32.2.

Phytochrome is implicated in the *Mougeotia* chloroplast rotation response because the system is most sensitive to red light and because the red light effect can be reversed by far-red light. Using microbeams of red and far-red light, Wolfgang Haupt and his coworkers in Germany showed that in *Mougeotia*, phytochrome is localized near the periphery of the cytoplasm in the vicinity of the plasma membrane. When a microbeam of red light was directed at the cell surface under a microscope, the edge of the chloroplast adjacent to the microbeam rotated 90°, even though the chloroplast itself was not illuminated.

EXPERIMENTS WITH PLANE POLARIZED LIGHT

The orientation of individual phytochrome molecules within the cell was probed by the use of plane-polarized light. Each phytochrome molecule has a preferred orientation for the absorption of plane-polarized light. The greatest absorption occurs when this light is parallel to the absorbing surface of the pigment. If we think of the pigment as a rod and the polarized light as a narrow slit of light, more light will be absorbed when the light slit is parallel to the long axis of the pigment than in any other orientation.

Haupt and his co-workers discovered that the ability of plane-polarized red light to cause chloroplast rotation depends on the plane of polarization relative to the long axis of the cell. Half of a cell was illuminated over the entire surface with polarized red light (R) vibrating *perpendicular* to the cell axis; the other half was illuminated with light vibrating *parallel* to the cell axis.

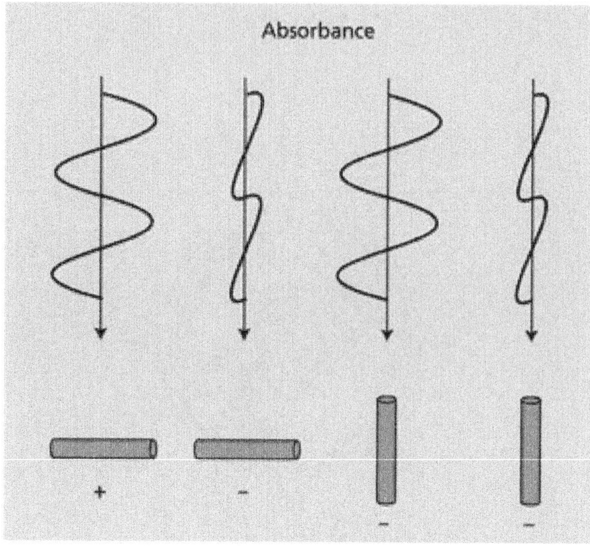

Figure 32.3.

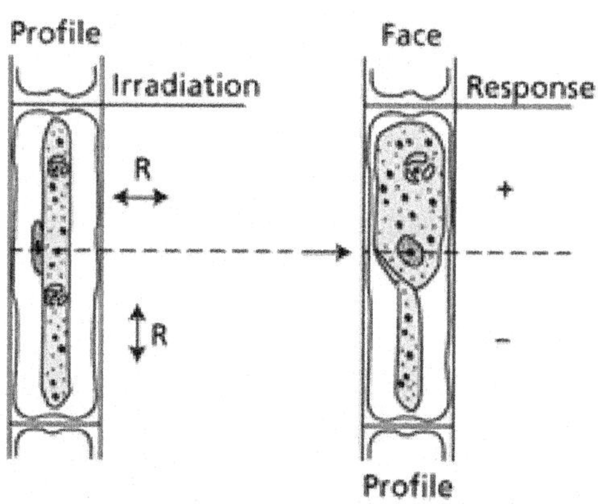

Figure 32.4.

The chloroplast rotated from the profile to the face position only in response to red light vibrating in the perpendicular direction, producing a twist in the chloroplast. On the basis of these experiments, it was concluded that phytochrome molecules have a defined orientation.

Two additional experiments demonstrated that the phytochrome orientation changes during the conversion from Pr to Pfr. The cytoplasmic layer was irradiated with a microbeam of plane-polarized red light vibrating either *parallel* (upper light treatment) or *perpendicular* (lower light treatment) to the cell's long axis.

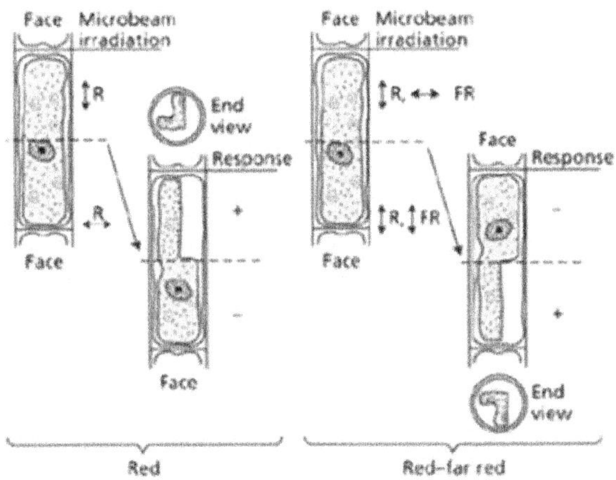

Figure 32.5.

The apparent contradiction is due to the cylindrical shape of the cell. Thus, the phytochrome molecules near the two edges will be oriented 90° relative to the phytochrome molecules at the center. In the second experiment, the cytoplasmic layer on the right side of the cell was irradiated with a microbeam of polarized red light vibrating *parallel* to the cell's long axis.

On the basis of the previous experiment, we can predict that red light vibrating parallel to the cell's long axis will induce chloroplast rotation. However, the red light treatment was immediately followed by a beam of far-red light. One half of the cell (top) was given a second irradiation of far-red light vibrating *perpendicular* to the cell axis; the other half of the cell (bottom) was given a second irradiation of far-red light vibrating *parallel* to the cell axis. Far-red light prevented red light–induced chloroplast rotation only when it was vibrating perpendicular to the cell axis. This result was explained by the proposal that phytochrome undergoes a 90° change in orientation upon photoconversion of Pr to Pfr and *vice versa*.

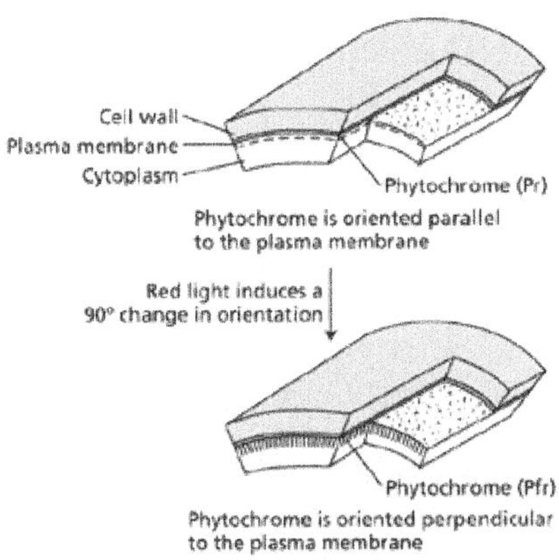

Figure 32.6.

A MODEL FOR CHLOROPLAST ROTATION

Finally, what is the mechanism of chloroplast rotation in *Mougeotia*? The effects already described are best accounted for by a model in which phytochrome is located either on the plasma membrane of *Mougeotia* or on the microtubules that form a cylindrical scaffold directly beneath the membrane. It is difficult to imagine how Pr and Pfr could maintain their respective orientation without being associated either with the membrane or with the cylinder of microtubules.

The nucleotide sequence that codes for *Mougeotia* phytochrome, consistent with those of other known phytochromes, suggests that the phytochrome is not an integral membrane protein, because it lacks hydrophobic transmembrane domains. On the other hand, the amino acid sequence at the carboxy-terminal end is similar to that of proteins that bind microtubules. So the evidence suggests that phytochrome is bound to the cytoskeleton adjacent to the plasma membrane.

During rotation, the chloroplast moves away from a localized region of Pfr and toward a region of Pr. *Mougeotia* chloroplasts move in response to unilateral light as a result of gradients of the active form of phytochrome, Pfr, that form an alternating pigment pattern around the periphery of the cell cylinder. The gradients of Pfr are rapidly translated into gradients of actin-myosin microfilament interactions, which generate the motive force for chloroplast rotation.

The primary mechanism of Pfr action remains obscure. Calcium ions from internal stores and calcium-activated calmodulin are known to participate in the response by interacting with the scaffold of microtubules, but the precise relationship between phytochrome and calcium is not yet understood. Studies using drugs that inhibit microtube and microfilament action do not support a role for microtubules in the mechanism of chloroplast movement. However, experiments with Cytochalasin B do indicate that microfilaments are involved.

PHYTOCHROME AND HIGH-IRRADIANCE RESPONSES

HIRs, such as the inhibition of stem or hypocotyl growth, have usually been studied in dark-grown, etiolated seedlings. Note that the main peak of activity is in the far-red rather than the red region of the spectrum, and there are peaks in the blue and UV-A regions as well. Because the absence of a peak in the red is unusual for a phytochome-mediated response, at first researchers believed that another pigment might be involved. In the late 1960s, Karl M. Hartmann in Freiburg, Germany proposed a model to explain the apparent discrepancy. Taking a cue from the enhancement effect of photosynthesis that was discovered by R. Emerson and co-workers, Hartmann speculated that the HIR might be the result of an interaction between the Pr and Pfr forms of phytochrome.

To probe such an interaction, Hartmann irradiated plants with not one but two wavelengths of light simultaneously, using two wavelengths that were inactive in the HIR for lettuce hypocotyl inhibition : 658 nm and 768 nm. Although physiologically inactive, the two wavelengths are still *photochemically* active in driving phytochrome conversions, each in the opposite direction. By carefully varying the irradiances of each wavelength, Hartmann was able to induce a range of ratios of Pfr to total phytochrome in the plant from near zero to higher values. He was able to show that the optimum ratio of Pfr to total phytochrome for the inhibition of lettuce hypocotyl growth is approximately 0.03. Remarkably, this is precisely the ratio that results when the plants are irradiated with a single wavelength at 720 nm, the far-red peak of the HIR.

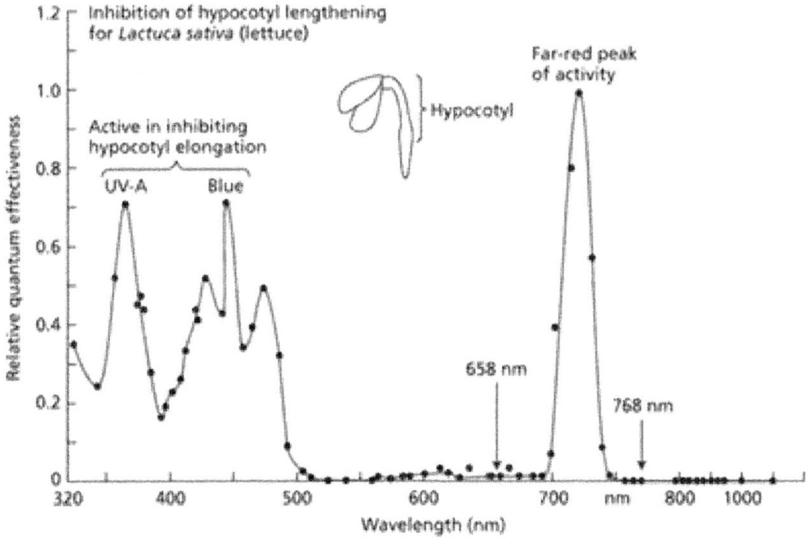

Figure 32.7 : HIR action spectrum for the inhibition of hypocotyl elongation of dark-grown lettuce seedlings. The peaks of activity for the inhibition of hypocotyl elongation occur in the UV-A, blue, and far-red regions of the spectrum.

THE ORIGINS OF PHYTOCHROME AS A BACTERIAL TWO-COMPONENT RECEPTOR

Bacteria often use **two-component signaling systems** to sense and respond to environmental signals such as nutrients or light. The first component is a **sensor protein**, consisting of an *input domain* that receives the input signal, linked to a *transmitter domain* that transmits the signal to the second component. The second component is the **response regulator protein**, composed of a *receiver domain* that accepts the signal from the transmitter of the sensor protein, linked to an *output domain* that initiates the cellular response.

These two components interact by a series of phosphorylation and dephosphorylation reactions. When it receives a signal, the sensor autophosphorylates a specific histidine residue in the transmitter domain. This phosphate is then transferred to a specific aspartate residue on the receiver domain of the response regulator. But what has all this to do with phytochrome?

Although they lack the specific histidine residue that becomes autophosphorylated, the C termini of phytochromes have some sequence similarity to the transmitter domains of bacterial sensor proteins. This sharing of sequences raises the interesting possibility that phytochrome evolved from the sensor protein of a bacterial two-component system.

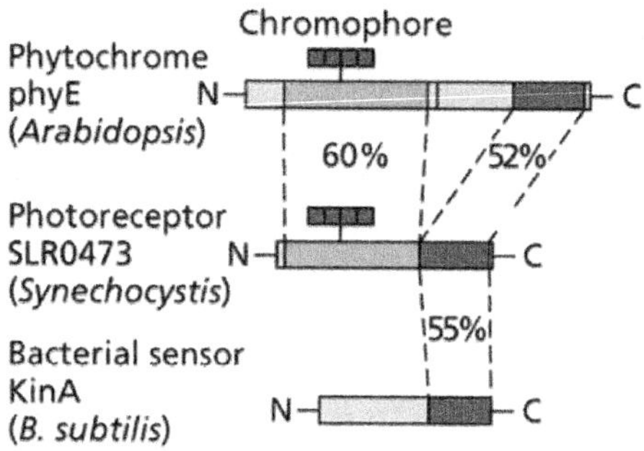

Figure 32.8 : Homologous regions of phytochrome E, cyanobacterial photoreceptor, and sensor protein from *Bacillus subtilis*. The percentages indicate the proportion of identical amino acids between pairs of sequences.

Provocative new evidence for the idea that phytochrome evolved from the histidine kinase of a bacterial two-component system has recently emerged from studies of cyanobacteria. The cyanobacterium *Fremyella* uses light to regulate the composition of its photo-synthetic pigments contained within the light-harvesting phycobilisomes, a response termed **chromatic adaptation.**

Like phytochromes, phycobiliproteins bind linear tetrapyrrole chromophores that are closely related in structure to the phytochrome chromophore, but there is no evidence for red/far-red reversible responses of the type seen in higher plants. Instead, chromatic adaptation occurs in response to growth in red light and green light.

When the gene for the photoreceptor for this response (*RcaE*) was cloned, it was found to encode a 74 kDa polypeptide, the N-terminal end of which is related to the N terminus of higher-plant phytochromes, while the C-terminal end is related to histidine kinases. However, the lack of the conserved chromophore-binding site and of red/far-red reversibility suggests that this protein is a distant relative of higher-plant phytochromes.

Publication of the complete sequence of the genome of the cyanobacterium *Synechocystis* in 1996 led to great excitement because this organism contains a gene that has between 50 and 60% amino acid identity with *Arabidopsis* phyE along its length, including a bona fide chromophore attachment site. In addition, the amino acid sequence of the C terminus of the *Synechocystis* pigment protein is related to the KinA sensor protein of *Bacillus subtilis*. This *Synechocystis* gene has been named *cph1* (*c*yanobacterial *ph*ytochrome *1*).

Recently it was shown that the purified Cph1 apoprotein can assemble with phycocyanobilin, a cyanobacterial linear tetrapyrrole, in vitro, and the resultant holoprotein exhibits red/far-red reversibility and absorption spectra similar to higher-plant phytochrome assembled in vitro with the same chromophore.

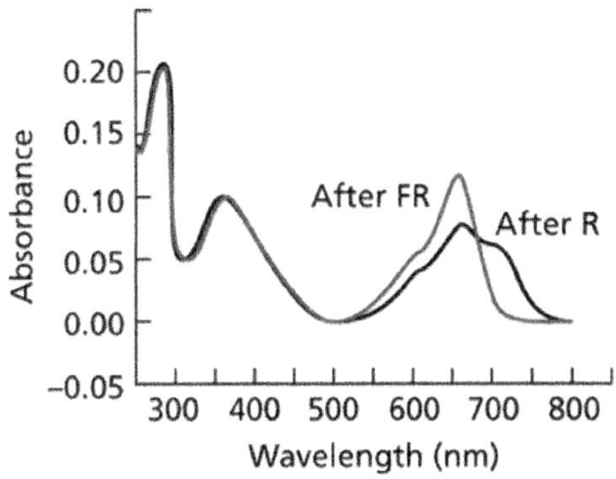

Figure 32.9 : Absorption spectra of cyanobacterial phytochrome after irradiation with red light.

Perhaps the most exciting finding of all was the identification, 10 base pairs downstream from the cyanobacterial phytochrome gene, of another gene having the features of a bacterial response regulator. Since the C-terminal end of the cyanobacterial phytochrome contains all the conserved features of a histidine kinase transmitter module, and the downstream response regulator resembles aspartate kinase receiver modules, it was now possible to test whether this phytochrome acts as part of a light-regulated transmitter–receiver pair, as had been proposed earlier.

Using the purified proteins, the Cph1 protein was shown to phosphorylate itself in a manner consistent with its having a histidine kinase activity, and further, this phosphate could be transferred to the response regulator. Surprisingly, Cph1 is far more active as a kinase in the Pr form than in the Pfr form, suggesting that the light signal is transduced through regulation of Pr abundance, rather than through Pfr abundance as in higher plants.

Taken together, the results from these studies suggest that plant phytochromes originally evolved from a class of photoreversible transmitter histidine kinases in cyanobacteria. Are plant phytochromes also kinases? In support of this idea, a phytochrome gene isolated from the moss *Ceratodon* contains an additional C-terminal sequence that has kinase homology.

In addition, highly purified preparations of higher-plant phytochrome have been found to contain a polycation-stimulated protein kinase activity that is also greater in the Pr form than in the Pfr form. However, the kinase activity associated with these phytochromes is of the serine–threonine type, and consistent with this observation, these proteins lack the conserved histidines of the bacterial sensor proteins. More recently, it was shown that recombinant oat phyA is a serine-threonine kinase with the same bio-chemical properties as the plant-derived photoreceptor. Moreover, R-light stimulated its autophosphorylation, supporting the hypothesis that Pfr is the active form. Polycations were shown to stimulate Pr phosphorylation more than Pfr, a result consistent with previous studies.

Thus, while higher-plant phytochromes share a homology with bacterial two-component signaling systems, the reaction mechanism and components may have diverged considerably. Nevertheless, the discovery of cyanobacterial phytochromes has opened up many exciting new areas of research that have already contributed much to our understanding of phytochrome evolution and function.

PROFILING GENE EXPRESSION IN PLANTS

In micro-array analysis short oligonucleotides or cDNA sequences representing thousands of gene sequences are spotted on a glass slide. RNA samples isolated from control and experimental plants are labelled using two different fluorescent dyes. Both the control and the experimental sample are hybridized to the micro-array and the slide is then scanned using two lasers that will excite one or the other dye. The light emitted from the dye is proportional to the amount of RNA that is bound to the gene-specific oligonucleotide or cDNA sequence and thus provides a relative measure of gene expression. To visualize the relative expression levels from both samples, images are often false coloured so that one signal is represented with green and the other signal as a red, when the signals overlap they are yellow. This technique has been used to examine changes in gene expression when plants are grown under red or far-red light to define phytochrome signaling pathways and have shown that multiple transcription factors are early targets of phytochrome control.

More recently, several informatics tools have been developed to further exploit extensive transcriptomics data-sets. These tools use algorithms to mine expression profiling data-sets to look for common promoter motifs, visualize cell- or tissue-specific patterns of gene expression and visualize complex networks.

With the advent of extremely high-throughput sequencing platforms like Illumina/Solexa and SOLiD, it is now possible to generate hundreds of millions of sequence reads to quantitate gene expression. These technologies when combined with full-genome sequencing offer unique opportunities to examine light signaling networks in both crop and non-traditional model species where access to micro-array technology has been a major obstacle to progress in the past.

TWO-HYBRID SCREENS AND CO-IMMUNOPRECIPITATION

Two-hybrid library screens are based on the finding that many transcriptional activators contain separable DNA binding and transcriptional activation domains. In a two-hybrid screen, a DNA binding domain of a transcription factor is fused to a test or "bait" protein such as phyA. Yeast cells carrying the bait protein are transformed with a library of plant cDNA products that are fused to a transcriptional activation domain ("prey"). If the bait protein interacts with the prey protein, then the activation domain will be brought into close proximity to the DNA binding site of the transcriptional activator, thus driving reporter gene expression that will pigment the yeast cells or provide an essential amino acid, permitting growth.

Co-immunoprecipitation experiments are often conducted to provide independent verification of the results of two-hybrid screens. These assays can be performed using *in vitro* synthesized versions of bait and prey proteins. The proteins are mixed together and immunoprecipitated using an antibody that is specific to either the bait or prey construct. If the two proteins co-precipitate, it is evidence they interact with each other.

PHYTOCHROME EFFECTS ON ION FLUXES

Phytochrome can rapidly alter the properties of membranes. We have already seen that low-fluence red light is required before the dark period to induce rapid leaflet closure during nyctinasty, and that fluxes of K^+ and Cl^- into and out of extensor and flexor cells mediate the response. However, the rapidity of leaf closure in the dark (lag time about 5 minutes) would seem to rule out mechanisms based on gene expression. Instead, rapid phytochrome-induced changes in membrane permeability and transport appear to be involved.

The role of the plasma membrane proton pump in *Samanea* pulvini (the structures at the base of the petiole that include the extensor and flexor cells) was studied by use of a proton-sensitive liquid membrane electrode. During phytochrome-mediated leaflet closure, the apoplastic pH of the flexor cells (the cells that swell during leaflet closure) decreased, while the apoplastic pH of the extensor cells (the cells that shrink during leaflet closure) increased. Thus, the plasma membrane H^+ pump of the flexor cells appears to be activated by darkness (provided that phytochrome is in the Pfr form), whereas the H^+ pump of the extensor cells appears to be deactivated under the same conditions.

Increased H^+ pump activity on the plasma membranes of the flexor cells should increase the membrane potential, thereby allowing more K^+ to enter the cells. Conversely, a reduction in the plasma membrane H^+ pump activity in the extensor cells would promote the loss of K^+ ions. Consistent with this model, the reverse pattern of apoplastic pH change was seen during leaflet opening.

Studies have also been carried out on phytochrome regulation of K^+ channels. Since it is difficult to study membrane channels in the intact pulvinus, Richard Crain and his colleagues at the University of Connecticut isolated protoplasts (cells without their cell walls) of both extensor and flexor cells from *Samanea* leaves. They determined the open or closed state of the K^+ channels indirectly by measuring the membrane potential changes that occurred when the extra-cellular K^+ concentration was raised from 20 to 200 m*M*. When the extra-cellular K^+ concentration was raised, K^+ entered the protoplasts and depolarized the membrane potential only if the K^+ channels were open. When the extensor and flexor protoplasts were transferred to constant darkness, the state of the K^+ channels exhibited a circadian rhythmicity during a 21-hour incubation period and the two cell types varied reciprocally, just as they do *in vivo*. That is, when the extensor K^+ channels were open, the flexor K^+ channels were closed, and *vice versa*. Thus, the circadian rhythm of leaf movements has its origins in the circadian rhythm of K^+ channel opening.

To test the effects of phytochrome on K^+ channels, researchers first treated protoplasts with white light, then exposed them to a brief pulse of either red light alone or red light followed by far-red light and transferred them to continuous darkness. Red light caused the K^+ channels of the flexor protoplasts to open, and far-red light reversed the effect of red light. The extensor protoplasts were closed by both treatments. In contrast to red light, blue light caused the K^+ channels of the extensors to open, and those of the flexors to close. Thus, phytochrome seems to regulate specifically the K^+ channels of the flexor cells. Blue light regulates primarily the extensor cells, although it can also affect the flexors at specific times of day.

On the basis of the evidence thus far, we can conclude that phytochrome brings about leaflet closure by regulating the activities of the primary proton pumps and/or the K^+ channels of the flexor and extensor cells. Although the effect is rapid, it is not instantaneous, and it is therefore unlikely to be due to a direct effect of phytochrome on the membrane. Instead, phytochrome acts indirectly via one or more signal transduction pathways, as in the case of the regulation of gene expression by phytochrome.

However, some effects of red and far-red light on the membrane potential are so rapid that phytochrome may also interact directly with the membrane. Such rapid modulation has been measured in individual cells and has been inferred from the effects of red and far-red light on the surface potential of roots and oat (*Avena*) coleoptiles. The lag between the production of Pfr and the onset of measurable potential changes varies from organism to organism; for example, it is about 1.7 s for depolarization (inside becomes less negative relative to outside the plasma membrane) in the giant alga *Nitella* and 4.5 s for hyperpolarization in *Avena*.

Changes in the bioelectric potential of cells imply changes in the flux of ions across the plasma membrane. The *Mougeotia* microbeam irradiation experiments of Haupt seemed to support the idea that phytochrome is localized on the plasma membrane, and membrane isolation studies carried out in the 1970s provided evidence that a small portion of the total phytochrome is tightly bound to various organellar membranes.

These findings led some workers to suggest that membrane-bound phytochrome represents the physiologically active fraction, and that all the effects of phytochrome on gene expression are initiated by changes in membrane permeability. On the basis of sequence analysis, however, it is now clear that phytochrome is a hydrophilic protein without membrane-spanning domains. The current view is that phytochrome in *Mougeotia* is associated with microtubules located directly beneath the plasma membrane.

If phytochrome exerts its effects on membranes from some distance, no matter how small, involvement of a "second messenger" would be implied. For example, rapid changes in cytosolic free calcium have been implicated as second messengers in several signal transduction pathways, and there is evidence that calcium plays a role in chloroplast rotation in *Mougeotia*. More recently, however, the role of calcium as a second messenger in phytochrome action has been questioned, so calcium-independent mechanisms may also be involved.

MICRO-ARRAY ANALYSIS OF SHADE AVOIDANCE

To identify components of the shade-avoidance response pathway, micro-array analysis was conducted in Arabidopsis, comparing plants grown in low and high R : FR light environments. These studies have identified both early- and late-acting genes that regulate responses to shading by other plants. Characterizations of two Arabidopsis transcription factors (*ATHB-2* and *PIL1*) also suggests that shade avoidance is initiated by rapid transcriptional changes following a decrease in the ratio of red to far-red light. These profiling studies strengthen the previous genetic studies that show that phyA acts to antagonize phyB-mediated responses under prolonged low R : FR light conditions through the transcriptional control of target genes that include components of the auxin transport and signaling pathways. Interestingly, it appears that the circadian clock is capable of regulating or "gating" physiological responses to R : FR through a modulation of gene expression. That is, changes in gene expression due to a change in the R :FR only occur at certain phases of the circadian rhythm.

Chapter 33

Blue-Light Responses : Morphogenesis and Stomatal Movements

BLUE-LIGHT SENSING AND LIGHT GRADIENTS

How do Plants Sense the Direction of the Light Signal?

Light gradients between lighted and shaded sides have been measured in coleoptiles of monocots and in hypocotyls of dicot seedlings irradiated with unilateral blue light. When a coleoptile is illuminated with 450-nm blue light, the ratio between the light that is incident to the surface of the illuminated side and the light that reaches the shaded side is 4 : 1 at the tip and the midregion of the coleoptile, and 8 : 1 at the base.

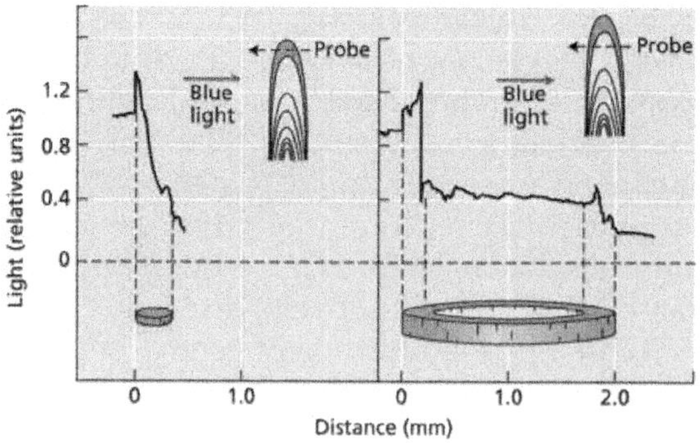

Figure 33.1 : Distribution of transmitted, 450 nm blue light in an etiolated corn coleoptile. The diagram in the upper right of each frame shows the area of the coleoptile being measured by a fiber-optic probe. A cross section of the tissue appears at the bottom of each frame. The trace above it shows the amount of light sensed by the probe at each point. A sensing mechanism that depended on light gradients would sense the difference in the amount of light between the lighted and shaded sides of the coleoptile, and this information would be transduced into an unequal auxin concentration and bending.

On the other hand, there is a *lens effect* in the sporangiophore of the mold *Phycomyces* irradiated with unilateral blue light, and as a result, the light measured at the distal cell surface of the sporangiophore is about twice the amount of light that is incident at the surface of the illuminated side. Light gradients and lens effects could play a role in how the bending organ senses the direction of the unilateral light.

GUARD CELL OSMOREGULATION AND A BLUE LIGHT-ACTIVATED METABOLIC SWITCH

Studies on photo-synthetic algae, such as *Chlorella,* have shown that irradiation by blue light increases respiratory rates of dark-grown cells, with an action spectrum that is typical of specific blue-light responses. These changes in respiratory rates are associated with a drastic shift in metabolism, which can also be observed in experiments using blue and red light.

Red light stimulates photo-synthetic carbon fixation in these algae, and the fixed carbon is incorporated into glucose and sucrose. The metabolic shift is observed when low photon fluxes of blue light are added to the red-light treatment; glucose and sucrose are no longer synthesized, and the fixed carbon is used in the bio-synthesis of organic acids, amino acids, and proteins.

Blue light is also a powerful regulator of guard cell metabolism. Guard cells irradiated with red light accumulate sucrose; if low photon fluxes of blue light are added to the red-light irradiation, sucrose stops accumulating, and guard cells accumulate K^+ and malate^{2-}. Thus, the blue light–stimulated shift to the synthesis of the organic acid malate in guard cells parallels the shift to the synthesis of organic acids and amino acids in photo-synthetic algae.

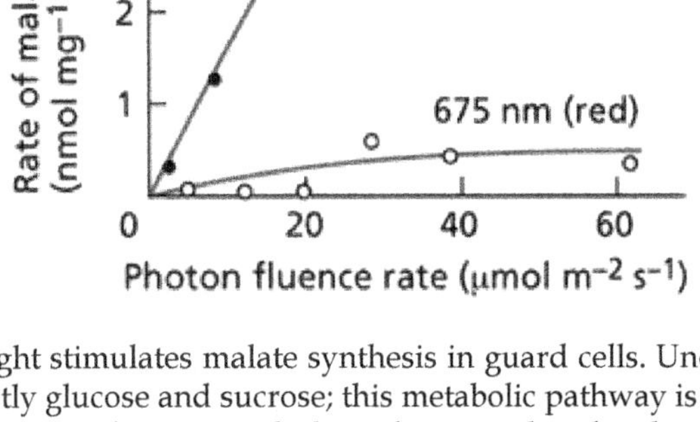

Figure 33.2 : Blue light stimulates malate synthesis in guard cells. Under red light, guard cells synthesize mostly glucose and sucrose; this metabolic pathway is deactivated by blue light, and metabolism switches to starch degradation and malate bio-synthesis.

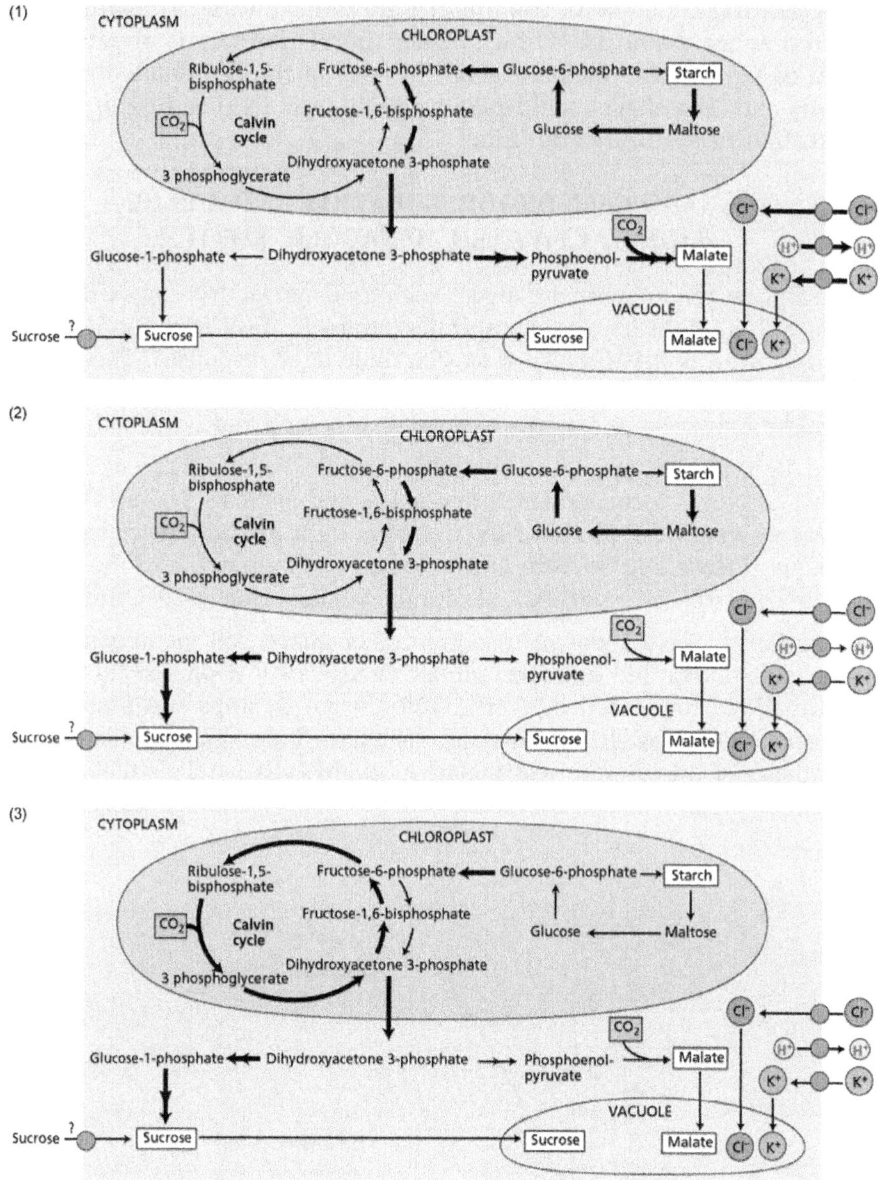

Figure 33.3 : Three distinct osmoregulatory pathways in guard cells. The dark arrows identify the major metabolic steps of each pathway that lead to the accumulation of osmotically active solutes in the guard cells. (1) Potassium and its counterions. Potassium and chloride are taken up in secondary transport processes driven by a proton gradient; malate is formed from the hydrolysis of starch. (2) Accumulation of sucrose from starch hydrolysis. (3) Accumulation of sucrose from photo-synthetic carbon fixation. The possible uptake of apoplastic sucrose is also indicated.

The experimental use of blue and red light has been useful for investigations of osmo-regulatory pathways associated with stomatal movements. Blue light also stimulates starch hydrolysis and malate bio-synthesis. Prolonged incubations of guard cells in detached epidermis under low fluence rates of blue light result in large stomatal apertures accompanied by extensive starch hydrolysis.

Guard cells also show a sucrose-dependent osmo-regulatory phase. Sucrose can originate from starch hydrolysis. On the other hand, guard cells from detached epidermis irradiated with red light under ambient CO_2 concentrations operate a third metabolic pathway, showing no detectable potassium uptake or starch hydrolysis. In these conditions, guard cells slowly accumulate sucrose, and this sucrose accumulation is blocked by DCMU (dichlorophenyldimethylurea), an inhibitor of photo-synthetic electron transport. A recent study of guard cell osmoregulation under red light illumination and CO_2-free air showed that under these conditions, in which photosynthesis is inhibited by the absence of CO_2, red light activates the potassium-dependent pathway. These studies underscore the remarkable metabolic flexibility of guard cells.

A study of the effects of calcium and abscisic acid on stomatal apertures under red and blue light found that increasing calcium or ABA concentrations inhibited blue light-stimulated stomatal opening in a concentration-dependent fashion, but had no effect on red light-stimulated opening. These contrasting responses to blue and red light can be explained by the effect of the treatments on guard cell osmoregulation. ABA and high calcium concentrations have been shown to inhibit proton pumping and potassium uptake, which are central to blue light-stimulated stomatal opening. Red light, on the other hand, stimulates guard cell photo-synthesis and sucrose accumulation, and this osmoregulatory pathway appears to be insensitive to ABA and calcium.

Unlike visual systems, the colour information in the blue light and red light treatments does not appear to have functional significance for photosynthetic organisms. Rather, their dramatic experimental effects appear to be a result of a differential stimulation of carotenoids and chlorophylls that are selectively excited by blue light applied under a saturating background of red light, or by red light alone, respectively.

Under solar radiation, regulatory pathways activated by the excitation of these pigments are used by the chloroplast to sense key features in their environment and to control their metabolism.

THE COLEOPTILE CHLOROPLAST

Guard cell chloroplasts are a conspicuous structural feature of guard cells, and their physiology and bio-chemistry have been studied for several decades. In contrast, the chloroplasts of phototropically sensitive coleoptiles have only recently been analyzed in detail. The coleoptile chloroplast had been suggested as a possible site for photo-reception in oat coleoptiles by Kenneth V. Thimann and his coworkers in 1960, but interest in the flavin hypothesis later in that decade shifted attention away from coleoptile chloroplasts and these organelles have been virtually ignored for several decades.

Recent studies have shown that the coleoptile chloroplast evolves oxygen, fixes CO_2 photo-synthetically, and operates the xanthophyll cycle. Dark-grown, etiolated coleoptile tips have violaxanthin and antheraxanthin but no zeaxanthin, and they accumulate zeaxanthin when irradiated.

When dark-grown coleoptiles are pretreated with red light for different lengths of time, their zeaxanthin content is proportional to the length of the red-light treatment. Exposure of coleoptiles with different zeaxanthin content to a brief pulse of blue light elicitis phototropic bending that is linearly related to the zeaxanthin content of the coleoptiles. Most interestingly, the regression line through the data points extrapolates to zero, suggesting that there is no phototropic response in the absence of zeaxanthin.

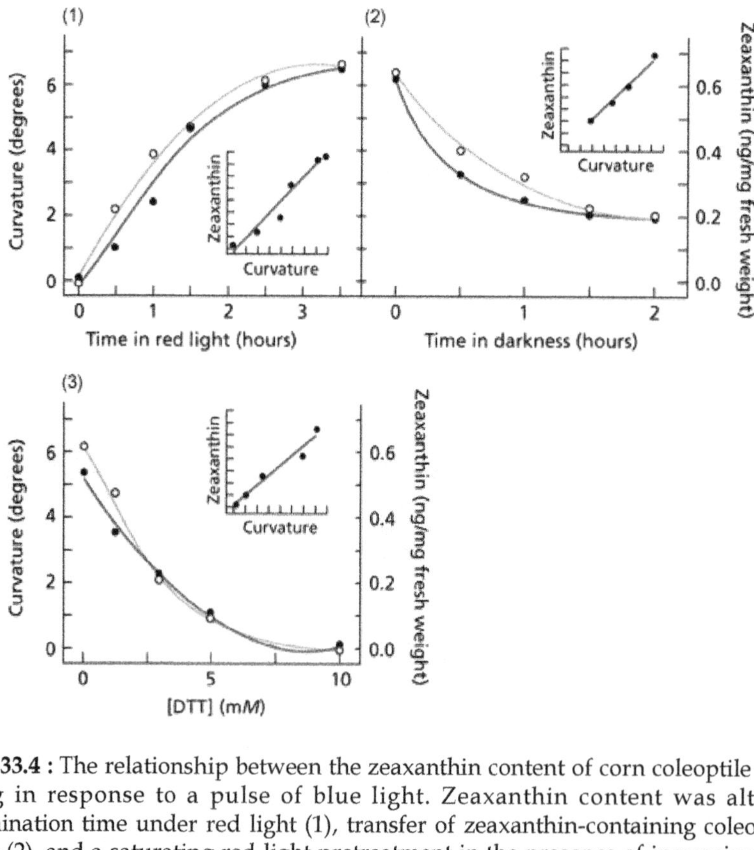

Figure 133.4 : The relationship between the zeaxanthin content of corn coleoptile tips and bending in response to a pulse of blue light. Zeaxanthin content was altered by preillumination time under red light (1), transfer of zeaxanthin-containing coleoptiles to the dark (2), and a saturating red-light pretreatment in the presence of increasing concentrations of DTT, an inhibitor of zeaxanthin formation (3).

Red light-treated coleoptiles with a high zeaxanthin content convert their zeaxanthin to violaxanthin when transferred to the dark. Phototropic bending in response to a pulse of blue light given at different time intervals after the transfer to the dark is linearly related to zeaxanthin content.

When corn coleoptiles are given a saturating red-light pretreatment and a blue-light pulse in the presence of increasing concentrations of DTT (an inhibitor of zeaxanthin formation), bending and zeaxanthin content decrease linearly as a function of DTT concentration.

Thus the relationship between the zeaxanthin content of coleoptile chloroplasts and blue light-dependent phototropic sensitivity is very similar to the relationship between the zeaxanthin content of the guard cell chloroplast and the blue-light sensitivity of stomatal opening.

Coleoptile and guard cell chloroplasts share some properties that are clearly distinct from mesophyll chloroplasts, including high rates of oxygen evolution and low rates of photo-synthetic carbon fixation, a high sensitivity of their xanthophyll cycle to low levels of incident photon fluxes, and a high starch content. The coleoptile and guard cell chloroplasts also show a specific blue light response with an action spectrum that closely matches the action spectra for blue light-stimulated stomatal opening and coleoptile phototropism.

These remarkable similarities suggest that these distinct characteristics are typical of chloroplasts specialized in signal transduction, as opposed to the classic role of the mesophyll chloroplast, specialized in photo-synthetic carbon fixation.

PHYTOCHROME-MEDIATED RESPONSES IN STOMATA

Many studies seeking to define a role of phytochrome in stomatal movements have yielded mostly negative results. Recent findings, however, have shown that phytochrome modulates stomatal movements in the orchid *Paphiopedilum* and in the zeaxanthin-less mutant of Arabidopsis *npq1*.

As discussed earlier, guard cells from *Paphiopedilum* lack chlorophyll and fail to show red light stimulated opening, but have a seemingly normal specific blue light response. Nevertheless, *Paphiopedilum* stomata also open in response to low fluence rates of red or green light. That opening can be reversed by far-red light, implicating phytochrome as the involved photoreceptor.

Far-red reversible stomatal opening can also be observed in Arabidopsis *npq1 mutants*. Earlier studies with *npq1* stomata have shown that, despite their lack of a specific blue-light response, they open normally in response to white light. However, if the experiments with white light are carried out under a far-red light background, the white light-stimulated opening is substantially reduced. Wild-type stomata are insensitive to far-red light. These findings suggest that in both *Paphiopedilum* and Arabidopsis *npq1*, phytochrome-dependent opening compensates for the absence of a major component of the stomatal response to light (guard cell photo-synthesis in *Paphiopedilum* and a specific blue-light response in *npq1*).

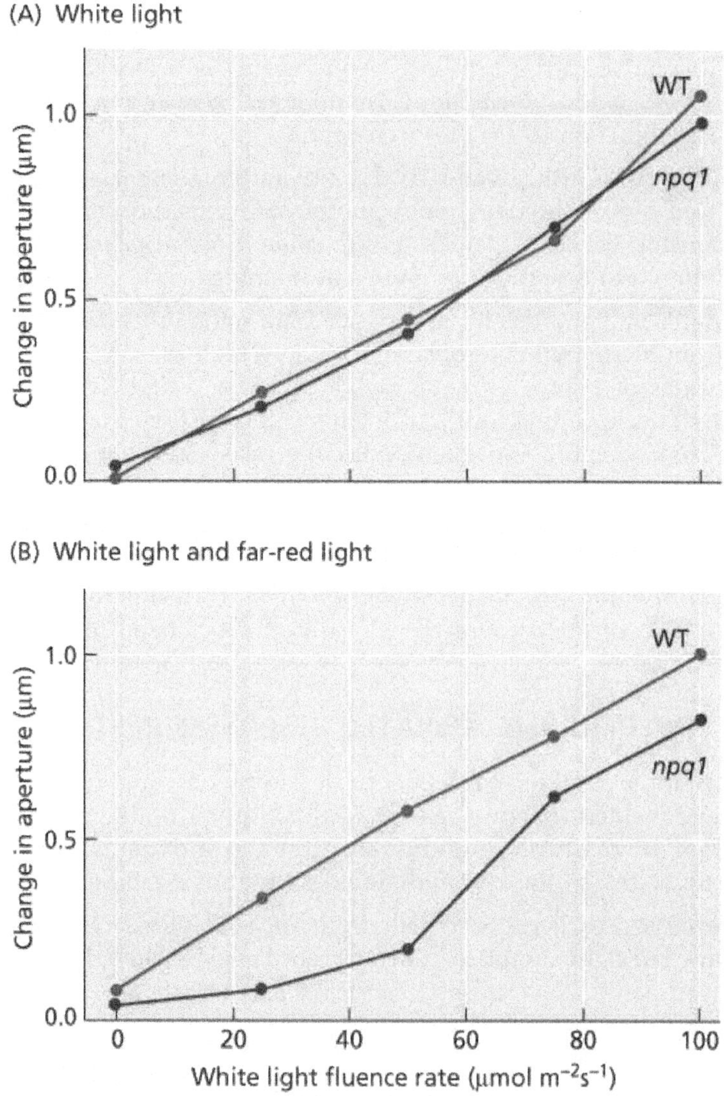

Figure 33.5 : Stomata from the zeaxanthin-less mutant of Arabidopsis *npq1* lack a blue light response but show a phytochrome response. (A) Despite the lack of blue light response in *npq1*, wild-type and *npq1* stomata show indistinguishable dose-response curves to white light. (B) In the presence of 100 μmol m^{-2} s^{-1} of far-red light, the dose-response curve of wild-type stomata to white light remains the same but the response of *npq1* stomata is substantially reduced. These results indicate that phytochrome-dependent opening in *npq1* compensates for the absence of a blue-light response.

Recent work with stomata from *cop1* mutants of Arabidopsis showed that they remain widely open in the dark. In addition, this study shows that the proteins CRY1, CRY2, PHOT1, and PHOT2 are positive modulators of the zeaxanthin-mediated stomatal response to blue light, whereas COP1 represses that modulation. Recall from that COP1 and cryptochrome interact with phytochrome, both *in vivo* and *in vitro*. Taken together, these results point to a novel control of stomatal movement involving phytochrome, COP1, CRY1, CRY2, PHOT1, and PHOT2.

Chapter 34

IN VITRO PROPAGATION OF *STACHYS TMOLEA BOISS.* (LAMIACEAE)

Bengi (BABA) ERDAG and A. Kamil YUREKLI*

ADU, Faculty of Science and Arts, Department of Biology, 09010 Kepez-Aydin/TURKEY

* Ege University. Faculty of Science. Department of Biology, 35100 Bornova-Izmir/TURKEY

ABSTRACT

In this report, micro-propagation of *Stachys tmolea* Boiss. (Lamiaceae) is described. Sterilized seeds of the species were germinated on two modified Murashige-Skoog media (mMS). Explants from seedlings formed calli on mMS (containing no plant growth regulators). mMS containing 2 mg/l NAA plus 2 mg/l BA. Heller's nutrient medium and Heller's medium containing 1 mg/l NAA plus 5 mg/l BA also gave calli. Callogenesis was better on mMS containing 2 mg/l NAA plus 2 mg/l BA and Heller's medium. Regeneration of calli was carried out on mMS containing different concentrations of NAA and BA. Multipled shoot formation was obtained on mMS (without growth regulator) and mMS containing 2 mg/l NAA plus 2 mg/l BA. Majority of shoots developed roots when they transferred on mMS containing 0.04 mg/l NAA plus 2 mg/l BA at the end of third subculture. They were then transferred to pots. (Abbreviation – mMS : Modified Murashige-Skoog medium; NAA : Naphtaleneacetic acid; BA : Benzyl Adenin)

Keywords

Stachys tmolea, Endemic plant, Callogenesis, Plant regeneration

INTRODUCTION

Extinction is the closest threat for endemic plants because they are distributed on very limited localities and mostly with relatively few representatives. Today's ecological problems such as pollution, enlargement of urban areas and destruction of natural habitats increase the risk of extinction for endemic plants.

A plant under extinction risk (more or less severe) can be protected with two main ways of conservation: *In situ* and *ex situ* studies. *In situ* conservation of plants may not be sufficient to protect them. For this reason, *in situ* studies on conservation of plant should be supported by *ex situ* investigations on *in vitro* propagation procedures for the species (Kaya *et al.*, 1998). Tissue culture techniques have been established as a useful approach for *ex situ* conservation of rare and threatened plant species (Cuenca *et al.*, 1999).

This report describes the micropropagation of *Stachys tmolea* Boiss. (Lamiaceae). endemic to Western Anatolia. The genus *Stachys* L., contains about 200 species, mostly herbaceous annual or perennials, rarely shrubs (Davis, 1982). In the literature, there is no record on propagation of *Stachys tmolea.* Therefore, it is hoped that this report will contribute to our knowledge about *in vitro* propagation of Turkish endemic plants.

MATERIAL AND METHODS

Seeds *of Stachys tmolea* were collected from Nif Mountain in Izmir province at an altitude of ca 850 m.. Seeds of the species at the herbarium of Ege University (EGE) were also used in the study.

The seeds were washed three times with distilled water and were retained in 4.5 % sodium hypochlorite solution and were transferred to ethanol (purity was 70 %). As a final step of surface sterilization, seeds were rinsed again with sterile distilled water (three times). Each step mentioned above were applied for 10 minutes. At the end of the sterilization, seeds were dried between sterile filter papers and were taken onto culture medium in laminar air flow sterile cabinet.

In order to micropropagate the species, we used different types of modified Murashige-Skoog's (Murashige and Skoog, 1962) nutrient medium (Table 1) and Heller's medium (Heller, 1953).

All media were supplemented with 30 g/l sucrose and 7 g/1 agar for solidification. The pH was adjusted to 5.8. All of the experimental tools and media were autoclaved under 1.4 atm. and 121° C conditions for 15 minutes.

For germination, modified MS (mMS) media (no. l and no. 2) were used as similar to the study of Saez *et al.* (1994). Seeds (two seeds/tube) were inoculated in test tubes (150 x 25 mm) containing 10 ml medium and they were then closed with their original lids. Seeds were also inoculated in flasks (100 ml) containing 50 ml medium. Each flask containing 6 seeds covered with 4 layers of aluminium foil.

Repeat of experiments were depended on amount of available seeds. All measurements were carried out between seventh day of inoculation and the end of sixth weeks. Results reflect evaluations that were realized at the end of sixth week that maximum germination occurred.

For callus induction, medium no. l (Table 1), medium no. 3, medium no. 6 and medium no. 7 were used (Table 2). Calli, dissected to smaller pieces(~l cm.) were retained on the medium to provide biomass increase. Subculturing was realized in intervals of 15 days and weights were measured at the end of second subculture and the value were calculated proportional (%) to first weight.

In the regeneration experiments, calli were cultured on media no. 2, no. 4 and no. 5 (Table 2 (Regenerated plantlets were transferred to media no. l and no. 5 for rooting after the third subculture.

The germination experiments were made in growth chamber (27 ± 2 C in dark). Conditions of the growth chamber were adjusted to l6 h light/ 8 h dark photoperiod by 2000 lux cold white lamb for callogenesis and regeneration.

RESULTS AND DISCUSSION

The lust step in proper application of tissue culture techniques is selection of suitable medium. We changed Ca concentration of media in two opposite ways. Akcam and Yurekli (1993) have obtained callus by changing Ca concentration in MS basal medium. Natrium was added to the medium. Although Na is not a macronutrients, it is known to be found in higher concentrations in C_4 plants (Brownell and Crossland, 1974). According to Cay and Yurekli (1990), majority of plants have C_4 type metabolism in higher altitudes by reason of increased ecological stresses. Amount of amonium and nitrate in the medium were increased to provocate metabolism of this saxicole plant. Because, provocation of saxicole plants depends on pre-existance of these substances (Lauchli and Bieleski 1983).

In the germination experiments, on medium no. l where total 50 seeds (two seeds/tube) were examined, 12±2 % germination was observed in tubes (150 x 25 mm) while germination rate was 48±4 % in flask (in 100 ml volume and six seeds/flask). The germination rate was 6±1 % on medium no. 2 in tubes (Table 3). Here, too, total 50 seeds were also used on this medium. Medium no. 2 was not experimented in flasks because of insufficient number of seeds.

Table 1. Differences of organic and inorganic contents of the media

INORGANICS	MS	mMS1	mMS2
NH_4NO_3 mg/l	1650	2000	2000
Na_3PO_4 mg/l	–	510	340
$CaCl_2.2H_2O$ mg/l	440	220	660
$MnSO_4.4H_2O$ mg/l	22,3	17	17
ORGANICS			
Tiamine HCI mg/l	0,1	0,5	0,5
Nicotinic acid mg/l	0,5	0,025	0,025
sucrose g/l	3	30	30
MS mMS	: Murashige-Skoog Medium : Modified MS Medium		

Table 2. Media used in the study and their hormone concentrations.

Medium number	Basal Nutrient Medium	NAA mg/l	BA mg/l
3	mMS no. 2	2	2
4	mMS no. 2	2	0.04
5	mMS no. 2	0.04	2
6	Heller's Medium	-	-
7	Heller's Medium	1	5

Table 3. Germination ratio in modified MS media.

MEDIUM	TYPE OF VESSEL	% GERMINATION
m MS no. 1	Tube	12 + 2
m MS no. 1	Flask	48 + 4
m MS no. 2	Tube	6 + 1

Table 4. Callus formation rate and callus characters in different media, used in the study.

ORIGINE OF CALLUS	MEDIUM	CALLUS FORMATION (%)	CALLUS CHARACTER Colour & Texture
S.	mMS no. 3	75	Green, soft
St.	mMS no. 3	48	Green, soft
Pit.	mMS no. 1	52	Brown, hard
Pit.	mMS no. 6	100	Green, hard
Pit.	mMS no. 7	73	Green, soft
S. : Seedling St. : Stem		Pit. : Plantlet	

It is obvious that germination was better in the medium no. 1. But seedlings were better developed in the medium no. 2. For this reason, mMS no. 2 was used for callus induction and regeneration.

Germination experiments have shown that better results can be obtained by using flasks 100 ml in volume than smaller ones. It is possible that this is due to increased secretion of phenolic compounds to the medium during germination activities and higher volume of flasks may decrease level of adverse effects by phenolics.

Table 5. % Biomass development in different media, used in the study.

MEDIUM	BIOMASS DEVELOPMENT (%)
mMS no. 4	245 + 54
mMS no. 6	100 + 7

Table 6. Regeneration features in different media, used in the study.

MEDIUM No.	FEATURES OF REGENERATION
mMS no. 2	Fast regeneration
mMS no. 4	Organogenezis
mMS no. 5	Main shoot dominancy

Figure 1. Callus biomass on medium no 4.

Figure 2. Main shoot dominancy on medium no. 5.

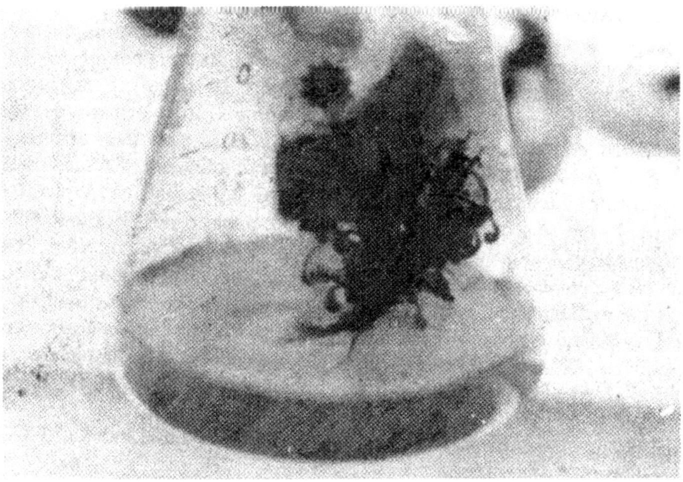

Figure 3. Root formation on medium no. 5

For callus induction, sterile seedlings from *in vitro* germinated seeds were used. The callus productivity of seedlings where they contact to the medium by their lower parts was 75 % after their transfer to medium no. 3. The calli were green and soft in texture (Table 4). On the other hand, stem explants showed 48 % callus productivity on the same medium. The calli were green and soft in texture, In the study of Brown and Charlwood (1986), preferential callus growth was also observed on media containing high NAA and BA. Calli of seedling (on base) were larger than calli of stem explants. This may be due to either different endogenous levels of these hormones in the tissues or to a difference in sensitivity (Trewavas and Cleland, 1983).

On the medium no. 1, no. 6 and no. 7 was observed callus induction. The callus productivity was 52 % in multiplid shoots where their contact points to the medium after their transfer to medium no. l. They were brown and hard in texture. The callus productivity of seedlings where they contact to the medium by their lower parts was 100 % after their transfer to medium no. 6. They were green and hard in texture. Green colour and soft texture were main features of calli from seedlings which observed on Heller's medium containing lmg/l NAA and 5 mg/l BA. The callus productivity was 73 %. Green coloured and soft calli showed the shoot regeneration ability while brown coloured and hard calli failed to regenerate shoot and died. In fact, in these media, callus induction is normally not expected because of insufficient (medium no. l and medium no. 6) and undesirable (medium no. 7) amount of growth regulators. But these observed inductions may be due to endogenous hormone levels of explant. During our efforts on this we could not find a chance to make hormonal analysis due to lack of required amount of plant sample.

Green and soft calli from mMS no. 3 were used in biomass induction experiments. All of calli (1 g in weight) were subcultured with intervals of 15 days. At the end of the 2nd subculture, biomass of some calli measured and then the biomass increase were calculated depending on the first weight. The highest value for the biomass increase was 245 % on medium no. 4 (mMS no. 2 containing 2 mg/l NAA plus 0,04 mg/l BA) while it was only 100 % on medium no. 6 (Heller's medium without no growth regulators) (Table 5). In Heller's medium without plant growth regulators, biomass increase can be explained by "habituation" phenomenon (Meins and Binns, 1978; Stoutemyer and Britt, 1969). Biomass increase on medium no. 4 is probably related with their endogenous hormon levels.

Calli were rapidly regenerated when they were transferred on medium no. 5 (mMS no. 2 containing 0.04 mg/I NAA plus 2 mg/l BA) (Table 6). A main shoot dominancy was observed as a result of the regeneration on medium no. 5 (Figure 2). In the study of Woo, was also observed similar results on *Robinia pseudoacacia L.* (Woo *et al.*, 1995). Stem explants were directly regenerated when they were transferred on medium no. 2. It is normally unexpected that explant was regenerated on medium without plant growth regulators. It is assumed that endogenous hormons of explant may stimulate shoot regeneration. At the end of the 2nd week the regeneration was started on the medium no. 4 (mMS no. 2 containing 2 mg/l NAA plus 0,04 mg/l BA) and organogenesis were also occurred during subculture applications. In tissue culture studies, shoot growth from callus is expected in relatively lower concentrations of NAA, but this expection is not universal phenomenon (Brown and Charlwood, 1986). The amount of endogenous hormones present in the explant could alter the morphogenetic responses as is evident in the case of pea meristems. Although pea nut meristems differantiate into whole plants in a wide range of concentrations of BA and NAA, a low level of BA (0,02 mg/l) and a high level of NAA (2 mg/l) has been found to be best suited for maximum plant regeneration (100 %) (Kartha, 1981). A multiplied shoot induction was also observed on media no. 1 and no. 3. This may be due to endogenous hormon of explants.

Root induction experiments were carried out on media no. 1 (mMS no. 1 hormon- free) and no. 5 (mMS no. 2 containing 0,04 mg/l NAA plus 2 mg/l BA). Obvious and rapid root induction was only seen on medium no. 5 after 3 rd subculture (Figure 3). On medium no. 5, rate of cytokinin was higher than auxin. Our success to obtain root induction on these media, containing smaller amount of auxin may be explained by an assumption of faster auxin metabolism of species, from higher altitudes. Alipinic and subalpinic are more effected by U. V. and blue light than the plants of lowlands. This may be reason for morphological and physiological variations. Klien (1967) suggested that ultravoilet illumination affects and ability of plant cells to use auxine efficiently and then occurs developmental modifications with growth regulators.

Rooted plantlets were transferred to natural condition by habituating. Cover of flask were opened in growth chamber for habituating them to septical conditions and distilled water were then added as 5 ml/day to avoid desiccation. After that plantlets were transferred to sand : soil mixture in plastic pots. Plantlets were found to be normal in appearance.

CONCLUSION

Finally it can be said that when *Stachys tmolea* is experimented in tissue culture studies:

1. In view of endemic plant's native conditions, medium components can be changed.
2. The larger culture dishes are more preferable than smaller ones.
3. Especially, endemic plants of high altitudes may have faster auxin metabolism. For this reason, previously given ratios of growth regulators for callus induction, regeneration and root induction can be neglected.

ACKNOWLEDGEMENTS

We are very grateful to TUBITAK for the financial support to our project as Ph.D. thesis (TBAG-1190). We also thank to Ege University, the Faculty of Science Department of Biology and Botanical Garden for their permission to use the facilities.

REFERENCES

1. Akcam, E., Yurekli, A. K. (1993). Callus culture studies on *Origanum sipyleum J. of Fac. Sci. Ege Univ.* 15(1); p.21-25.
2. Brown, J. T. and Charlwood, B. V. (Differentiation and monoterpene biosynthesis in plant cell culture. In *Secondary Metabolism in Plant Cell Cultures.* (Morris, P., Scragg, A- H., Stafford, A., and Fowler, M. W., eds). Cambridge University Press, Cambridge (1986).
3. Brownell, P. F. and Crossland, C. J. (1974). Growth responses to sodium by Bryophyllum tobiflorum under conditions inducing Crassulacean Acid metabolism. *Plant Physiol.* 54; p. 416-417.
4. Cuenca S.; Amo-Marco, J. B. and Parra R. (1999). Micro-propagation from inflorescence stems of the Spanish endemic plant Centaurea paui Loscos ex Wilk (Compositae). *Plant Cell Reports.* 18 (7-8); p.674-679.
5. Cay, N. and Yurekli, A. K. (1990). Anatomical and physiological determination of photosynthetic metabolism types of some plants of Spil (Manisa) Mountain, Xth National Congress of Biology.
6. Davis, P. H. (1965-1988). Flora of Turkey and the East Aegean Island..Volume 1-10. Edinburg.
7. Heller, R. (1953). Recherches sur la nutrition minerale des tissue vegetaux cultives *in vitro. Ann. Sci. Nat. Bot.. Biol. Veg.,* 14(1); p. 223.
9. Kaya, Z.; Kiln, E. and Giiner, A. (1998) National plan for *In situ* conservation of plant genetic diversity of Turkey, Ministry of Environment, Department of Plant Conservation and Erosion Control. Ankara.
10. Lauchli, A. and Bieleski, R. L. (1983). Inorganic plant nutrition. Springer Verlag, Berlin pp.449.
11. Meins, F., Jr. and Binns, A. N. (1978). Epigenetic clonal variation in the requirements of plant cells for cytokinins. In *The clonal basis for development,* S. Subtelny and I. M. Sussex, eds. Newyork: Academic press, p.185-201.
12. Murashige, T. and Skoog, F. (1962).A revised medium for rapid growth and bioassays with tobacco tissue cultures. *Plant Physiology.* 15; p. 473-497.

13. Saez, F.; Sanchez, P.and Piqueras, A. (1994). Micro-propagation of *Thymus piperella. Plant Cell & Tissue and Organ Culture.* 39(3); p.269-272.

14. Stoutemyer, V. T. and Britt, O. K. (1969) Growth and habituation in tissue cultures of English ivy, Hedera helix. *American journal Botany* 56; p. 222-26.

15. Trewavas, A. J. and Cleland, R. E.(1983). Is plant development regulated by changes in concentration of growth substances or by changes in the sensitivity to growth substances? *Trends in Biochemical Science.* 8; p.354-357.

16. Woo, J. H.; Choi, M. S. and Park, Y. G. (1995). Plant regeneration from callus cultures of Black locust *(Robinia pseudoacacia* L.). *Journal of Korean Forestry Society.* 84(2); p.145-150.

Chapter 35

STRESS-RESPONSIVE EXPRESSION, SUBCELLULAR LOCALIZATION AND PROTEIN–PROTEIN INTERACTIONS OF THE RICE METACASPASE FAMILY

Lei Huang, Huijuan Zhang, Yongbo Hong, Shixia Liu, Dayong Li and Fengming Song*

National Key Laboratory for Rice Biology, Institute of Biotechnology, Zhejiang University, Hangzhou 310058, China; E-Mails: leihero2008@163.com (L.H.); zhanghj82@zju.edu.cn (H.Z.); yongbohong@126.com (Y.H.); liurui90913@163.com (S.L.); dyli@zju.edu.cn (D.L.)

* Author to whom correspondence should be addressed; E-Mail: fmsong@zju.edu.cn; Tel.: +86-571-8898-2269; Fax: +86-571-8898-2271. Academic Editor: Ann Cuypers

ABSTRACT

Metacaspases, a class of cysteine-dependent proteases like caspases in animals, are important regulators of programmed cell death (PCD) during development and stress responses in plants. The present study was focused on comprehensive analyses of expression patterns of the rice metacaspase (*OsMC*) genes in response to abiotic and biotic stresses and stress-related hormones. Results indicate that members of the OsMC family displayed differential expression patterns in response to abiotic (e.g., drought, salt, cold, and heat) and biotic (e.g., infection by *Magnaporthe oryzae*, *Xanthomonas oryzae* pv. *oryzae* and *Rhizoctonia solani*) stresses and stress-related hormones such as abscisic acid, salicylic acid, jasmonic acid, and 1-amino cyclopropane-1-carboxylic acid (a precursor of ethylene), although the responsiveness to these stresses or hormones varies to some extent. Subcellular localization analyses revealed that OsMC1 was solely localized and OsMC2 was mainly localized in the nucleus. Whereas OsMC3, OsMC4, and OsMC7 were evenly distributed in the cells, OsMC5, OsMC6, and OsMC8 were localized in cytoplasm. OsMC1 interacted with OsLSD1 and OsLSD3 while OsMC3 only interacted with OsLSD1 and that the zinc finger domain in OsMC1 is responsible for the interaction activity. The systematic expression and biochemical analyses of the OsMC family provide valuable information for further functional studies on the biological roles of OsMCs in PCD that is related to abiotic and biotic stress responses.

Keywords

Rice; metacaspase; expression patterns; abiotic and biotic stress; subcellular localization; protein–protein interaction.

1. INTRODUCTION

Programmed cell death (PCD) is an important life process that orchestrates cell suicide and keeps the proper metabolism function [1,2]. During PCD, a kind of enzymes, known as caspases, are often activated and thus initiate the cell death program [3–6]. In mammals, caspases are a family of cysteine-dependent aspartate-directed proteases, which cleave a variety of intracellular polypeptides and cause the stereotypic morphological and biochemical changes of the cells [7]. However, higher plants do not have close homologues of caspases, but possess a phylogenetically distant family of caspase-like proteins, called metacaspases [8]. Recent genome-wide characterizations in various plant species have revealed that the plant metacaspases are presented as a multigene family. For example, there are nine members in Arabidopsis [9], 8 or 9 members in rice [10,11] and 6 members in grapevine [12]. Phylogenetic analyses indicate that the plant metacaspases can be divided into two categories, type I and type II, according to the protein structure. The type I metacaspases possess a N-terminal extension prodomain ranging from 80 to 120 amino acids in length and two CxxC-type zinc finger domains, whereas type II metacaspases do not have such structures [9]. Both types of metacaspases have a p20 and a p10 subunits in the C-terminal regions, but the linker region between the two subunits in the type I is shorter than the type II metacaspase [11]. Recent biochemical studies demonstrated that two aspartate residues at the p10 subunit in a tobacco type II metacaspase contribute to the substrate-binding pocket [13] and identified a number of physiological substrates for some of the plant metacaspases [14,15].

Although the activity of some plant metacaspases is post-translationally regulated [16–25], transcriptional regulation of the gene expression is a major mechanism that modulates the activity of metacaspases in plants. A number of studies have indicated that the expression of the metacaspase genes can be regulated by developmental cues [11,12] and induced by a wide range of abiotic and biotic stresses [26–31]. For example, Arabidopsis *AtMC1* and *AtMC3* [32], tomato *LeMCA1* [27], pepper *CaMC9* [28], wheat *TaMCA4* [29] and *Nicotiana benthamiana NbMCA1* [30] were previously reported to be upregulated by infection from fungal or bacterial pathogens. Recent functional studies using overexpression or knockout/knockdown approaches have demonstrated that the metacaspases play critical roles in regulating PCD involved in developmental processes and stress responses. Two type II metacaspases, Arabidopsis AtMC9 and Norway spruce mcII-Pa, were found to regulate PCD that are required for the formation of xylem vessel elements [33] and embryogenesis [22,34], respectively. In Arabidopsis, AtMC1 and AtMC3, two type I metacaspases, were shown to positively and negatively regulate PCD, respectively [35], while overexpression of *AtMC8* increased oxidative stress-induced PCD [26]. Knockout of *AtMCP2d*, coding for a type II metacaspase, suppressed PCD induced by fumonisin B1, oxidative stress and *Pseudomonas syringae* [24]. Additionally, the pepper CaMC9 [28], wheat TaMC4 [29] and *N. benthamian* NbMCA1 [30] are positive regulators of PCD and defense response. It was recently found that the functions of Arabidopsis AtMC1 and Norway spruce mcII-Pa are linked to autophagy [36,37].

Subcellular localization and interaction relationships with other proteins are critical to the mode of action for plants' metacaspases. For example, the Arabidopsis AtMC9 was found to be present in apoplast, nucleus, and cytoplasm and its subcellular localization can be changed during late autolysis process [14,33]. In regulation of PCD, AtMC1 is a positive regulator via interacting with AtLSD1, a negative regulator of cell death [38], whereas AtMC2 is a negative regulator with weak interaction with AtLSD1 [35].

The rice metacaspase (OsMC) family contains eight members [10,11]. Among them, three (OsMC1-3) belong to type I and the remaining five (OsMC4–8) are members of type II [11]. However, very little is known about the biological function of the rice metacaspases so far. As a first step toward understanding the function of rice metacaspases in abiotic and biotic stress-related PCD, we performed a comprehensive analysis of gene expression in response to abiotic and biotic stresses as well as to stress-related hormones. We also analyzed the subcellular localizations of all rice metacaspases and examined the possible interactions of type I with OsLSDs, negative regulators of PCD [38]. Our results presented in this study provide valuable information for further functional studies on the biological roles of rice metacaspases in PCD that is related to abiotic and biotic stress responses.

2. RESULTS

2.1. Expression Patterns of OsMCs in Response to Abiotic Stresses

Abiotic stresses such as dehydration (e.g., drought and salt) and extreme temperature (e.g., cold and heat) are the main deleterious factors affecting on plant growth/development and ultimately crop yield [39]. To explore the involvement of *OsMCs* in abiotic stresses, the expression patterns of *OsMCs* in rice plants after drought, salt, cold, and heat stresses were analyzed.

2.1.1. Expression Patterns in Response to Drought and Salt Stresses

We first analyzed the expression patterns of *OsMCs* in rice leaves and roots under drought and salt stresses (Figure 1). To assess the accuracy of qRT-PCR

[40] and confirm the drought and salt stresses applied to the experimental rice plants, we examined the expression changes of *SNAC1* (*Stress-responsive NAC 1*) in leaves and *Wsi18* (*Water stress-induced 18*) in roots, which are drought- and salt-responsive in leaf and root tissues, respectively [41,42], after drought and salt stresses. As shown in Figure 1A, the expression of *SNAC1* in leaf tissues of plants treated with drought or salt stress was significantly induced while the expression of *OsWsi18* in root tissues of plants treated with drought or salt stress was also upregulated, as compared with corresponding controls in rice plants without stress treatment. Meanwhile, the expression levels of *SNAC1* and *OsWsi18* calculated with three different reference genes such as *Actin, GAPDH,* and *Ubiquitin* [43] exhibited similar patterns (Figure 1A). These indicated that the drought and salt treatments in our stress experiments were satisfactory for further analyses of the expression patterns of *OsMCs* in response to drought and salt stresses. In leaf tissues of rice plants treated with drought or salt stress, the expression levels of all *OsMC* genes were significantly downregulated at 3 h after treatment, leading to >3-fold of reduction, as compared with that in control plants without stress treatment (Figure 1B). However, the expressions of *OsMCs* in root tissues of rice plants

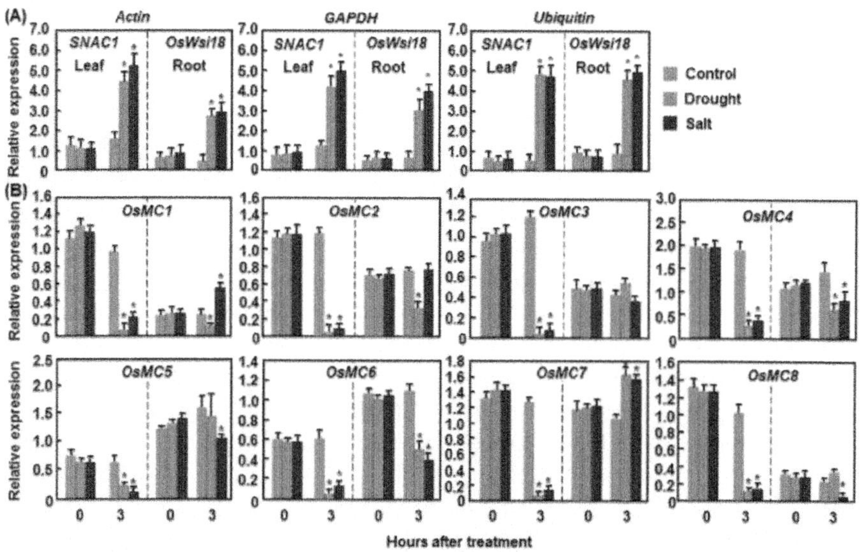

Figure 1. Expression patterns of *OsMCs* in rice leaf and root tissues after treatment with drought and salt stresses. Drought stress was applied to two-week-old seedlings by transferring to three layers of filter papers for fast dehydration. Salt stress was applied to two-week-old seedlings by drenching with 150 mM NaCl solution. Leaf and root samples were collected at indicated time points for qRT-PCR analyses. (**A**) Expression patterns of *SNAC1* and *OsWsi18* calculated with different reference genes; and (**B**) Expression patterns of *OsMCs*. Relative expression is shown as folds of transcript level of different reference genes (**A**) or *Actin* gene (**B**). Left and right parts in each graph divided by dashed lines represent the expression levels in leaf and root tissues, respectively. Data presented are the means ± SD from thee independent experiments and * above the columns indicate significant difference at $p < 0.05$ level.

treated with drought and salt stresses exhibited quite diverse patterns as compared with the patterns in leaf tissues. In drought stress-treated rice plants, only the expression of *OsMC7* was slightly induced, while the expression of *OsMC3*, *OsMC5*, and *OsMC8* was not affected (Figure 1B). By contrast, the expression levels of *OsMC1*, *OsMC2*, *OsMC4*, and *OsMC6* were significantly decreased, leading to >1-fold of reduction, as compared with those in the control plants (Figure 1B). In salt stress-treated rice plants, the expression of *OsMC2* and *OsMC3* was not affected, whereas the expression levels of *OsMC1*and *OsMC7* were increased by approximately 1-fold (Figure 1B). By contrast, the expression levels of *OsMC4*, *OsMC5*, *OsMC6*, and *OsMC8* were significantly decreased, as compared with those in control plants (Figure 1B).

2.1.2. Expression Patterns in Response to Cold and Heat Stresses

We next analyzed the expression patterns of *OsMCs* in rice leaves under cold and heat stresses (Figure 2). Similarly, we examined the expression changes of *OsSRO1c* (*SIMILAR TO RCD ONE 1c*) and *OsHCI1* (*Heat and cold induced 1*), which were reported to be strongly induced by cold and heat stresses, respectively [44,45], after cold and heat stresses to confirm the stress treatment applied to the experimental rice plants. As shown in Figure 2A, the expression of *OsSRO1c* and *OsHCI1* in leaf tissues of plants treated with cold and heat stresses was significantly induced, respectively, at 12 h after treatment, as compared with corresponding controls without stress treatment. The expression levels of *OsSRO1c* and *OsHCI1* calculated with three different reference genes such as *Actin*, *GAPDH*, and *Ubiquitin* [43] exhibited similar patterns (Figure 2A). The data indicated that the cold and heat treatments were effective to the rice plants in our stress experiments, which were satisfied for analyses of the expression patterns of *OsMCs* in response to cold and heat stresses. In cold stress-treated plants, the expression level of *OsMC4* was significantly induced, resulting in >4-fold of increase at 12 h after treatment, while the expression levels of *OsMC3*, *OsMC5*, and *OsMC7* were decreased, as compared with those in the control plants (Figure 2B). The expression of *OsMC1*, *OsMC2*, *OsMC6*, and *OsMC8* was not affected by cold stress (Figure 2B). In heat stress-treated plants, the expression levels of *OsMC4* and *OsMC7* were significantly decreased, giving >1-fold of reduction; whereas the expression levels of *OsMC1*, *OsMC3*, *OsMC5*, and *OsMC6* were slightly induced (Figure 2B). By contrast, the expression of *OsMC2* and *OsMC8* was not affected by heat stress (Figure 2B).

Collectively, these data indicate that the expression of *OsMCs* display differential expression patterns upon different abiotic stresses and such differential expression may imply their involvement in different networks and have different biological functions.

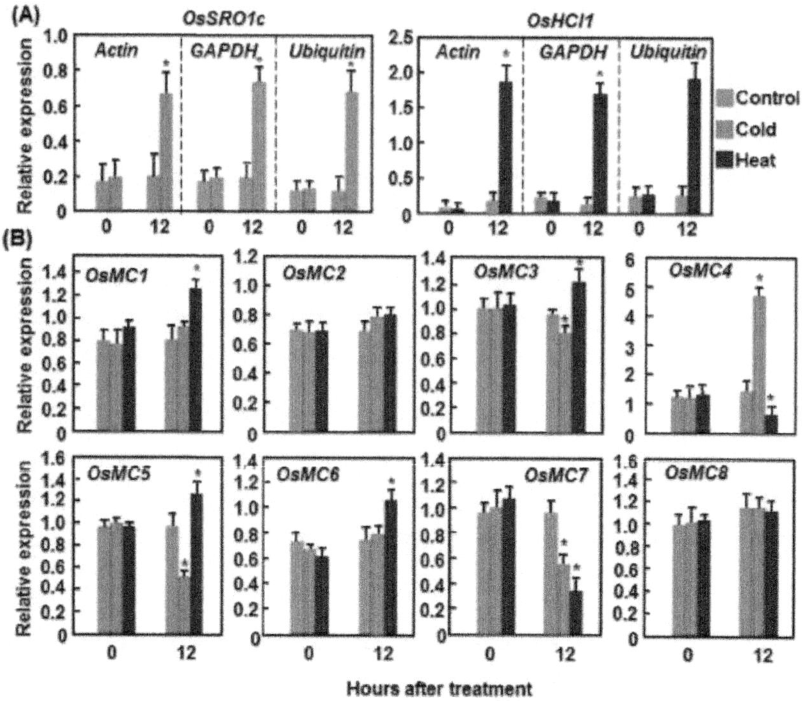

Figure 2. Expression patterns of *OsMCs* in response to cold and heat stresses. Cold and heat stresses were applied by placing two-week-old seedlings in growth chambers with temperatures set at 4 and 42 °C, respectively. Leaf samples were collected at indicated time points for qRT-PCR analyses. (**A**) Expression patterns of *OsSRO1c* and *OsHCI1* calculated with different reference genes; and (**B**) Expression patterns of *OsMCs*. Relative expression is shown as folds of transcript levels of three different reference genes (**A**) or *Actin* gene (**B**). Data presented are the means ± SD from thee independent experiments and * above the columns indicate significant difference at $p < 0.05$ level.

2.2. Expression Patterns of OsMCs in Response to Pathogens

To explore the possible involvement of *OsMCs* in response to biotic stress, we analyzed the expression patterns of the *OsMCs* genes in rice after infection with different types of pathogens. Rice blast disease caused by *Magnaporthe oryzae* and sheath blight disease caused by *Rhizoctonia solani* are two of the most important fungal diseases of rice, whereas the bacterial blight disease caused by *Xanthomonas oryzae* pv. *oryzae* (*Xoo*) is a major rice bacterial disease. Among the thee diseases chosen, *M. oryzae* and *Xoo* are generally (hemi)biotrophoic pathogens while *R. solani* is a typical necrotrophic fungus.

2.2.1. Expression Patterns in Response to M. oryzae

In rice-*M. grisea* interactions, a pair of near isogenic lines (NILs) H8R and H8S interact differentially with strain 85-14-B1 of *M. oryzae*, resulting in incompatible and compatible interactions, respectively [46]. We first confirmed the induction of defense response in incompatible and compatible interactions between rice and *M. oryzae* by analyzing the expression changes of known defense-related genes in rice plants after inoculation. As shown in Figure 3A, the expression levels of *OsPR1a* (*Pathogenesis-related 1a*), a well-known defense-related gene that is induced by *M. oryzae* [47], in both of the incompatible and compatible interactions were significantly increased at 48 h after inoculation. Meanwhile, typical disease symptom appeared in leaves of H8S plants while small restricted disease spots was observed on leaves of H8R plants at 3 days after inoculation. Similar expression patterns of *OsPR1a* were observed when calculated with three different reference genes such as *Actin*, *GAPDH*, and *Ubiquitin* [43] (Figure 3A). These data demonstrate the effectiveness and reliability of our pathogen inoculation experiments. As compared with those in the mock-inoculated plants, the expression levels of *OsMC1*, *OsMC3*, and *OsMC6* were significantly induced in leaves of H8R plants at 48 h after inoculation, leading to 2~3-fold of increases, while the expression levels of *OsMC1* and *OsMC3* in H8S plants was decreased (Figure 3B). Similar expression patterns and dynamics were observed for *OsMC5* and *OsMC7* (Figure 3B). The expression of both *OsMC5* and *OsMC7* in incompatible and compatible interactions was induced at 24 h after inoculation;

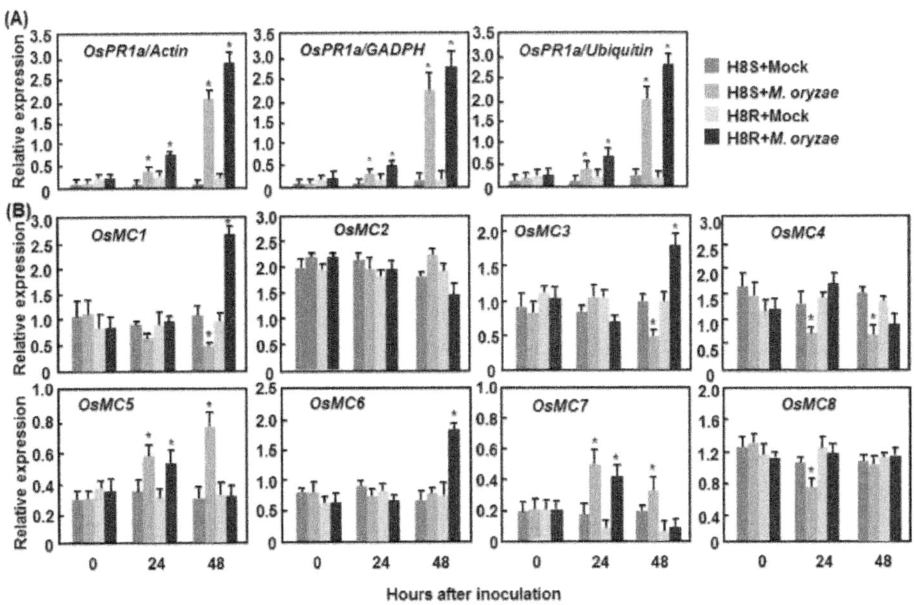

Figure 3. Expression patterns of *OsMCs* in incompatible and compatible interactions between rice and *Magnaporthe oryzae*. Two-week-old seedlings were inoculated by spraying with spore suspension of *M. oryzae* and leaf samples were collected at indicated time points after inoculation for qRT-PCR analyses. (**A**) Expression patterns of *OsPR1a* calculated with different reference genes; and (**B**) Expression patterns of *OsMCs*. Relative expression is shown as folds of transcript levels of three different reference genes (**A**) or *Actin* gene (**B**). Data presented are the means ± SD from thee independent experiments and * above the columns indicate significant difference at $p < 0.05$ level.

however, increased expression at 48 h after inoculation was only observed in compatible interaction (Figure 3B). Additionally, the expression of *OsMC4* and *OsMC8* was decreased only in compatible interaction (Figure 3B). The expression of *OsMC2* was not affected by infection of *M. grisea* in both of the incompatible and compatible interactions (Figure 3B).

2.2.2. Expression Patterns in Response to Xoo

In rice-*Xoo* interactions, a pair of NILs IR-BB10 and IR24 interact differentially with *Xoo* strain PXO99^A, resulting in incompatible and compatible interactions, respectively [48]. Also, we examined the expression changes of defense-related genes as a sign of activation of defense response in incompatible and compatible interactions between rice and *Xoo* before analyzing the expression patterns of *OsMCs*. As shown in Figure 4A, the expression levels of *OsPR1a*, a defense-related gene that is induced by *Xoo* [49], both the incompatible and compatible interactions were significantly increased at 48 h after inoculation. Meanwhile, typical disease symptoms appeared in the inoculated leaves of IR-BB10 and IR24 plant at 7 days after inoculation, but the IR-BB10 plants showed a disease resistance

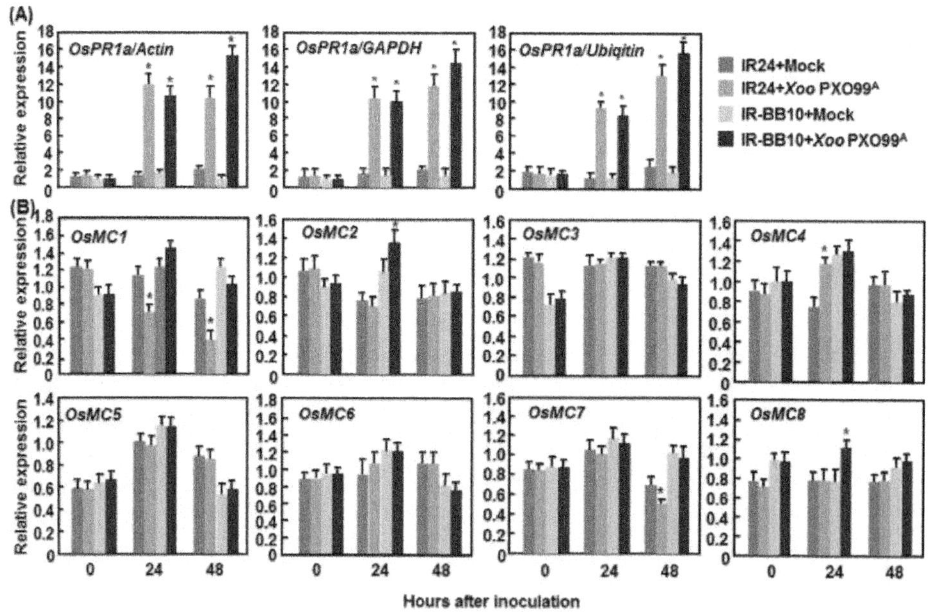

Figure 4. Expression patterns of *OsMCs* in the incompatible and compatible interactions between rice and *Xanthomonas oryzae* pv. *oryzae*. Eight-week-old seedlings were inoculated with *X. oryzae* pv. *oryzae* strain PXO99^A by leaf clipping method and leaf samples were collected at indicated time points after inoculation for qRT-PCR analyses. (**A**) Expression patterns of *OsPR1a* calculated with different reference genes; and (**B**) Expression patterns of *OsMCs*. Relative expression is shown as folds of transcript levels of three different reference genes (**A**) or *Actin* gene (**B**). Data presented are the means ± SD from thee independent experiments and * above the columns indicate significant difference at $p < 0.05$ level.

phenotype with shorter lesions than the IR24 plants. Similar expression patterns of *OsPR1a* were observed when calculated with three different reference genes such as *Actin, GAPDH,* and *Ubiquitin* [43] (Figure 4A). These data demonstrate the effectiveness and reliability of our pathogen inoculation experiments. Surprisingly, the expression of most of the *OsMC* genes was not affected significantly in both ofthe IR-BB10 and IR24 plants after *Xoo* infection (Figure 4B). However, the expression levels of *OsMC1* at 24 and 48 h and *OsMC7* at 48 h were significantly decreased while the expression level of *OsMC4* at 24 h was induced markedly in IR24 plants after *Xoo* infection, but the levels in IR-BB10 plants were not affected (Figure 4B). By contrast, the expression of *OsMC2* and *OsMC8* in IR-BB10 plants was induced significantly at 24 h after *Xoo* infection (Figure 4B).

2.2.3. Expression Patterns in Response to R. solani

In rice-*R. solani* interaction, the expression level of *OsPR10* (*Pathogenesis-related 10*), a defense-related gene that is induced by *R. solani* [50], was significantly increased by 2 folds at 36 h after inoculation, as compared with that in the mock-inoculated plants (Figure 5A). Similar expression patterns of *OsPR10* were

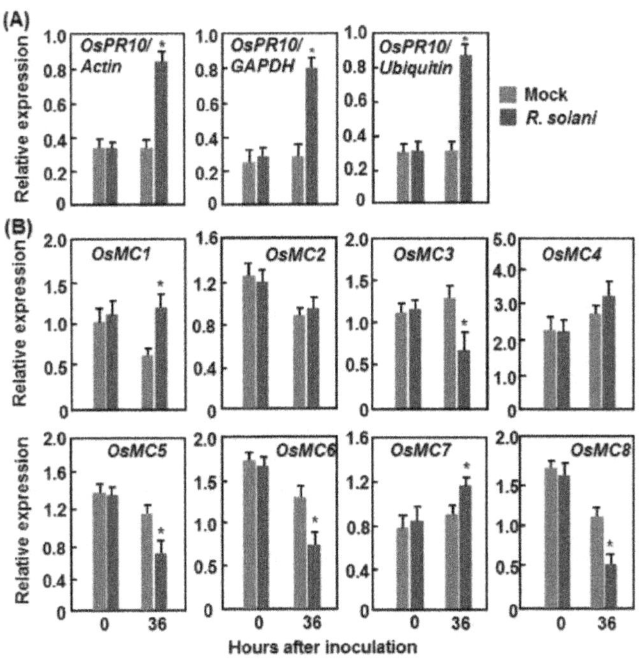

Figure 5. Expression patterns of *OsMCs* in response to *Rhizoctonia solani*. Six-week-old plants were inoculated by attaching mycelial of *R. solani* strain GD-118 onto sheath and leaf samples were collected at indicated time points after inoculation for qRT-PCR analyses. (**A**) Expression patterns of *OsPR10* calculated with different reference genes; and (**B**) Expression patterns of *OsMCs*. Relative expression is shown as folds of transcript levels of three different reference genes (**A**) or *Actin* gene (**B**). Data presented are the means ± SD from three independent experiments and * above the columns indicate significant difference at $p < 0.05$ level.

observed when calculated with three different reference genes such as *Actin*, *GAPDH*, and *Ubiquitin* [43] (Figure 5A). Therefore, the effectiveness and reliability of our pathogen inoculation experiments were satisfied for further analyses of expression of *OsMCs* in response to *R. solani*. Among the *OsMCs*, the expression of *OsMC1* and *OsMC7* was induced significantly by *R. solani* as compared with that in the mock-inoculation plants (Figure 5B). By contrast, the expression of *OsMC3*, *OsMC5*, *OsMC6*, and *OsMC8* was downregulated by *R. solani* while the expression of *OsMC2* and *OsMC4* was not affected by *R. solani* (Figure 5B).

Together, these data indicate that the expression of *OsMCs* exhibit differential expression patterns in rice upon infections from different fungal and bacterial pathogens and such differential expression may imply their involvement in defense response to pathogens.

2.3. Expression Patterns of OsMCs in Response to Stress-Related Hormones

Abscisic acid (ABA), salicylic acid (SA), jasmonic acid (JA), and ethylene (ET) are well-known stress-related hormones in plants and are involved in abiotic and biotic stress responses [51–54]. The differential expression patterns of *OsMCs* in response to multiple abiotic and biotic stresses led us to examine the responsiveness of *OsMCs* to these stress-related hormones. As shown in Figure 6, the *OsMC* genes responded differentially as revealed by their expression changes in plants

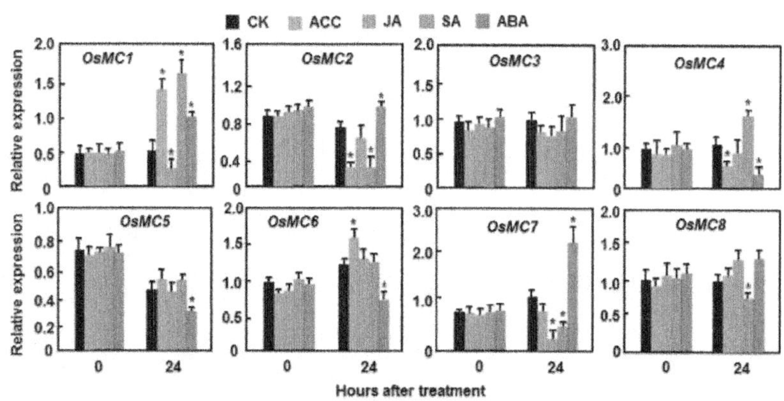

Figure 6. Expression patterns of *OsMCs* in response to stress-related hormones. Two-week-old seedlings were sprayed with 100 μM SA, 100 μM MeJA, 100 μM ACC, 100 μM ABA or similar volumes of solution as controls. Leaf samples were collected at indicated time points and relative expression of the genes is shown as folds of transcript level of the actin gene. Data presented are the means ± SD from thee independent experiments and * above the columns indicate significant difference at $p < 0.05$ level.

after treatment with ABA, SA, JA, or 1-amino cyclopropane-1-carboxylic acid (ACC, a precursor of ET). In ACC-treated plants, dramatic expression changes of *OsMC1*, *OsMC2*, *OsMC4* and *OsMC6* were observed (Figure 6). The expression levels of *OsMC1* and *OsMC6* were increased while the expression levels of *OsMC2* and *OsMC4* were decreased at 24 h after treatment (Figure 6). The expression of *OsMC3*, *OsMC5*, *OsMC7*, and *OsMC8* was not affected by ACC treatment (Figure 6). After JA treatment, the expression levels of *OsMC1* and *OsMC7* were significantly decreased while the expression of *OsMC2*, *OsMC3*, *OsMC4*, *OsMC5*, *OsMC6*, and *OsMC8* was not affected (Figure 6). In response to SA treatment, the expression levels of *OsMC1* and *OsMC4* were upregulated by 3- and 1.5-fold, respectively, whereas the expression levels of *OsMC2*, *OsMC7*, and *OsMC8* were downregulated significantly by >1.5-fold (Figure 6). The expression of *OsMC3*, *OsMC5*, and *OsMC6* was not affected by SA treatment (Figure 6). After ABA treatment, the expression levels of *OsMC1*, *OsMC2*, and *OsMC7* were increased while the expression levels of *OsMC4*, *OsMC5*, and *OsMC6* were decreased (Figure 6). The expression of *OsMC3* and *OsMC8* was not affected by ABA treatment (Figure 6). Taken together, the *OsMC* genes respond to multiple stress-related hormones with differential expression patterns, which may imply their relationships with the stress hormone-mediated signaling for their biological functions.

2.4. Putative Cis-Elements in Promoter Regions of the OsMC Genes

To gain insights into the transcriptional regulation of the *OsMC* genes, putative *cis*-elements in the promoter regions were examined. A number of well-known *cis*-elements that are involved in growth/development and stress response were detected in the 1 kb upstream regions of the *OsMC* genes (Table 1). One MYB-related (S000176), one MYC-related (S000407) and one WRKY-related (S000447) *cis*-elements, which are known to be involved in abiotic and biotic stress responses [55,56], were found in the promoter regions of all *OsMC* genes. Two other WRKY-related (S000390 and S000457) *cis*-elements were also detected in the promoter regions of the *OsMC* genes except the *OsMC6* (Table 1). The promoter regions of *OsMC3*, *OsMC4*, *OsMC5*, *OsMC6*, and *OsMC8* further contain GCC-box (S000430), known recognition site for ERFs [57]. Three ABA-responsive *cis*-elements, ABRELATERD1 (S000414), ABREOSRAB21 (S000012), and ABRERATCAL (S000408), were found in the promoter regions of *OsMC2*, *OsMC3*, *OsMC6*, and *OsMC8*; however, the promoter regions of *OsMC1*, *OsMC4*, *OsMC5*, and *OsMC7* do not contain known ABA-responsive *cis*-element (Table 1). Additionally, both of the *OsMC1* and *OsMC5* promoter regions harbor one NtBBF1motif (S000273), which is known to be auxin-responsive [58].

Table 1. Putative *cis*-elements in promoter regions of the *OsMC* genes.

Regulator	*Cis*-Element	Code	Numbers of *cis*-Elements							
			OsMC1	*OsMC2*	*OsMC3*	*OsMC4*	*OsMC5*	*OsMC6*	*OsMC7*	*OsMC8*
ABA	ABRELATERD1	S000414	–	2	1	–	–	2	–	1
	ABREOSRAB21	S000012	–	–	2	–	–	–	–	–
	ABRERATCAL	S000408	–	3	–	–	–	2	–	1
MYB	MYBCORE	S000176	2	4	5	1	1	2	1	1
MYC	MYCCONSENSUSAT	S000407	4	14	8	4	4	8	6	8
WRKY	WBOXATNPR1	S000390	1	1	1	–	4	–	–	3
	WBOXNTERF3	S000457	2	1	1	1	2	–	1	6
	WRKY710S	S000447	3	4	3	4	6	1	1	10
ERF	GCCCORE	S000430	–	–	3	1	2	2	–	1
Auxin	NTBBF1ARROLB	S000273	1	–	–	–	1	–	–	–

2.5. Subcellular Localization of OsMCs

To explore the subcellular localization of the OsMC proteins, we first searched putative nuclear localization signal (NLS) in the OsMC proteins by PSORT analysis. Putative NLSs were identified in OsMC1 and OsMC7, which contain NLS of RRRH and PVKGRRH sequences, respectively. No NLS was detected in OsMC2, OsMC3, OsMC4, OsMC5, OsMC6, and OsMC8 proteins. We then examined experimentally the subcellular localization of OsMC proteins though transient expression of GFP:OsMC fusions in leaves of *Nicotiana benthamiana* plants that expressed a red nuclear marker RFP–H2B protein [59]. Confocal laser scanning microscope observation at 48 h after agroinfiltration revealed that the GFP:OsMC1 fusion was solely localized to the nucleus, co-localized with the known nucleus marker RFP–H2B protein and the GFP:OsMC2 fusion was mainly localized in the nucleus with a very weak signal in the cytoplasm (Figure 7). The GFP:OsMC3, GFP:OsMC4 and GFP:OsMC7 fusions were found to distribute both in the cytoplasm and nucleus whereas the *GFP:OsMC5*, *GFP:OsMC6*, and GFP:OsMC8 fusions were exclusively localized in the cytoplasm without any localization in the nucleus (Figure 7). The GFP alone was detected in both the cytoplasm and nucleus. These results indicate that the OsMC proteins exhibit differential subcellular localization, which may be associated with their specific biological functions in the cells.

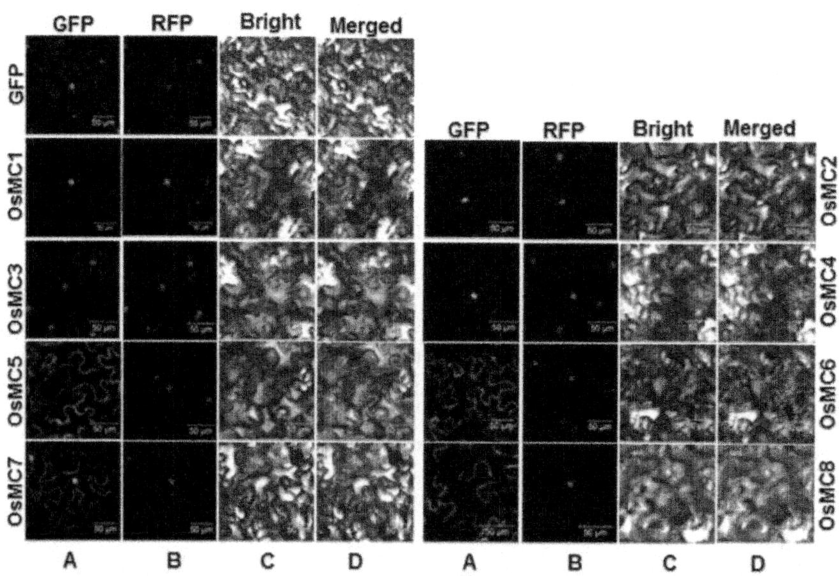

Figure 7. Subcellular localization of the OsMC proteins when transiently expressed in leaves of *Nicotiana benthamiana*. Agrobacteria harboring pFGC-OsMC1/2/3/4/5/6/7/8 or pFGC-Egfp were infiltrated into leaves of *N. benthamiana* plants expressing a red nucleus marker protein RFP-H2B and leaf samples were collected at 24 h after agroinfiltration. Images were observed and taken under a confocal laser scanning microscope in dark field for green fluorescence (**A**); red fluorescence (**B**); white field for cell morphology (**C**) and in combination (**D**), respectively. Bar = 50 μM.

2.6. Interaction Relationships between Type I OsMCs and OsLSDs

It was previously reported that the Arabidopsis type I metacaspase AtMC1 interacted with AtLSD1 [35], a negative regulator of cell death [38]. To examine whether thee rice type I metacaspases, OsMC1, OsMC2, and OsMC3, can also interact with OsLSDs. In Arabidopsis, there are one AtLSD1 and two LSD One Like proteins (AtLOL1 and AtLOL2) [38,60]. Five putative AtLSD1 homologs were identified in rice genome [10,61] and the nomenclature for these proteins are somewhat confused in literatures. To avoid confusion in nomenclature for these genes, we generated a phylogenetic tree including the rice OsLSDs and OsLOLs and Arabidopsis AtLSD1 and AtLOLs and assigned unique names to them (Figure 8A). OsLOL1 (OsLSD1) and OsLOL2 were previously reported and found to function in PCD, disease resistance and growth [62–64]. In our yeast two hybrid experiments, a positive control (pGADT7-T + pGBKT7-53) and a negative control (pGADT7-T + pGBKT7-Lam) were always included to rule out possible false interactions (Figure 8B). As shown in Figure 8C, significant interactions between OsMC1 and OsLSD1/OsLSD3 and between OsMC3 and OsLSD1 were detected; no interaction in other pairwise combinations between OsMC1/2/3 and OsLSDs/OsLOLs was observed. Notably, OsMC2 did not interact with any of OsLSDs while OsLSD2 did not interact with all three type I metacaspases OsMC1, OsMC2, and OsMC3 (Figure 8C). We further examined which region of OsMC1 is responsible for the interaction activity with OsLSD1 and OsLSD3. As shown in Figure 8D, the N-terminal region (1-84 aa), which contains the conserved zinc finger domain, did interact with both of OsLSD1 and OsLSD3, but the C-terminal region of OsMC1 (85–368 aa), which lacked the conserved zinc finger domain, did not interact with OsLSD1 and OsLSD3, indicating that the zinc finger domain in OsMC1 is critical for its interaction activity with OsLSDs. Furthermore, we also examined whether these type I OsMCs interact with each other or itself. Self-interaction in OsMC1 was observed (Figure 8E). However, OsMC1, OsMC2, and OsMC3 did not interact to each other and OsMC2 and OsMC3 did not show self-interaction in yeast (Figure 8E).

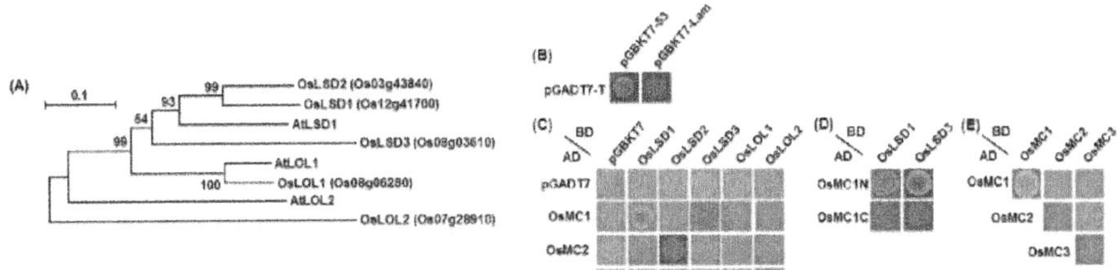

Figure 8. Interactions between type I OsMCs and OsLSDs. (**A**) Phylogenetic tree of rice OsLSDs/OsLOLs with Arabidopsis AtLSD1/AtLOLs. Phylogenetic tree was constructed by Neighbor-joining method using MEGA program; The bar represents 0.1 amino acid substitution per site; (**B**) Positive (pGADT7-T + pGBKT7-53) and negative (pGADT7-T + pGBKT7-Lam) controls; (**C**) Interactions of OsMC1, OsMC2 and OsMC3 with OsLSD1 and OsLSD3 but not with OsLSD2, OsLOL1 and OsLOL2; (**D**) Interactions of N- and C-terminals of OsMC1 with OsLSD1 and OsLSD3; and (**E**) Interactions among OsMC1, OsMC2 and OsMC3. Yeasts harboring the indicated plasmid combinations were grown on selective medium SD/Trp Leu His Ade and P-galactosidase activity showing positive interactions was examined by addition of X-a-gal. Repeated experiments showed similar results. AD, pGADT7 vector; BD, pGBKT7 vector.

3. DISCUSSION

Although the rice OsMC family has been previously characterized [10,11], none of the family members has been studied for the biological function so far. The results presented in this study indicate that members of the OsMC family respond differentially to multiple abiotic and biotic stresses as well as to some well-known stress-related hormones. However, the responsiveness of the members of the OsMC family to these stresses or hormones varies to some extent. For example, *OsMC1* was found to be induced by all of the stresses or hormones tested; while other *OsMCs* responded to less stresses or hormones (Figures 1-6). Such differential stress- or hormone-related responsiveness may imply the specific involvement of the members of the *OsMC* family in different abiotic and biotic stress responses.

It was previously reported that the Arabidopsis *AtMC1* and *AtMC3,* tomato *LeMCA1,* pepper *CaMC9,* wheat *TaMCA4* and *N. benthamiana NbMCA1* were significantly induced by infection from fungal or bacterial pathogens [27–30,32], indicating the importance of transcriptional regulation of metacaspase genes for the biological functions in plants. In the present study, we also found that infection with *M. oryzae, Xoo,* or *R. solani,* thee different pathogens with distinct infection styles, did cause transcriptional reprogramming in expression of the *OsMC* genes. For example, more than half of the *OsMCs* displayed differential expression in either compatible or incompatible interaction between rice and *M. oryzae.* Among these *OsMCs, OsMC5,* a type II metacaspase, was induced by the blast fungus in both compatible and incompatible interactions (Figure 3) but not by SA and JA (Figure 6), two well-known signaling molecules in defense responses. This is in accordance with a previous observation from RNA-Seq analysis of rice samples infected with *M. grisea* [65]. On the other hand, some of the *OsMCs* exhibited differential expression patterns in different rice-pathogen interactions. For example, *OsMC1* was found to be induced by the blast fungus in the incompatible interaction of rice-*M. oryzae* (Figure 3) and by SA (Figure 6); however, its expression was downregulated in the incompatible interaction of rice-*Xoo* (Figure 4). Further, *OsMC7* showed a higher level of expression in compatible rice-*M. oryzae* interaction (Figure 3) but was downregulated by SA and JA (Figure 6), whereas its expression was not affected dramatically in interactions of rice-*Xoo* and rice-*R. solani* (Figures 4 and 5). The differential expression of *OsMC1* and *OsMC7* in response to different pathogens may be attributed to different mechanisms that are used by *M. grisea, Xoo,* and *R. solani* during their infection processes [66]. This is similar to the observations that *OsEDR1* was found to play opposite roles in the rice-*M. oryzae* and rice-*Xoo* interactions [67].

Diverse abiotic stresses have been found to induce PCD and the abiotic stress-induced PCD significantly affects plant growth and development [68]. It was found that abiotic stress-induced PCD is often accompanied with accumulation of reactive oxygen species (ROS) [68]. In fact, ROS and other abiotic factors such as ultraviolet light and ozone were reported to induce expression of metacaspase genes in Arabidopsis and maize [26,31]. In the present study, we found that, like the responsiveness to pathogens, members of the *OsMC* family also respond differentially to dehydration (drought and salt) and extreme temperature (cold and heat) stresses. Interestingly, expression of most *OsMC* genes was downregulated by drought and salt stresses (Figures 1 and 2). Some of the *OsMC* genes exhibited overlapping expression patterns under different abiotic stress conditions, indicating their possible involvement in diverse abiotic stress responses. For example, the expression of the *OsMC* genes in leaf tissues was significantly downregulated by drought and salt stresses (Figure 1). However, the downregulated expression patterns of *OsMCs* in leaf tissues are not well correlated with the patterns in the root tissues (Figure 1). This is in agreement with the observations that some of the transcriptional factor genes exhibited distinct expression patterns in leaf and root tissues under abiotic stress conditions [69,70]. Furthermore, the drought and salt stress-caused differential expression patterns of some *OsMCs* are correlated with the patterns in rice plants treated with ABA (Figures 1, 2 and 6), which is believed to play critical roles in response to drought and salt stresses [51,52]. For example, the expression of *OsMC4* was suppressed by drought and salt stresses (Figure 1) and also was suppressed by ABA (Figure 6). On the other hand, opposite expression patterns of *OsMCs* under abiotic stress conditions were also observed. For example, the expression of *OsMC4* was induced by cold stress but suppressed by heat stress, whereas the expression of *OsMC5* was suppressed by cold stress but induced by heat stress (Figure 2), indicating opposite roles for *OsMC4* and *OsMC5* in extreme temperature stresses.

Notably, the range of changes in the expression of *OsMCs* in response to abiotic and biotic stress varied greatly. For example, expression of *OsMC1*, *OsMC3*, and *OsMC6* was induced by 2–3-fold in leaves inoculated with an incompatible strain of *M. oryzae* (Figure 3), whereas the expression changes of some *OsMCs* in response to pathogens or abiotic stress were less than 2-fold (Figures 1–5). Similar results were also observed for the expression changes of *MC* genes in other plants; for example, 3–4-fold induction for *CamC9* in pepper leaves inoculated with *Xanthomonas campestris* pv. *vesicatoria* [28], ~5-fold upregulation for *TaMC4* in wheat leaves challenged with the avirulent race of *Puccinia striiformis* f. sp. *tritici* [29], ~1.5-fold induction for *NbMCA1* in *N. benthamiana* leaves infected with *Colletotrichum destructivum* [30] and ~1-fold change for *TaeMCAII* in wheat under heat stress [71]. The difference in the expression changes among the *OsMC* genes in response to different pathogens may be explained by the possibility that both of the transcriptional and post-transcriptional regulations are required for the functions of *OsMCs* in given abiotic ad biotic stress responses [16,17].

Several features on the subcellular localization and protein-protein interactions provide valuable information on the mode of action of OsMCs. In the present study, we examined the subcellular localization of all members in the OsMC family and our results indicate that the members of this family exhibit diverse subcellular localization. The OsMC1 protein contains a predicted NLS with sequence of RRRH and was found to be localized in nucleus (Figure 7). This is further supported by the interaction between OsMC1 and OsLSD1 (Figure 8), whose homolog in *Pisum sativa*, PsLSD1, was shown to be a nucleus-localized protein [72]. Further experiments are required to examine the *in vivo* OsMC1-OsLSD1 interaction and whether such interaction takes place in the nucleus. OsMC7, along with OsMC3 and OsMC4, were evenly distributed in the cells (Figure 7), although OsMC7 contains a putative NLS sequence. Interestingly, OsMC5, OsMC6, and OsMC8 were found to be exclusively localized in the cytoplasm without any localization in the nucleus (Figure 7). However, it was previously reported that the Arabidopsis AtMC9 is present in apoplast, nucleus and cytoplasm [14] and its subcellular localization can be changed from an even cytoplasmic localization in living cells to patches or aggregates of different sizes in cells during late autolysis [33]. Thus, whether dynamics in subcellular localization upon developmental and stress signals are the case for the OsMC proteins needs to be examined further. On the other hand, the type I metacaspase, OsMC1 interacted with OsLSD1 and OsLSD3 while OsMC3 only interacted with OsLSD1 (Figure 8). The N-terminal LSD1-type zinc finger domain in OsMC1 is responsible for the interaction activity with OsLSD1 and OsLSD3 (Figure 8). This is in agreement with the interactions between AtMC1 and AtLSD1 [35]. Furthermore, OsMC1 was found to interact with itself while *OsMC2 and OsMC3* did not (Figure 8), indicating possible different modes of action for these type I metacaspases.

4. MATERIALS AND METHODS

4.1. Plant Materials and Growth

Rice (*Oryza sativa* L.) cv. Yuanfengzao and two NIL pairs (provided by Dr. Zuhua He, Shanghai Institute of plant physiology and ecology, Chinese Academy of Sciences, Shanghai, China) were used in this study for different purposes. The cv. Yuanfengzao was used for analyses of gene expression in response to abiotic stress, hormone treatments, and infection by *R. solani*. The pair of NILs H8S and H8R was used for analysis of gene expression in compatible and incompatible interactions between rice and *M. oryzae* race ZB1 (provided by Mr. Rongyao Chai, Zhejiang Academy of Agricultural Sciences, Hangzhou, China) while the pair of NILs IR-BB10 and IR24 (provided by International Rice Research Institute, Los Banos, the Philippines) was used for analysis of gene expression in compatible and incompatible interactions between rice and *X. oryzae* pv. *oryzae* (*Xoo*) strain PXO99^{A} (provided by Dr. Jean Leach, Colorado State University, Fort Collins, CO, USA). The rice plants were grown in a growth room under a 14-h light/10-h dark cycle at 26 °C.

4.2. Treatments with Abiotic Stress and Hormones

Two-week-old cv. Yuanfengzao seedlings were treated with varied abiotic stresses or hormones. Drought stress was applied by transferring the hydroponically cultivated seedlings to thee layered filter papers for fast dehydration. Salt stress treatment was achieved by drenching 150 mM NaCl (Sinopharm Chemical Reagents Co., Shanghai, China) solution to the rice seedlings. For cold stress, seedlings were transferred to a growth chamber at 4 °C. Heat stress was applied by transferring the seedlings to a chamber with temperature at 42 °C. Leaf samples were harvested for all four abiotic stress treatments and root samples were only collected for drought and salt treatments. For hormone treatment, seedlings were sprayed with 100 μM ABA, 100 μM methyl jasmonate (MeJA) (Sigma-Aldrich, St. Louis, MO, USA), 100 μM ACC (Sigma-Aldrich, St. Louis, MO, USA), 150 μM SA (Sigma-Aldrich, St. Louis, MO, USA) in a solution containing 0.1% ethanol (Sinopharm Chemical Reagents Co., Shanghai, China) and 0.02% Tween-20 (Sinopharm Chemical Reagents Co., Shanghai, China) or with the solution as controls. All samples were stored at -80 °C until use.

4.3. Inoculation with Different Pathogens

M. oryzae strain 85-14B1 was cultivated on complete medium at 25 °C for 10 days and spores were collected to prepare inoculum. Two-week-old H8S and H8R seedlings at thee leaf stages were inoculated spraying with 5×10^5 conidia/mL spore suspension containing 0.02% Tween-20 or with similar volume of solution containing only 0.02% Tween-20 as mock-inoculation controls [46]. Inoculated plants were kept under moist conditions in the dark at room temperature for 24 h and then moved to a growth chamber (12 h 28 °C light/12 h 24 °C dark). *Xoo* strain PXO99^{A} was grown in NA broth at 28 °C with shaking and cells were collected by centrifugation and diluted in distilled water to OD600 = 0.8. Eight-week-old IR-BB10 and IR24 plants were inoculated using leaf clipping method [73]. *R. solani* strain GD-118 (provide by Dr. Weiliang Chen, Zhejiang University, Hangzhou, China) was grown in 250 mL of potato dextrose broth medium and incubated on a 28 °C shaker for 3 days and the culture was centrifuged to collect the mycelia. Six-week-old cv. Yuanfengzao plants were inoculated by closely attaching mycelial balls (8 mm in diameter) on the sheath with aluminium foil [74]. Mock-inoculation control plants were only inoculated by attaching aluminium foil onto sheath without mycelial balls. Samples were collected from pathogen-inoculated and mock-inoculated plants at indicated time points and stored at –80 °C until use.

4.4. qRT-PCR Analysis of Gene Expression

Total RNAs were extracted from frozen samples using TRIzol reagent (Invitrogen, Shanghai, China) and then treated with RNase-free DNase (TaKaRa, Dalian, China). First-strand cDNA was synthesized from 1 μg total RNA using AMV reserves transcriptase (TaKaRa, Dalian, China) according to the manufacturer's instructions. According to the previously suggested guidelines [40,75], thee different reference genes were used to assess the accuracy of qRT-PCR data and no template controls were always included in qRT-PCR experiments. Each qRT-PCR reaction contained 12.5 μL of SYBR premix Ex Taq (TaKaRa, Dalian, China), 1 μL of cDNA samples and 10 μM gene-specific primers (Table 2) in a final volume of 25 μL and was performed on a CFX96 Real-time System (Bio-Rad, Hercules, CA, USA). The relative gene expression data were calculated using the $2^{-\Delta\Delta Ct}$ method. Rice *Actin* (accession number KC140129), *GAPDH* (accession number AK062215) and *Ubiquitin* (accession number AK059011) genes were used as internal reference controls. Thee independent biological replicates were conducted for all experiments.

Table 2. Primers used in this study.

Primers	Sequences (5'-3')	Size (bp)
	Cloning of cDNAs	
OsMC1-F OsMC1-R	ATGGATCACTTCGGCGGACG TTACAGGACGAACGGCTTGC	1107
OsMC2-F OsMC2-R	ATGGCGAGCGCGAGGCCGCC TCACAAGAGGAAGGGCTTCC	1110
OsMC3-F OsMC3-R	ATGGGCTGCAACTGCCTCGT TCACAGGAGAAACGGTTTCC	1203
OsMC4-F OsMC4-R	ATGGGGCGGAAGAGAGCGGT TTAGCATATGAAAGCCACGT	1230
OsMC5-F OsMC5-R	ATGGGGGGCCGGAAGCGCGC TCAGCATATGAAGGCCACAC	1263
OsMC6-F OsMC6-R	ATGGGCCGCAAGCGCGCGCT TCAGCATATAAAAGACACAT	1254
OsMC7-F OsMC7-R	ATGGAGAGGGGTCAGAAGAA TCAGAGCGCCGTCATGGCCT	1026
OsMC8-F OsMC8-R	ATGGCGGTCGTCAGCGGCGG TCACAGGATAAACTGCTCCT	909
OsLSD1-F OsLSD1-R	ATGTGCATTGCTGAACCAAT TCAGCTGCTGGGCTTCTGGT	606
OsLSD2-F OsLSD2-R	ATGGTGGCTTCAAGAGCTCCA CTATCCTAGACTGAAAAGCA	444
OsLSD3-F OsLSD3-R	ATGCAGAGCCAGATCGTGT TTACTTTTTACCACCAGTTGTA	519
OsLSD4-F OsLSD4-R	ATGCAGAGCCAGATCGTGTG CTATTTCCCAGTTGTAACTCCA	561
OsLSD5-F OsLSD5-R	ATGCAGGACCAGCTGATCTG TCATCTTTTCCATGAGGTGAC	444

(Contd...)

Primers	Sequences (5'-3')	Size (bp)
	Subcellular Localization Assays	
OsMC1-GFP-F OsMC1-GFP-R	CGCGGATCC ATGGATCACTTCGGCGGACG TGCTCTAGA TTACAGGACGAACGGCTTGC	1107
OsMC2-GFP-F OsMC2-GFP-R	CGCGGATCC ATGGCGAGCGCGAGGCCGCC TGCTCTAGA TCACAAGAGGAAGGGCTTCC	1110
OsMC3-GFP-F OsMC3-GFP-R	CGCGGATCC ATGGGCTGCAACTGCCTCGT TGCTCTAGA TCACAGGAGAAACGGTTTCC	1203
OsMC4-GFP-F OsMC4-GFP-R	CGCGGATCC ATGGGGCGGAAGAGAGCGGT TGCTCTAGA TTAGCATATGAAAGCCACGT	1230
OsMC5-GFP-F OsMC5-GFP-R	CGCGGATCC ATGGGGGGCCGGAAGCGCGC TGCTCTAGA TCAGCATATGAAGGCCACAC	1263
OsMC6-GFP-F OsMC6-GFP-R	CGCGGATCC ATGGGCCGCAAGCGCGCGCT TGCTCTAGA TCAGCATATAAAAGACACAT	1254
OsMC7-GFP-F OsMC7-GFP-R	CGCGGATCC ATGGAGAGGGGTCAGAAGAA TGCTCTAGA TCAGAGCGCCGTCATGGCCT	1026
OsMC8-GFP-F OsMC8-GFP-R	CGCGGATCC ATGGCGGTCGTCAGCGGCGG TCCCCCGGG TCACAGGATAAACTGCTCCT	909
	qRT-PCR Assays	
OsMC1-RT-F OsMC1-RT-R	GCTTCATCAAGGCGGTGGAGTG AAGTTGGCGACCTTGCGGATG	142
OsMC2-RT-F OsMC2-RT-R	CGACCCGTACAGGGTGCCGA GCACAGCGCCTCGTCG- TAGC	166
OsMC3-RT-F OsMC3-RT-R	GGCTCCTTCGTCCGCAAGAT CACAGGAGAAACGGTTTCCTGT	101
OsMC4-RT-F OsMC4-RT-R	TCGACGTTCGTGGAGATGCTC ATTCACGAGCCGCCTGATCTT	126
OsMC5-RT-F OsMC5-RT-R	GTGCCAGACCGACCAGACAT CCGCTCTTCTCCGACAGGAT	102
OsMC1-RT-F OsMC1-RT-R	GCTTCATCAAGGCGGTGGAGTG AAGTTGGCGACCTTGCGGATG	142
OsMC6-RT-F OsMC6-RT-R	CCACACCGCAGGGTTCTTCAT GTCCAGGCTGCTGAGTGTATCC	147
OsMC7-RT-F OsMC7-RT-R	ATACAGACCGTGCTGGCGTC AGGAATGGCGTCTCGGCGTT	143
OsMC8-RT-F OsMC8-RT-R	TCCGGCAAGTGCCTCGTAAC CAATGCGGTCGGTCACAGGAT	150
	Yeast Two-Hybrid Assays	
OsMC1-BD-F OsMC1-BD-R	CGGAATTC ATGGATCACTTCGGCGGACGT CGCGGATCC TTACAGGACGAACGGCTTGCG	1107
OsMC2-BD-F OsMC2-BD-R	CCGGAATTC ATGGCGAGCGCGAGGCCG CGCGGATCC TCACAAGAGGAAGGGCTTC	1110
OsMC3-BD-F OsMC3-BD-R	CCGGAATTC ATGGGCTGCAACTGCCTCGTC CGCGGATCC TCACAGGAGAAACGGTTTCC	1203

OsMC1-BDN-R	TGCTCTAGA CTTGCCGCGGGAGCCCGGGA	252
OsMC1-BDC-F	CGCGGATCC AAGCGCGCCGTCCTGATCGGC	855
OsLSD1-AD-F OsLSD1-AD-R	CCGGAATTC ATGTGCATTGCTGAACCAAT CGCGGATCC TCAGCTGCTGGGCTTCTGGT	606
OsLSD2-AD-F OsLSD2-AD-R	CCGGAATTC ATGGTGGCTTCAAGAGCTCCA CGCGGATCC CTATCCTAGACTGAAAAGCA	444
OsLSD3-AD-F OsLSD3-AD-R	CCGGAATTC ATGCAGAGCCAGATCGTGT CGCGGATCC TTACTTTTTACCACCAGTTGTA	519
OsLSD4-AD-F OsLSD4-AD-R	CCGGAATTC ATGCAGAGCCAGATCGTGTG CGCGGATCC CTATTTCCCAGTTGTAACTCCA	561
OsLSD5-AD-F OsLSD5-AD-R	CCGGAATTC ATGCAGGACCAGCTGATCTG CGCGGATCC TCATCTTTTCCATGAGGTGAC	444

4.5. Subcellular Localization of the OsMC Proteins

The coding sequences of the *OsMC* genes was amplified using gene-specific primers (Table 2) and cloned into pMD-19T (Takara, Dalian, China) by T/A cloning, yielding plasmids pMD-19T-OsMC1/2/3/4/5/6/7/8. After confirmation by sequencing, the coding sequences were amplified from plasmids pMD-19T-OsMC1/2/3/4/5/6/7/8 with gene-specific primers containing restriction enzyme sites and cloned into pFGC-Egfp at the corresponding restriction enzyme sites, yielding pFGC-OsMC1/2/3/4/5/6/7/8. The recombinant plasmids pFGC-OsMC1/2/3/4/5/6/7/8 and the pFGC-Egfp empty vector were transformed into *Agrobacterium tumefacies* strain GV3101 by electroporation using GENE PULSER II Electroporation System (Bio-Rad Laboratories, Hercules, CA, USA). Agrobacteria harboring pFGC-OsMC1/2/3/4/5/6/7/8 and pFGC-Egfp were grown in YEP medium (50 μg/mL rifampicin, 50 μg/mL kanamycin and 25 μg/mL gentamicin) for 24 h with continuous shaking at 28 °C, collected by centrifugation and resuspended in infiltration buffer (10 mM MgCl2, 10 mM MES, 200 μM acetosyringone, pH 5.7). Agrobacteria carrying pFGC-OsMC1/2/3/4/5/6/7/8 or pFGC-Egfp empty vector were infiltrated into leaves of 4-week-old *N. benthamiana* plants expressing a red nuclear marker RFP-Histone 2B protein [56] using 1 mL needless syringes and the agroinfiltrated plants were grown in a growth room at 25 °C for 48 h. Fluorescence signals were excited at 488 nm and detected using a 500–530 nm emission filter preformed with Zeiss LSM 780 confocal laser scanning microscope (Carl Zeiss, Jena, Germany).

4.6. Yeast Two-Hybrid Interaction Assays

Protein-protein interactions between OsMCs and OsLSDs were examined though yeast two-hybrid assay using Matchmaker Gold Yeast Two-Hybrid System according to the manufacturer's instructions. The coding sequences of OsMC1/2/3 were amplified using gene-specific primers (Table 2) from pMD-19T-OsMC1/2/3 and inserted into pGBKT7, yielding pGBKT7-OsMC1/2/3 bait vectors or inserted into pGADT7, yielding pGADT7-OsMC1/2/3 prey vectors. The sequences for the N-terminal (1–84 aa) and C-terminal (85–368 aa) regions of the *OsMC1* gene were also amplified from pMD-19T-OsMC1 with gene-specific primers (Table 2) and cloned into pGBKT7, yielding pGBKT7-OsMC1N and pGBKT7-OsMC1C vectors. The coding sequences of the *OsLSD* genes was amplified using gene-specific primers (Table 2) and cloned into pM D-19T (Takara, Dalian, China), yielding plasmids pMD-19T-OsLSD1/2/3 and pMD-19T-OsLOL1/2. After confirmation by sequencing, the coding sequences were amplified from plasmids pMD-19T-OsLSD1/2/3 and pMD-19T-OsLOL1/2 with gene-specific primers containing restriction enzyme sites and cloned into pGADT7, yielding pGADT7-OsLSD1/2/3 and pGADT7-OsLOL1/2 prey vectors. The bait and prey vectors in different combinations were co-transformed into Y2HGold yeast cells and confirmed by colony PCR. The transformed yeasts were cultivated on SD/Trp Leu His Ade medium for 3 days at 30 °C and added with X-a-gal (5-bromo-4chloro-3-indolyl-a-D-galactopyranoside) solution to examine the activity of P-galactosidase. Interactions were evaluated based on the growth situation of the transformed yeast cells on the SD/Trp Leu His Ade medium and the production of blue pigment after the addition of X-a-Gal. Yeasts co-transformed with pGBKT7-53 or pGBKT7-Lam and pGADT7-T were as positive and negative controls, respectively.

4.7. Bioinformatics Analysis of Cis-Element in the Promoters of the OsMC Genes

Approximately 1000 bp sequences upstream of the transcriptional start site of each *OsMC* gene were downloaded from the rice genome sequence database at the Rice Genome Annotation Project (http://rice.plantbiology.msu.edu). Putative *cis*-elements in the promoter regions were identified by searching against the PLACE *cis*-element database at http://www.dna.affrc.go.jp/PLACE/signalscan.html.

4.8. Accession Numbers

Sequences of the *OsMC* genes used in this study were obtained from the Rice Genome Annotation Database (http://rice.plantbiology.msu.edu/) under the following identifiers: *OsMC1* (LOC_Os03g27120), OsMC2 (LOC_Os03g27210), *OsMC3* (LOC_Os03g27170), *OsMC4* (LOC_Os05g41660), *OsMC5* (LOC_Os05g41670), *OsMC6* (LOC_Os01g58580), *OsMC7* (LOC_Os 11g04010), *OsMC8* (LOC_Os03g27190), *OsLOL1* (LOC_Os08g06280), *OsLOL2* (LOC_Os07g28910), *OsLSD1* (LOC_Os12g41700), *OsLSD2* (LOC_Os03g43840) and *OsLSD3* (LOC_Os08g03610).

5. CONCLUSIONS

As the first step toward understanding of the functions of *OsMCs,* the present study was focused on the possible involvement of *OsMCs* in abiotic and biotic stress responses though comprehensive analysis of the expression changes in rice plants treated with different abiotic stresses, three different pathogens and four well-known stress-related hormones. Our data indicate that members of the *OsMC* family respond differentially to multiple abiotic and biotic stresses as well as to stress-related hormones, as summarized in Table 3. Based on the expression patterns, *OsMC1, OsMC7,* and *OsMC8* may have functions in interactions with *M oryzae, Xoo,* and *R. solani,* while *OsMC3, OsMC5,* and *OsMC6* seem to be involved in interaction with fungal pathogens such as *M. oryzae* and *R. solani. OsMC4* responds in compatible interactions with *M oryzae* and *Xoo* while *OsMC2* specifically responds in incompatible rice-*Xoo* interaction. *OsMC4, OsMC6,* and *OsMC7* in root tissues respond with similar patterns under drought and salt stresses. *OsMC2* is specifically downregulated by drought stress while *OsMC5* and *OsMC8* are specifically downregulated by salt stress. *OsMC1* showed opposite expression patterns in response to drought and salt stresses, whereas *OsMC3* did not show expression change in these two stresses. *OsMC3, OsMC4,* and *OsMC5* displayed opposite expression patterns in response to heat and cold stresses, respectively, while expression of *OsMC7* is downregulated by both heat and cold stresses. *OsMC1* and *OsMC6* are upregulated by heat stress but not by cold stress. Importantly, our data revealed that the OsMC proteins have different subcellular localizations and type I metacaspases OsMC1 and OsMC3 can interact with OsLSD1 or OsLSD3. Results from this systematic analysis of the *OsMC* family not only provide evidence for the possible involvement of *OsMCs* in response to abiotic and biotic stresses, but also provide clues for further functional analysis of *OsMCs* genes in stress tolerance and pathogen resistance in rice.

Table 3. Summary on the expression patterns of OsMC*s* in response to pathogens, abiotic stresses and hormones.

Genes	*M. oryzae*		*Xoo*		*R.*	Abiotic Stress				Hormones			
	Compatible	Incompatible	Compatible	Incompatible	*solani*	Drought	Salt	Heat	Cold	ACC	JA	SA	ABA
OsMC1	↓	↑	↓	--	↑	↓	↑	↑	--	↑	↓	↑	↑
OsMC2	--	--	--	↑	--	↓	--	--	--	↓	--	↓	↑
OsMC3	↓	↑	--	--	↓	--	--	↑	↓	--	--	--	--
OsMC4	↓	--	↑	--	--	↓	↓	↓	↑	↓	--	↑	↓
OsMC5	↑	↑	--	--	↓	--	↓	↑	↓	--	--	--	↓
OsMC6	--	↑	--	--	↓	↓	↓	↑	--	↑	--	--	↓
OsMC7	↑	↑	↓	--	↑	↑	↑	↓	↓	--	↓	↓	↑
OsMC8	↓	--	--	↑	↓	--	↓	--	--	--	--	↓	--

↑, represents significant upregulation; ↓, indicates significant downregulation; -- indicates no significant change. Expression patterns in root under drought and salt stresses are listed.

ACKNOWLEDGMENTS

We thank Zuhua He, Institute of Plant Physiology and Ecology, Shanghai Institutes for Biological Sciences, Chinese Academy of Sciences, for providing the rice materials and Rongyao Chai, Zhejiang Academy of Agricultural Science, for providing the strain of *M. oryzae*. This work was supported by the National Natural Science Foundation of China (No. 31272028), National R & D Project of Transgenic Crops of the Ministry of Agriculture of China (No. 2011ZX08009-003-001), the National High-Tech R & D Program (No. 2012AA101505) and National Research Foundation for the Doctoral Program of Higher Education of China (No. 20120101110070).

AUTHOR CONTRIBUTIONS

Lei Huang and Fengming Song designed the study. Lei Huang, Huijuan Zhang, Yongbo Hong, Shixia Liu, and Dayong Li performed the experiments. Fengming Song performed data interpretation and wrote the manuscript with Lei Huang.

Conflicts of Interest

The authors declare no conflict of interest.

REFERENCES

1. Kroemer, G.; El-Deiry, W.S.; Golstein, P.; Peter, M.E.; Vaux, D.; Vandenabeele, P.; Zhivotovsky, B.; Blagosklonny, M.V.; Malorni, W.; Knight, R.A.; *et al.* Classification of cell death: Recommendations of the nomenclature committee on cell death. *Cell Death Differ.* **2005**, *12*, 1463–1467.
2. Kerr, J.F.; Wyllie, A.H.; Currie, A.R. Apoptosis: A basic biological phenomenon with wide-ranging implications in tissue kinetics. *Br. J. Cancer* **1972**, *26*, 239–257.
3. Cohen, G.M. Caspases: The executioners of apoptosis. *Biochem. J.* **1997**, *326*, 1–16.
4. Kitanaka, C.; Kuchino, Y. Caspase-independent programmed cell death with necrotic morphology. *Cell Death Differ.* **1999**, *6*, 508–515.
5. Kroemer, G.; Martin, S.J. Caspase-independent cell death. *Nat. Med.* **2005**, *11*, 725–730.
6. Aravind, L.; Koonin, E.V. Classification of the caspase-hemoglobinase fold: Detection of new families and implications for the origin of the eukaryotic separins. *Proteins* **2002**, *46*, 355–367.
7. Earnshaw, W.C.; Martins, L.M.; Kaufmann, S.H. Mammalian caspases: Structure, activation, substrates, and functions during apoptosis. *Annu. Rev. Biochem.* **1999**, *68*, 383–424.
8. Uren, A.G.; O'Rourke, K.; Aravind, L.; Pisabarro, M.T.; Seshagiri, S.; Koonin, E.V.; Dixit, V.M. Identification of paracaspases and metacaspases: Two ancient families of caspase-like proteins, one of which plays a key role in MALT lymphoma. *Mol. Cell.* **2000**, *6*, 961–967.
9. Tsiatsiani, L.; van Breusegem, F.; Gallois, P.; Zavialov, A.; Lam, E.; Bozhkov, P.V. Metacaspases. *Cell Death Differ.* **2011**, *18*, 1279–1288.
10. Liu, Q.; Xue, Q. Molecular phylogeny, evolution, and functional divergence of the LSD1-like gene family: Inference from the rice genome. *J. Mol. Evol.* **2007**, *64*, 354–363.
11. Wang, L.K.; Zhang, H. Genome wide survey and characterization of metacaspase gene family in rice (*Oryza sativa*). *J. Genet.* **2014**, *93*, 93–102.
12. Zhang, C.; Gong, P.; Wei, R.; Li, S.; Zhang, X.; Yu, Y.; Wang, Y. The metacaspase gene family of *Vitis vinifera* L.: Characterization and differential expression during ovule abortion in stenospermocarpic seedless grapes. *Gene* **2013**, *528*, 267–276.
13. Acosta-Maspons, A.; Sepulveda-Garcia, E.; Sanchez-Baldoquin, L.; Marrero-Gutierrez, J.; Pons, T.; Rocha-Sosa, M.; Gonzalez, L. Two aspartate residues at the putative p10 subunit of a type II metacaspase from *Nicotiana tabacum* L. may contribute to the substrate-binding pocket. *Planta* **2014**, *239*, 147–160.
14. Tsiatsiani, L.; Timmerman, E.; de Bock, P.J.; Vercammen, D.; Stael, S.; van de Cotte, B.; Staes, A.; Goethals, M.; Beunens, T.; van Damme, P.; *et al.* The *Arabidopsis* metacaspase9 degradome. *Plant Cell* **2013**, *25*, 2831–2847.
15. Wrzaczek, M.; Vainonen, J.P.; Stael, S.; Tsiatsiani, L.; Help-Rinta-Rahko, H.; Gauthier, A.; Kaufholdt, D.; Bollhoner, B.; Lamminmaki, A.; Gevaert, K.; *et al.* GRIM REAPER peptide binds to receptor kinase PRK5 to trigger cell death in *Arabidopsis*. *EMBO J.* **2015**, *34*, 55–66.

16. Zhang, Y.; Lam, E. Sheathing the swords of death: Post-translational modulation of plant metacaspases. *Plant Signal. Behav.* **2011**, *6*, 2051–2056.

17. Lam, E.; Zhang, Y. Regulating the reapers: Activating metacaspases for programmed cell death. *Trends Plant Sci.* **2012**, *17*, 487–494.

18. Vercammen, D.; van de Cotte, B.; de Jaeger, G.; Eeckhout, D.; Casteels, P.; Vandepoele, K.; Vandenberghe, I.; van Beeumen, J.; Inze, D.; van Breusegem, F. Type II metacaspases Atmc4 and Atmc9 of *Arabidopsis thaliana* cleave substrates after arginine and lysine. *J. Biol. Chem.* **2004**, *279*, 45329–45336.

19. Vercammen, D.; Belenghi, B.; van de Cotte, B.; Beunens, T.; Gavigan, J.A.; de Rycke, R.; Brackenier, A.; Inze, D.; Harris, J.L.; van Breusegem, F. Serpin1 of *Arabidopsis thaliana* is a suicide inhibitor for metacaspase 9. *J. Mol. Biol.* **2006**, *364*, 625–636.

20. Belenghi, B.; Romero-Puertas, M.C.; Vercammen, D.; Brackenier, A.; Inze, D.; Delledonne, M.; van Breusegem, F. Metacaspase activity of *Arabidopsis thaliana* is regulated by *S*-nitrosylation of a critical cysteine residue. *J. Biol. Chem.* **2007**, *282*, 1352–1358.

21. Wen, S.; Ma, Q.M.; Zhang, Y.L.; Yang, J.P.; Zhao, G.H.; Fu, D.Q.; Luo, Y.B.; Qu, G.Q. Biochemical evidence of key residues for the activation and autoprocessing of tomato type II metacaspase. *FEBS Lett.* **2013**, *587*, 2517–2522.

22. Bozhkov, P.V.; Suarez, M.F.; Filonova, L.H.; Daniel, G.; Zamyatnin, A.A., Jr.; Rodriguez-Nieto, S.; Zhivotovsky, B.; Smertenko, A. Cysteine protease mc II-Pa executes programmed cell death during plant embryogenesis. *Proc. Natl. Acad. Sci. USA* **2005**, *102*, 14463–14468.

23. Watanabe, N.; Lam, E. Two *Arabidopsis* metacaspases AtMCP1b and AtMCP2b are arginine/lysine-specific cysteine proteases and activate apoptosis-like cell death in yeast. *J. Biol. Chem.* **2005**, *280*, 14691–14699.

24. Watanabe, N.; Lam, E. Arabidopsis metacaspase 2d is a positive mediator of cell death induced during biotic and abiotic stresses. *Plant J.* **2011**, *66*, 969–982.

25. Piszczek, E.; Dudkiewicz, M.; Mielecki, M. Biochemical and bioinformatic characterization of type II metacaspase protein (TaeMCA II) from wheat. *Plant Mol. Biol. Rep.* **2012**, *30*, 1338–1347.

26. He, R.; Drury, G.E.; Rotari, V.I.; Gordon, A.; Willer, M.; Farzaneh, T.; Woltering, E.J.; Gallois, P. Metacaspase-8 modulates programmed cell death induced by ultraviolet light and H2O2 in *Arabidopsis*. *J. Biol. Chem.* **2008**, *283*, 774–783.

27. Hoeberichts, F.A.; ten Have, A.; Woltering, E.J. A tomato metacaspase gene is upregulated during programmed cell death in *Botrytis cinerea*-infected leaves. *Planta* **2003**, *217*, 517–522.

28. Kim, S.M.; Bae, C.; Oh, S.K.; Choi, D. A pepper (*Capsicum annuum* L.) metacaspase 9 (*Camc9*) plays a role in pathogen-induced cell death in plants. *Mol. Plant Pathol.* **2013**, *14*, 557–566.

29. Wang, X.D.; Wang, X.J.; Feng, H.; Tang, C.L.; Bai, P.F.; Wei, G.R.; Huang, L.L.; Kang, Z.S. *TaMCA4*, a novel wheat metacaspase gene functions in programmed cell death induced by the fungal pathogen *Puccinia striiformis* f. sp. *tritici*. *Mol. Plant Microbe Interact.* **2012**, *25*, 755–764.

30. Hao, L.; Goodwin, P.H.; Hsiang, T. Expression of a metacaspase gene of *Nicotiana benthamiana* after inoculation with *Colletotrichum destructivum* or *Pseudomonas syringae* pv. *tomato*, and the effect of silencing the gene on the host response. *Plant Cell Rep.* **2007**, *26*, 1879–1888.

31. Ahmad, R.; Zuily-Fodil, Y.; Passaquet, C.; Bethenod, O.; Roche, R.; Repellin, A. Ozone and aging up-regulate type II metacaspase gene expression and global metacaspase activity in the leaves of field-grown maize (*Zea. mays* L.) plants. *Chemosphere* **2012**, *87*, 789–795.

32. Watanabe, N.; Lam, E. Recent advance in the study of caspase-like proteases and Bax inhibitor-1 in plants: Their possible roles as regulator of programmed cell death. *Mol. Plant Pathol.* **2004**, *5*, 65–70.

33. Bollhoner, B.; Zhang, B.; Stael, S.; Denance, N.; Overmyer, K.; Goffner, D.; van Breusegem, F.; Tuominen, H. Post mortem function of Atmc9 in xylem vessel elements. *New Phytol.* **2013**, *200*, 498–510.

34. Suarez, M.F.; Filonova, L.H.; Smertenko, A.; Savenkov, E.I.; Clapham, D.H.; von Arnold, S.; Zhivotovsky, B.; Bozhkov, P.V. Metacaspase-dependent programmed cell death is essential for plant embryogenesis. *Curr. Biol.* **2004**, *14*, R339–R340.

35. Coll, N.S.; Vercammen, D.; Smidler, A.; Clover, C.; van Breusegem, F.; Dangl, J.L.; Epple, P. *Arabidopsis* type I metacaspases control cell death. *Science* **2010**, *330*, 1393–1397.

36. Minina, E.A.; Filonova, L.H.; Fukada, K.; Savenkov, E.I.; Gogvadze, V.; Clapham, D.; Sanchez-Vera, V.; Suarez, M.F.; Zhivotovsky, B.; Daniel, G.; *et al.* Autophagy and metacaspase determine the mode of cell death in plants. *J. Cell. Biol.* **2013**, *203*, 917–927.

37. Coll, N.S.; Smidler, A.; Puigvert, M.; Popa, C.; Valls, M.; Dangl, J.L. The plant metacaspase AtMC1 in pathogen-triggered programmed cell death and aging: functional linkage with autophagy. *Cell Death Differ.* **2014**, *21*, 1399–1408.

38. Dietrich, R.A.; Richberg, M.H.; Schmidt, R.; Dean, C.; Dangl, J.L. A novel zinc finger protein is encoded by the Arabidopsis *LSD1* gene and functions as a negative regulator of plant cell death. *Cell* **1997**, *88*, 685–694.

39. Vinocur, B.; Altman, A. Recent advances in engineering plant tolerance to abiotic stress: Achievements and limitations. *Curr. Opin. Biotechnol.* **2005**, *16*, 123–132.

40. Remans, T.; Keunen, E.; Bex, G.J.; Smeets, K.; Vangronsveld, J.; Cuypers, A. Reliable gene expression analysis by reverse transcription-quantitative PCR: Reporting and minimizing the uncertainty in data accuracy. *Plant Cell* **2014**, *26*, 3829–3837.

41. Hu, H.H.; Dai, M.Q.; Yao, J.L.; Xiao, B.Z.; Li, X.H.; Zhang, Q.F.; Xiong, L.Z. Overexpressing a NAM, ATAF, and CUC (NAC) transcription factor enhances drought resistance and salt tolerance in rice. *Proc. Natl. Acad. Sci. USA* **2006**, *103*, 12987–12992.

42. Yi, N.; Oh, S.J.; Kim, Y.S.; Jang, H.J.; Park, S.H.; Jeong, J.S.; Song, S.I.; Do Choi, Y.; Kim, J.K. Analysis of the *Wsi18*, a stress-inducible promoter that is active in the whole grain of transgenic rice. *Transgenic Res.* **2011**, *20*, 153–163.

43. Jain, M.; Nijhawan, A.; Tyagi, A.K.; Khurana, J.P. Validation of housekeeping genes as internal control for studying gene expression in rice by quantitative real-time PCR. *Biochem. Biophys. Res. Commun.* **2006**, *345*, 646–651.

44. You, J.; Zong, W.; Du, H.; Hu, H.H.; Xiong, L.Z. A special member of the rice SRO family, OsSRO1c, mediates responses to multiple abiotic stresses though interaction with various transcription factors. *Plant Mol. Biol.* **2014**, *84*, 693–705.

45. Lim, S.D.; Cho, H.Y.; Park, Y.C.; Ham, D.J.; Lee, J.K.; Jang, C.S. The rice RING finger E3 ligase, OsHCI1, drives nuclear export of multiple substrate proteins and its heterogeneous overexpression enhances acquired thermotolerance. *J. Exp. Bot.* **2013**, *64*, 2899–2914.

46. Sun, L.J.; Zhang, H.J.; Li, D.Y.; Huang, L.; Hong, Y.B.; Ding, X.S.; Nelson, R.S.; Zhou, X.P.; Song, F.M. Functions of rice NAC transcriptional factors,ONAC122 and ONAC131, in defense responses against *Magnaporthe grisea*. *Plant Mol. Biol.* **2013**, *81*, 41–56.

47. Agrawal, G.K.; Rakwal, R.; Jwa, N.S.; Agrawal, V.P. Signalling molecules and blast pathogen attack activates rice *OsPR1a* and *OsPR1b* genes: A model illustrating components participating during defence/stress response. *Plant Physiol. Biochem.* **2001**, *39*, 1095–1103.

48. Young, S.A.; Guo, A.; Guikema, J.A.; White, F.F.; Leach, J.E. Rice cationic peroxidase accumulates in xylem vessels during incompatible interactions with *Xanthomonas oryzae* pv. *oryzae*. *Plant Physiol.* **1995**, *107*, 1333–1341.

49. Mitsuhara, I.; Iwai, T.; Seo, S.; Yanagawa, Y.; Kawahigasi, H.; Hirose, S.; Ohkawa, Y.; Ohashi, Y. Characteristic expression of twelve rice *PR1* family genes in response to pathogen infection, wounding, and defense-related signal compounds (121/180). *Mol. Genet. Genomics* **2008**, *279*, 415–427.

50. Zhao, C.J.; Wang, A.R.; Shi, Y.J.; Wang, L.Q.; Liu, W.D.; Wang, Z.H.; Lu, G.D. Identification of defense-related genes in rice responding to challenge by *Rhizoctonia solani*. *Theor. Appl. Genet.* **2008**, *116*, 501–516.

51. Zhang, J.H.; Jia, W.S.; Yang, J.C.; Ismail, A.M. Role of ABA in integrating plant responses to drought and salt stresses. *Field Crop. Res.* **2006**, *97*, 111–119.

52. Lee, S.C.; Luan, S. ABA signal transduction at the crossroad of biotic and abiotic stress responses. *Plant Cell Environ.* **2012**, *35*, 53–60.

53. Halim, V.A.; Vess, A.; Scheel, D.; Rosahl, S. The role of salicylic acid and jasmonic acid in pathogen defence. *Plant Biol.* **2006**, *8*, 307–313.

54. Gontia-Misha, I.; Sasidharan, S.; Tiwari, S. Recent developments in use of 1-aminocyclopropane-1-carboxylate (ACC) deaminase for conferring tolerance to biotic and abiotic stress. *Biotechnol. Lett.* **2014**, *36*, 889–898.

55. Abe, H.; Urao, T.; Ito, T.; Seki, M.; Shinozaki, K.; Yamaguchi-Shinozaki, K. Arabidopsis AtMYC2 (bHLH) and AtMYB2 (MYB) function as transcriptional activators in abscisic acid signaling. *Plant Cell* **2003**, *15*, 63–78.

56. Eulgem, T.; Rushton, P.J.; Robatzek, S.; Somssich, I.E. The WRKYsuperfamily of plant transcription factors. *Trends Plant Sci.* **2000**, *5*, 199–206.

57. Chakravarthy, S.; Tuori, R.P.; D'Ascenzo, M.D.; Fobert, P.R.; Despres, C.; Martin, G.B. The tomato transcription factor Pti4 regulates defense-related gene expression via GCC box and non-GCC box *cis* elements. *Plant Cell* **2003**, *15*, 3033–3050.

58. Baumann, K.; de Paolis, A.; Costantino, P.; Gualberti, G. The DNA binding site of the Dof protein NtBBF1 is essential for tissue-specific and auxin-regulated expression of the *rolB* oncogene in plants. *Plant Cell* **1999**, *11*, 323–333.

59. Chakrabarty, R.; Banerjee, R.; Chung, S.M.; Farman, M.; Citovsky, V.; Hogenhout, S.A.; Tzfira, T.; Goodin, M. pSITE vectors for stable integration or transient expression of autofluorescent protein fusions in plants: Probing *Nicotiana. benthamiana*-virus interactions. *Mol. Plant Microbe Interact.* **2007**, *20*, 740–750.

60. Epple, P.; Mack, A.A.; Morris, V.R.F.; Dangl, J.L. Antagonistic control of oxidative stress-induced cell death in *Arabidopsis* by two related, plant-specific zinc finger proteins. *Proc. Natl. Acad. Sci. USA* **2003**, *100*, 6831–6836.

61. Wang, L.J.; Tian, Y.C.; He, C.Z. Cloning of a novel rice gene OsLSD1 and bioinformatic analysis of LSD1-like gene family from Arabidopsis and rice. *Prog. Biochem. Biophys.* **2005**, *32*, 268–274.

62. Wang, L.J.; Pei, Z.Y.; Tian, Y.C.; He, C.Z. OsLSD1, a rice zinc finger protein, regulates programmed cell death and callus differentiation. *Mol. Plant Microbe Interact.* **2005**, *18*, 375–384.

63. Xu, C.X.; He, C.Z. The rice *OsLOL2* gene encodes a zinc finger protein involved in rice growth and disease resistance. *Mol. Genet. Genomics* **2007**, *278*, 85–94.

64. Wu, J.H.; Zhu, C.F.; Pang, J.H.; Zhang, X.R.; Yang, C.L.; Xia, G.X.; Tian, Y.C.; He, C.Z. OsLOL1, a C2C2-type zinc finger protein, interacts with OsbZIP58 to promote seed germination though the modulation of gibberellin biosynthesis in *Oryza sativa*. *Plant J.* **2014**, *80*, 1118–1130.

65. Kawahara, Y.; Oono, Y.; Kanamori, H.; Matsumoto, T.; Itoh, T.; Minami, E. Simultaneous RNA-seq analysis of a mixed transcriptome of rice and blast fungus interaction. *PLoS ONE* **2012**, *7*, e49423

66. Liu, W.; Liu, J.; Triplett, L.; Leach, J.E.; Wang, G.L. Novel insights into rice innate immunity against bacterial and fungal pathogens. *Annu. Rev. Phytopathol.* **2014**, *52*, 213–241.

67. Shen, X.L.; Liu, H.B.; Yuan, B.; Li, X.H.; Xu, C.G.; Wang, S.P. OsEDR1 negatively regulates rice bacterial resistance via activation of ethylene biosynthesis. *Plant Cell Environ.* **2011**, *34*, 179–191.

68. Petrov, V.; Hille, J.; Mueller-Roeber, B.; Gechev, T.S. ROS-mediated abiotic stress-induced programmed cell death in plants. *Front. Plant Sci.* **2015**, *6*, 69.

69. Jeong, J.S.; Kim, Y.S.; Baek, K.H.; Jung, H.; Ha, S.H.; Do Choi, Y.; Kim, M.; Reuzeau, C.; Kim, J.K. Root-specific expression of *OsNAC10* improves drought tolerance and grain yield in rice under field drought conditions. *Plant Physiol.* **2010**, *153*, 185–197.

70. Zheng, X.; Chen, B.; Lu, G.; Han, B. Overexpression of a NAC transcription factor enhances rice drought and salt tolerance. *Biochem. Biophys. Res. Commun.* **2009**, *379*, 985–989.

71. Piszczek, E.; Dudkiewicz, M.; Sobczak, M. Molecular cloning and phylogenetic analysis of cereal type II metacaspase cDNA from wheat. *Biol. Plant.* **2011**, *55*, 614–624.

72. He, S.P.; Huang, K.W.; Zhang, X.; Yu, X.C.; Huang, P.; An, C.C. The LSD1-type zinc finger motifs of *Pisum sativa* LSD1 are a novel nuclear localization signal and interact with importin α. *PLoS ONE* **2011**, *6*, e22131.

73. Chen, H.L.; Wang, S.P.; Zhang, Q.F. New gene for bacterial blight resistance in rice located on chomosome 12 identified from Minghui 63, an elite restorer line. *Phytopathology* **2002**, *92*, 750–754.

74. Jia, Y.; Liu, G.; Park, D.S.; Yang, Y. Inoculation and scoring methods for rice sheath blight disease. *Methods Mol. Biol.* **2013**, *956*, 257–268.

75. Bustin, S.; Beaulieu, J.-F.; Huggett, J.; Jaggi, R.; Kibenge, F.; Olsvik, P.; Penning, L.; Toegel, S. MIQE precis: Practical implementation of minimum standard guidelines for fluorescence-based quantitative real-time PCR experiments. *BMC Mol. Biol.* **2010**, *11*, 74.

This page left intentionally blank.

INDEX

F

G

H

I

J

K

L

M

N

O

P

Q

R

S

T

U

V

W

Z

This page left intentionally blank.